Drones and Geographical Information Technologies in Agroecology and Organic Farming

Contributions to Technological Sovereignty

Editors

Massimo De Marchi
Climate Justice, Jean Monnet Center of Excellence
Department of Civil Environmental Architectural Engineering
University of Padova, Padova, Italia

Alberto Diantini
Postdoc Researcher, Department of Historical and
Geographic Sciences and the Ancient World
University of Padova, Padova, Italy

Salvatore Eugenio Pappalardo
Laboratory GIScience and Drones 4 Good
Department of Civil Environmental Architectural Engineering
University of Padova, Padova, Italy

GIScience

STeDe Sustainable Territorial Development

Climate Justice
Jean Monnet
Centre of Excellence

With the support of the
Erasmus+ Programme
of the European Union

CRC Press
Taylor & Francis Group
Boca Raton London New York

CRC Press is an imprint of the
Taylor & Francis Group, an **informa** business
A SCIENCE PUBLISHERS BOOK

Cover credits: Francesca Peroni and Daniele Codato

First edition published 2022
by CRC Press
6000 Broken Sound Parkway NW, Suite 300, Boca Raton, FL 33487-2742

and by CRC Press
4 Park Square, Milton Park, Abingdon, Oxon, OX14 4RN

Library of Congress Cataloging-in-Publication Data (applied for)

ISBN: 978-0-367-14638-2 (hbk)
ISBN: 978-1-032-15355-1 (pbk)
ISBN: 978-0-429-05284-2 (ebk)

DOI: 10.1201/9780429052842

Typeset in Times New Roman
by Innovative Processors

Acknowledgement

This book has been made available open access thanks to the funding of three initiatives implemented at the Department of Civil Environmental and Architectural Engineering at the University of Padova:

(1) the Advanced Master on "GIScience and Unmanned System for the integrated management of the territory and the natural resources - with majors"
(2) the International Joint Master Programme on "Climate Change and Diversity - Sustainable Territorial Development" (CCD - STeDe)
(3) the "Climate Justice Jean Monnet Center of Excellence" with the support of Erasmus + Programme of the European Union, call for proposals EAC/A02/2019 – Jean Monnet Activities; Decision number 620401; Project number: 620401-EPP-1-2020-1-IT-EPPJMO-CoE.

Disclaimer

Preface

The United Nations decade of ecosystem restoration (2021-2030), the Glashow Climate Pact (November 2021) reaffirms the role of Nature Based Solutions in the fight against climate change and in building shared adaptation solutions. The Glashow Climate Pact highlights the importance of ensuring the integrity of all ecosystems, the protection of biodiversity "recognized by some cultures as Mother Earth, the importance for some of the concept of 'climate justice', when taking action to address climate change".

In April 2020 Boaventura de Souza Santos published the "Cruel Pedagogy of Virus" focusing on how the COVID pandemic/syndemic has arrived at the end of six decades of uneven development and highlights the global predatory capitalism and patriarchy embodied in many development discourses, consolidating social exclusion, resource extraction, human and nature domination, environmental injustice, and accumulation by dispossession.

Deconstructing development, sustainable development, sustainable growth asks for recognizing practices of critical development, alternative development, alternatives to development, post-development to embrace what Max-Neef called "the development at human scale".

Change starts from new practices, challenging the menu of globalizing universalizing development theories and initiatives to inhabit pluriverses of words and worlds.

Agroecology, as young science that is about to turn a century, can contribute in various ways to the current challenges of facing environmental and climate emergency, halting biodiversity loss, pursuing just food systems.

The indigenous, peasant, and environmental movements of active citizenship, inspired by agroecology, promote food sovereignty, just food systems, the collaboration between food producers and consumers, the renewed alliance between natural, agricultural and urban ecosystems, technological sovereignty, innovation attentive to human rights.

This book explores the challenges posed by the new geographic information technologies in agroecology and organic farming. It discusses the differences among technology-laden conventional farming systems and the role of technologies in strengthening the potential of agroecology and organic farming. In conventional thinking, the use of new technologies is an almost exclusive domain of precision agriculture. Traditions and links with the past are typical western urban images of agroecology compared with modern industrial agriculture, based on mechanization and evolving technology use. The many agriculture 4.0 and sustainable agricultures are still adopting a productive paradigm rooted in yield and profit of farm (as firm), innovation is something universally coming from specialized centers, local knowledge is negligible.

There is a profound connection between social and technological innovation and the multiscale dimension of innovation, especially in the place-specific agroecosystem. Farmers and citizens are themselves innovators; they should have the agency to govern technologies and to develop appropriate place-based institutional-technological innovation.

Technology can not be a commodity, it is common. Traditional agricultural systems are not statics: 9000 years of agriculture in Mexico or several thousand years of Amazon polyculture have required knowledge and ability to care for complex territories (agroecosystems) granting the reproduction of human societies and the evolution of ecosystems.

In the perspective of "technologies for all" there is a basket of promising open applications consolidating agroecology and its plural dimensions of innovation based on knowledge-intensive approaches, knowledge sharing, co-creation of knowledge, common goods and heritages of humanity at different scales.

We want to recall the Kamunguishi Declaration issued by Zapara nationality, a disappearing Amazon population having their oral heritage and cultural manifestation recognized by UNESCO in the list of intangible heritage. Kamunguishi is *the house of the forest for continuous rebirth*:

the world is ony one *(Nukaki)*
the world is forest *(Naku)*
we are forest!

Massimo De Marchi
Alberto Diantini
Salvatore Eugenio Pappalardo

Contents

Part III: Landscapes and Ecosystem Services, Technologies for Agroecological Transitions

Part IV: Conclusions and Perspectives

CHAPTER

1

Agroecology and Sustainable Food Systems: Inquiring Technological Approaches

Massimo De Marchi[1]*, Salvatore Eugenio Pappalardo[2] and Alberto Diantini[3]

[1] Director of the Advanced Master on 'GIScience and Unmanned System for the Integrated Management of the Territory and the Natural Resources - with Majors', responsible of International Master Degree on Sustainable Territorial Development, Climate Change Diversity Cooperation (STeDe-CCD), Department of Civil Environmental Architectural Engineering, University of Padova

[2] Laboratory GIScience and Drones 4 Good, University of Padova

[3] Research Programme Climate Change, Territory, Diversity – Department of Civil Environmental Architectural Engineering – Postdoc Researcher at the Department of Historical and Geographic Sciences and the Ancient World, University of Padova

1.1. Introduction

The awareness of impacts of conventional industrial farming and the exceeding of multiple-planet boundaries (Campbell *et al.*, 2017; Montgomery, 2007; Sánchez-Bayoa and Wyckhuys, 2019) has been paired, in the last 30 years, by the faith in technology as a central pillar of innovation for agricultural transition to sustainability.

In this conventional thinking, agroecology is normally not associated with the use of new technologies and is an almost exclusive domain of precision agriculture. Traditions and links with the past are typical Western urban images of agroecology compared with conventional agriculture, based on mechanization and evolving technology use.

In the introductory chapter of this book, we start with reflection on the agroecological transition to map the multiplicity of labels for sustainability in agriculture, combined with the exploration of different interpretations of

*Corresponding author: massimo.de-marchi@unipd.it

sustainability and the link with technologies and innovation. The approach adopted is the 'technology for all' as a dynamic combination of available tools adapted to specific locations and cultures of myriads of agroecological small farms, going beyond the universalizing closed menu of technological supply for standardizing conventional large farms. The keyword is exploring the 'basket of options' suitable for the multiplicity of small farmers, herders, fisher-folk, peasants, indigenous people, and urban dwellers interested in growing directly their food, suitable for youth and elders, for women and men in cooperation among humans and non-humans.

1.2. Agroecological Transitions

Agroecology's origins, developments, and trends can be summarized by some key concepts following some fundamental contributions: analysis of agroecosystems looking for interaction between place, time, flows, decisions (Conway,1987); a new paradigm of research and development for world agriculture (Altieri, 1989); resource management science for poor farmers in marginal environments (Altieri, 2002); ecology of the food system (Francis *et al.*, 2003); science focusing on multi-scalarities and interdisciplinarity (Dalgaard *et al.*, 2003); combination of science, movement, practice (Wezel *et al.*, 2009), and a transdisciplinary, participatory, action-oriented approach (Mendez *et al.*, 2016).

Agroecology is not just a speculative exploratory science but is committed to change through the design of sustainable agroecosystems (Gliessman, 2007; Malezieux, 2012; Wezel *et al.*, 2014; Wezel *et al.*, 2020). Gliessman (2007, 2014, 2016) summarizes five possible levels of agroecological transition from conventional industrial farming to farming for just food systems.

The first level requires increase in the efficiency of industrial/conventional practices in order to reduce the use and consumption of expensive, scarce or environmentally-damaging inputs. This basic level of efficiency is well represented by precision agriculture or the different declinations of sustainable or smart agriculture, but it is far from a real transition.

The proliferation of multiple labels to describe innovation pluralism of sustainable agriculture often conceals a weak sustainability approach harbored in the paradigms of yield, granted by modernized industrial farming, optimizing chemical and biotechnological energy inputs by the new technological-controlled supply (HLPE, 2019; Klerkxa and Rose, 2020).

The so-called precision agriculture continues to rely on mechanization, fossil fuels, and chemicals, but uses them more efficiently so that instead of spraying an entire field, the chemical inputs are released only in the rows: the idea is to avoid excessive or not useful treatments and to concentrate the operation only when and where necessary. In effect, precision farming was developed as an approach apt to mitigating the environmental impacts of intensive farming implemented in large surfaces with external material and energy inputs (Zhang *et al.*, 2002; Gebbers and Adamchuk, 2010). GPS, satellite images, GIS, and drones help conventional farming in localizing in detail where to supply water, pesticides, and fertilizers. Prescription maps define the right place and moment for interventions of machinery fleets (Wolf and Buttel, 1996; Falkenberg, 2015; European Parliament, 2016; European Parliament, 2017; Altieri *et al.*, 2017; HLPE, 2019; Klerkxa and Rose, 2020). The yield goal remains the key objective, integrating a more efficient use of resources toward economic-environmental sustainability. Precision farming allows extractive agriculture to enter the sustainability era. Sustainable intensification, climate smart agriculture, nutrition-sensitive agriculture, sustainable Food value chain and other various Agriculture 4.0 declinations are often considered, in the mainstreaming discourses of agricultural policies and practices, the abundant innovative offer of salvation tools for the planet and prosperity. The basic idea is to use industrial practices more efficiently in order to minimize the environmental impacts – this is not a change of model, but a way to protect the yield-universalizing paradigm with belts of 'more efficient', 'less impactful', and 'sustainable'.

There is a level two, where the keyword is 'substitution', that is, replacement of industrial/conventional inputs and practices with sustainable alternatives. It is what organic, biological or ecological agriculture does according to the regulations, for example, of the European Union or the United States. These regulations define all the inputs allowed to guarantee products without traces of industrial/conventional phyto-sanitary products. But agricultural activity can be implemented in fields without trees or living fences. The risk, as Miguel Angel Altieri recalls, is the consolidation of a capitalist market for organic production with a new concentration of distribution and sale of 'ecological inputs' (Altieri, 2002; Guthman, 2004; Altieri, 2017).

In many cases, the change of the conventional production model to organic agriculture is maintained inside the industrial paradigm of the yield, simply with a change of the external input supply from chemical to organic, or better, to all products admitted by regulations.

Industrial organic farming relies on fossil fuel; for example, increase in tilling and soil labouring as an alternative to chemical weed control, and on the dependence of external inputs. Cycles are not closed on the farms and the approach is still inside the typical capitalist markets maintaining two limitations. On one side, the farm is dependent on the market fluctuations of external biological inputs; on the other, the organic food production is conditioned by a price system regulated by a market of commodities not recognizing the right to food and right to decent work of farmers (Guthman, 2004; Altieri, 2017). Industrial organic production maintains the stratification of small and big farms with many of the social and environmental injustices of conventional farming: the paradigm of yield and the basic objectives of producing commodities for the market are not questioned. Industrial organic production can cohabit with mechanization, technology, and a conventional farming machinery landscape without asking for a higher level of transition. Small and transformative organic farming opened a reflection on the convergence between the organic world, and agroecology (Migliorini and

Wezel, 2017), and on the transition to organic Agriculture 3.0 (Rahman *et al.*, 2017). Compared to agroecology, organic farming is still more technical and normative, and highly regulated by certification schemes; the logic of changes are driven by an alternative scientific and philosophical northern and Western view; the approach is still in the food chain with a vision on food health and food security; production systems rely on low external input substitution regulated by allowed and forbidden substances; and despite the relevance in the change of northern conventional agriculture, organic farming needs a redesign inspired by agroecology (Rahman *et al.*, 2017; Migliorini and Wezel, 2017; Altieri, 2017).

Only level three is the bifurcation point for a true agroecological transition: this level requires the re-design of agroecosystems to adopt functions based on ecological processes. This agroecological transition begins on the farm and in the landscape, but needs to be scaled up to be effective. In level three, agroecology meets landscape ecology and requires ecological infrastructure; thus the differentiation and complexification of the ecosystem happen not just in the field, even if it starts from the field and the farm (Gliessman, 2007; Malezieux, 2012; Wezel *et al.*, 2014; Perfecto *et al.*, 2009). The machinery landscape of conventional agriculture or industrial organic farming must introduce hedges, trees and forests, wetlands, and soil covered by leaves or dead vegetation. Agroecology stresses policies and equity; the creativity frames agroecological principles with the prominence of southern intercultural view and local indigenous knowledge; the agroecosystem is the point of reference for the management of relations among species, and material, and energy flows; food sovereignty and food networks inspire the approach; additionally, maybe agroecology could need formalization (Rahman *et al.*, 2017; Migliorini and Wezel, 2017). In many parts of the world, the agroecosystems, based on agroecology, are still supplying a plurality of services, and level three of the agroecological transition is already active and ready to jump to the next two levels.

With the transition to level four, a more direct connection is re-established between those who grow food and those who consume it. This level is fundamental, both to consolidate the existing agroecological farms resisting the universalizing paradigm of yields and to welcome the new agroecological farms walking the transition paths in order to leave behind the conventional/industrial agricultural approach (Gliessman, 2007, 2014, 2016; HLPE, 2019; Wezel *et al.*, 2020). The new food networks, connecting farmers and citizens, are based on direct relations and new tools of PGS (participatory guarantee systems). The trust among those who grow food and who consume it is not granted by a third-party certification body (as in conventional agriculture and organic farming) but by direct contact and accessible direct network of reciprocal commitments (Home *et al.*, 2017; FAO, 2018d; Montefrio and Johnson, 2019).

Foundations created at the scale of agroecosystems and landscape (level three) and new connections between food, farmers, and citizens (level four) can culminate in level five, building a new sustainable global food system, which strengthens the resilience of ecosystems over a basis of equity, participation, and justice (Gliessman, 2016; Wezel *et al.*, 2018; Anderson *et al.*, 2019; Côte *et al.*, 2019). Level five asks for a strong commitment by governments in adopting agroecological and food sovereignty policies (Jansen, 2015; Pimbert, 2018), and concretely acting for the scaling up of agroecology (Bellon and Ollivier, 2018).

Considering the multi-scale approach (farm, landscape, region, and world) and the three main dimension of agroecology (research, farmer practices, and social change), it is possible to define the combination of scale/actors involved at different levels of agroecological transition (Gliessman, 2007, 2014, 2016; HLPE, 2019; Wezel *et al.*, 2020). Levels one and two (efficiency and substitution) are implemented mainly at the farm level with a direct commitment of farmers and researchers and a minimal contribution of social actors limited to a final decision at the moment of buying food. Level three (re-design) creates the connections among farms and landscapes with the research sector supporting tools for evaluating social and ecological interactions; farmers are key actors of this change and citizens can support farmers' commitments. At level four, the interactions among growers and eaters require the adoption of a food network approach with a primary engagement of citizens and farmers and the contribution of applied interdisciplinary research monitoring effective changes. At this level, agroecology operates on multiple scales: local, regional, and national. A global equitable food system, level five, requires a strong commitment of citizens in pressing decision-makers for agroecology scaling up and the maintenance of appropriate institutions. In this context, farmers adopting agroecology should offer an inspiring example for the change of the agricultural system and research could act as a supportive platform for monitoring the effectiveness of this transition process.

Beyond the responsibility of research, farmers, and citizens, the scaling up of agroecology requires the engagement of institutions, both at local and global levels; reflections are undergoing, so the debate; and the need is to spread exemplary policy practices developed at the national, regional or municipal scale.

1.3. Sustainability and Sustainable Food Systems

The Mexican agroecologist and ethnobotanist, Efraim Hernandez Xolocotzi, analyzing the complexity of indigenous agroecosystem, recognized how sustainability was based in a solid co-evolution of social and environmental dimensions, resulting in the interactions of ecological, technological, and socio-economic place-based components (Hernandez Xolocotzi, 1977; Díaz León and Cruz León, 1998). Modern farming systems abandoned the connection with the ecological roots, allowing market-driven socio-economic components to become the paradigm of management in food systems. In this perspective, sustainability should recognize and rebuild the ecological services of agroecosystems, managing energy and material flows, starting from the natural nitrogen fixation and the co-operation with soil mycorrhizae (Gliessman, 2007, 2014). It does not mean avoidance of any input arriving from outside the system, but use of material and energy flow from natural or contiguous ecosystems; for example, integrating the management of urban organic waste in urban farming. Renewable energy sources should substitute non-renewable energy, without forgetting energy and material efficiency. Sustainability in the agroecosystem requires management and co-existence with different species, desired and unwanted species, and avoiding the control paradigm in the integrated management of soil fertility and vegetation health by maintaining the higher level of biological and ecological services naturally available in the agroecosystem (Altieri, 2012; Malezieux, 2012; Gliessman, 2014).

Agroecology looks at the agroecosystem by focusing on the principles of ecology, the cultural texture, the socio-economic dynamics, and the uniqueness of the place. Agroecology has a multi-scale look, not only at the localized agroecosystems but also at the problems of food production, the way of doing agriculture in different contexts, the environmental management and resource enhancement, and the cultural knowledge as a whole to design better systems: from farm to global food system (Francis *et al.*, 2003).

Awareness of food can be a starting point to reflect on the relationship between people and agroecosystems, considering that territories are open systems where the influence of society is not given only by the ecological components, but also by the decisions, the ability to develop co-operative and conflictive behaviours among people driven by desires and visions. So, what we have in mind is larger than what we have on the table (Francis *et al.*, 2003).

Human action shapes ecosystems in a direction that can be sustainable or lead to potential degradation. The current globalized system is not fully aware of knowledge about food and ecosystems due to the separation of the place of consumption from the place of production. The global society of the biosphere is creating uncertainty and instability among the people of ecosystems, generating ecological refugees, enlarging the space of collection of food, the substitution among different food chains, and the control of food commodities (Gadgil, 1994). Spaces on concern and collection of resources being far from the spaces of living require the development of a multi-scale awareness, maybe for people living in cities; thinking about food may create a connection with the ecosystem dimension (Gadgil, 1994). This separation leads to a lack of awareness about the implication of food consumed. Food choices are based more on price, global market availability, consuming well-advertised products, and forgetting all the connections between food and health. The latter cannot be reduced to consumer health; there are many concealed dimensions – health of farmers and food processing workers, the ecosystem health and society as a whole, environmental quality, and social impacts. Rebuilding this becomes a fundamental aspect, considering we have a global system producing various types of quality and lower quality food and not granting the right to foods, especially to people who more need it (Francis *et al.*, 2003; De Schutter, 2011; FAO, 2015).

The vision of conventional farming based on increasing yield and food quantity separating food chain from externalities (soil erosion, water, and air pollution, biodiversity loss, etc.) should let the floor to the agroecological approach, where there is no separation between society and nature, where people are part of the ecosystem. Then, if people are part of the ecosystem, the logic is based on co-existence, not on separation and extraction (Francis *et al.*, 2003; Declaration of the International Forum for Agroecology, Nyéléni, 2015; FAO, 2016a; FAO, 2016a). Food awareness goes beyond productivity to take into account the issue of complexity and justice: humans live in a complex open system, interacting with many ecosystems and species, with nature and societies, and should develop food-efficient systems, taking into account accessibility and sovereignty. In this context, the central theme is the consumer who co-operates in closing the circle, in promoting efficiency and justice inside the food network, and in co-operating in a participatory dialogue involving all the different parties – the researchers, the producers, who transform the food, who trade it, and who consume it. The agroecological cycle closes only if there is a responsible consumer: the 'consumer-actor' makes choices and influences them at various levels. It is necessary that agroecology knows how to inform and involve consumers, sharing information on what is happening in food production, in rural landscapes, in terms of production methods and proposals. So only with more information on how food is produced, processed, and circulated, it is possible to increase the awareness in consumer choices and on the overall choices of the different actors in the world food system (Francis *et al.*, 2003; HLPE, 2014; HLPE, 2017a; HLPE, 2017b).

The world sustainable food system is based on many small agroecosystems, which are capable of adapting to local and cultural contexts. In it, the food needs do not prevail on producing commodities for the global market, but food production is concerned with the desires and priorities of the populations, who therefore respond to social needs at different scales. The focus is on the food networks and not the food chains, complex food networks connecting farms and tables and caring for how food is produced, exchanged, distributed, and how it reaches the different tables with networks not only dependent on large-scale distribution chains (Gliessman, 2014; FAO, 2015).

Despite different ways of managing data and statistics (Ricciardi *et al.*, 2018; HLPE, 2019), traditional agroecosystems managed by small farmers provide about two-thirds of the world's food. On the other side, many large conventional industrial farms produce commodities with other purposes: livestock feeding and energy supply. So, the priority is the international market and the prices when food is no longer a right, but a commodity. A sustainable world food system equitably distributes food, reduces waste, ensures the important role of agricultural land to grant justice, accessibility, and sovereignty (Patel, 2009; Gliessman, 2014; Grey and Patel, 2014; Figueroa-Helland *et al.*, 2018).

Urban agriculture of the future will not be an agriculture 'fishing' in the world market but it will be based on social and proximity farming in order to close the cycle inside the city and create new urban agroecosystems (Gliessman, 2014; Altieri and Nichols, 2019; Almeida and Bizao, 2017; Rentig, 2017).

1.4. Technology for All: Innovation Narratives and Agroecology

Agroecology is a way of managing ecosystems that combine human and non-human needs with higher intensity of knowledge. Traditional agricultural systems are not static (Altieri, 2012): 9,000 years of agriculture in Mexico (Díaz León and Cruz León, 1998) or several thousand years of Amazon polyculture (Brugger *et al*., 2016; Maezumi *et al*., 2018; Neves and Heckemberger, 2019) have required knowledge and ability to care for complex territories (agroecosystems) granting the reproduction of human societies and the evolution of ecosystems.

Long-lasting sustainable agroecosystems show six characteristics (Altieri *et al*., 2012): permanence of productivity; risk reduction and resiliency; integration of economic viability, social equity, and cultural diversity; conservation and enhancement of natural resources, biodiversity, and ecosystem services; wise management of natural cycles and reducing dependency on non-renewable resources; and prevention of environmental land degradation. As Declaration of the International Forum for Agroecology (2015) summarizes: agroecology 'cultivates' biodiversity, respects Mother Earth, and is economically viable; and farmers should be socially rewarded not only for the production of food, but for all thee environmental services they create and maintain.

Agroecosystems, combining farming systems with complex livelihood structures, are rooted in the self-reliance of communities and the ownership of multiple sovereignties – spatial, food, technological, energy blended with sophisticated agroecological knowledge systems (Tomich *et al*., 2011; Altieri, 2012; Paracchini *et al*., 2020).

Agroecology, as a new paradigm changing the unsustainable ways of doing agriculture, can inspire the development of appropriate technologies which are able to grasp the productive potential of the agroecosystem, guaranteeing sustainable subsistence for all (Altieri, 1989). Many discourses on sustainable farming overestimate the role of technology, forgetting the articulated and multifaceted sustainability of existing agroecological systems, when analyzed merely from the yields' lens.

The 2019 HLPE's report, planned to explore the role of agroecology in sustainable food systems, resulted in a final document registering the struggle in the international food policy arena to frame agroecology into a continuous spectrum with 'other innovative approaches for sustainable agriculture and food systems that enhance food security and nutrition' (Anderson and Maughan, 2021).

The risk of the agroecology captured by the innovation imperative, as highlighted by the Nyeleni Declaration, requires distinguishing on one side, the different concepts of innovation, and on the other, the key difference of agroecology from 'other innovative approaches' of the wide 'silos' of Agriculture 4.0. The world is fascinated by the agroecology label depicted by actors with completely different visions: international companies interested in marketing products and institutions that perhaps use an agroecological cloak, with a reach cloakroom of synonyms, to use 'junk agroecology' to circulate the older yield paradigm (Alonso-Fradejas *et al.*, 2020).

Anderson and Maughan (2021) offer an overview of the 'innovation imperative' comparing two polarities: the dominant supportive approach adopted in sustainable agricultures versus the critical vision to innovation related to agroecology. The authors focus on three sub-frames of the main innovation structure: the measurement sub-frame, the technology sub-frame, and the rights sub-frame. Many Agriculture 4.0 and sustainable agricultures are still adopting a productive paradigm rooted in yield and profit of farm (as firm); innovation is something universally measurable following scientific-technical standards and where local knowledge is negligible. In this 'measurement sub-frame' the innovation approach of agroecology affirms the key role of site-specific local knowledge based on a different plurality in the way of knowing and measuring; place-based evaluation of innovation should be adopted and sustainability expresses the holistic multi-dimensional approach (López-Ridaura *et al.*, 2002; Ripoll-Bosch *et al.*, 2012; Valdez-Vazqueza *et al.*, 2017; de Oliveira Côrtesa *et al.*, 2019). About the technological sub-frame (Anderson and Maughan, 2021), agroecology sees a profound connection between social and technological innovation and the multi-scale dimension of innovation, especially in the place-specific agroecosystem. Farmers and citizens are themselves innovators: they should have the agency to govern technologies and to develop appropriate place-based institutional-technological innovations. New technologies are neither neutral nor good by definition, since they can create negative impacts. In this light, the adoption of a precautionary approach is the key attitude to deal with new technologies (Raghavan *et al.*, 2016; Gkisakis *et al.*, 2017; Gkisakis and Damianakis, 2020; Niggli *et al.*, 2016; Bellon Maurel and Huyghe, 2017; Daly *et al.*, 2019). Conversely, the different declinations of Farming 4.0 trust in the identification of innovation with new technologies driven by specialized agencies with the conventional top-down market approach of technology transfer. The narrative is based on some classical rhetoric figure of the green revolution: 'feeding the world', youth priorities, social change, and benefits for farmers. The prominence is on the quantity of novelty against the prudence of the precautionary approach (Anderson and Maughan, 2021).

In the right sub-frame (Anderson and Maughan, 2021), the two polarities see, from the side of industrial sustainable farming, the regimes of intellectual properties protecting innovation, the consumer right to choose the suitable technology, and, in other words, the farmers are entitled to any choice of agricultural products and innovation schemes – this right system should grant the priority right to innovate. A critical view to this approach is based on human rights as a priority framework; so rights should not be granted just to a few private or public institutions specializing in technology production and transfer, but all people are innovators, and the human rights framework protects the 'agency of people in all spheres of life' (Anderson and Maughan, 2021). The right of 'most affected' in the agroecosystems and in the food network should be prioritized. Agroecology, being a knowledge-intensive approach, advocates knowledge sharing, co-creation of knowledge, and intellectual commons. Agroecological knowledge is not fragmented and cannot be sold separately in the market of business intelligence: it is knowledge in the public domain, common good, and heritage of humanity at different scales. Production opportunities must not be taken away from a place to be placed in an international market, but local actors have to benefit first and foremost – 'thousand years of knowledge of the ecosystems through trial and errors up to agriculture' (Díaz León and Cruz León, 1998): isn't this innovation?

1.5. Geographical Information Tools and Knowledge: A Basket of Options

This book is organized into three parts and 14 chapters. It discusses the role of geographical technologies information and knowledge in strengthening the potential of agroecology.

The first part analyzes how technologies of geographic information offer tools to farmers and citizens in the quest for rights of nature and food sovereignty.

Chapter 2, by Massimo De Marchi and Alberto Diantini, offers an outlook on the relations among geographical information, science, and agroecology to disclose the 'power of maps' in agro-ecological transformative scaling up. Geography and cartography have a long and consolidated epistemological and empirical experience about the key role of maps, starting from the pre-digital era, in changing the world through the empowerment of weak and marginalized actors in cities and rural contexts. The chapters explore some key elements of 'mapping for change': from 'material' participatory cartography to immaterial participatory GIS and Volunteered Geography. Despite the low interactions, mostly informal, in the last decades among the science of geographical information and agroecology, there are many areas of common interests and mutual interaction and co-operation.

Alice Morandi, in Chapter 3, deals with the role of livestock in the quest for sustainable agricultural development. Livestock challenges the sustainability in agricultural development for the constellation of impacts, not only on the environment, but also on food security and sovereignty. The chapter explores how livestock can contribute to the Sustainable Development Goals of Agenda 2030 and on the other, how the livestock sector deals with the SDGs' strategic framework. The spatial decision support systems (SDSS), which are GIS tools to support spatial-explicit decision making, can support the livestock sector in simulating, and then implementing, multiple sustainability paths. The chapter presents Gleam-i (global livestock environmental assessment model), a webGIS tool elaborated by the FAO to develop policy assessments of livestock decisions. Gleam-i is applied in a case study of climate change mitigation in the Colombian poultry supply chains, showing the possibility to prevent impacts and to increase food security and food sovereignty. The Gleam-i is a promising open-source tool available on the web to a wide public. We hope the chapter creates interest among policymakers, civil society organizations, and the livestock sector to test alternative policies for climate justice and food sovereignty.

The second part of the book deals with technologies at farm levels; the three chapters provide practical experiences about a positive application of technologies in agroecology: the role of positioning systems, the diffusion of hyperspectral imagery, and the proximal sensing of drones. The chapters of this session have a common thread: the first part of each is more devoted to the description and presentation of the technology and the second part presents the applications in agroecology.

The chapter prepared by Angela Gatti and Alessio Zanoli focuses on the revolution in position precision provided by the availability of GNSS (global navigation satellite systems). The availability of GNSS technology in the consumer market has familiarized citizens with the user segment of the system relying upon more components: the space and the ground control segments. Recursive triangulations among satellites, earth stations, professional or consumer devices allow the refinement of position between metric or centimetric precision. About the use of GSS application in farming, in the last couple of decades, we witnessed a sort of metonymic discourse capturing the precision of position by precision farming, creating a sort of exclusive ownership of this technology. The social globalized imaginary is adsorbed by self-driving tractors spying the right quantity of chemical input controlled by a GNSS. However, imaginary should go beyond the discursive boundaries of Agriculture 4.0. The authors show many applications in agroecology and organic farming: from soil sampling to harvest and biomass monitoring, and the interesting application in livestock management. The chapter closes with a review of the emerging issue of low-cost GNSS, based on cheaply available devices (smartphone and u-blox) revolutionizing the accessibility of this technology and moving toward the democratization of GNSS tools for food sovereignty and agroecological transition.

András Jung and Michael Vohland prepared Chapter 5, which is devoted to hyperspectral remote sensing and field spectroscopy. Compared to normal multispectral satellite images (*see* Chapter 10), with a limited number of spectral bands, hyperspectral sensors collect huge data cubes supplying new generations of imageries with hundreds of spectral bands. These large amounts of spectral data require important processes (and machine resources) for data analysis, but can supply important information on soil constituents or vegetation with fine details of species, and the possibility to distinguish particular phenological or pathological conditions. The authors integrate the presentation of hyperspectral imageries with field spectroscopy and summarize some possible applications in agroecology and organic farming. Hyperspectral imageries and field spectroscopy are a promising technology, even if expensive in terms of equipment and data processing. They can represent an interesting case of reflection in scaling up agroecology and in the transition of public agricultural services in facilitating and sharing advanced technologies for small agroecology farming systems.

Drones for Good is the topic of Chapter 6, written by Salvatore Eugenio Pappalardo and Diego Andrade. UAS (unmanned aerial systems) or UAV (unmanned aerial vehicles) probably represents the icon of GeoICT applied in agroecosystems for a long time trapped in conventional industrial precision farming. The paradigm of Drones for Good and the use of drones not only outside the military domain but also outside the industrialist approach, open many opportunities in agroecological transition and community empowerment. The authors explore possible paths between agroecology and unmanned systems and present different technologies, starting from cameras worn by birds, kites, and balloons: these 'grandparents' of modern drones can disclose a lot of new opportunities. After describing the different UAV platforms (fixed wings and multi-rotors) and sensors, the authors review distinct approaches and methodologies of using UAV in agroecology. Based on their experiences in different contexts (in Ecuador and Italy), Pappalardo and Andrade share interesting case studies of UAV applications for agrobiodiversity conservation and community-based agroecosystems, from farms to landscapes, showing the role of UAV technologies in implementing the multi-scale paradigm of agroecology.

The third part of the book deals with technologies for agroecological transition at the landscape scale, integrating food sovereignty and ecosystem services.

GIS and webGIS are the topics of Chapter 7, opening the third part of the book. Luca Battistella, Federico Gianoli, Marco Minghini and Gregory Duveiller offer an outlook on the different types, trends, and constitutive characterization of web mapping: the collaborative approach in data supplying, validation, and sharing. After a comparison of two business models (proprietary versus open source), the authors describe the geospatial web components, making possible the transition of GIS technology from the desktop to the web. The evolution of web mapping and webGIS shows a large variety of services and tools with different levels of complexity and usability, increasing the inclusion of different categories of social actors, experiencing platforms without coding and handling intuitive tools, like story maps. The implementation, in many jurisdictions, of the right for environmental information, has been supported by the development of SDI (spatial data infrastructures) based on geoportals and geocatalogs, spreading the availability of open data. An example of SDI is the BIOPAMA Regional Reference Information System for Biodiversity and Protected Areas Management. The webGIS in agroecology has a strong potential at different scales of the food system. Despite the limited number of cases, there is a growing increase of applications, especially in the connections among farmers and citizens in making visible food networks, and agroecological approaches in caring for food sovereignty and rights of nature.

Antony Moore and Marion Johnson accompany us to know the experiences of agroecology in Aotearoa, New Zealand, inside the project He Ahuwhenua Taketake (indigenous agroecology). Three case studies of Maori and Moriori farms, based on collective land ownership (trusts) are presented. GIS is used to support a geodesign process, integrating local knowledge with technical-scientific contributions for indigenous agroecological management. The chapter describes the process of data collection and management (from survey to geodatabase) to produce overview maps and facilitating dialogue among indigenous people and institutions. Maori (mainland New Zealand) and Moriori (Chatham Islands) farmers perceive their agroecological practices in the holistic perspective of the '*ki uta ki tai*': from the mountains to the sea. Native plants are central for *mahinga kai* (food) and *rongoa* (traditional medicine) for humans and animals, in a network of relationships among different beings, the *Papatūanuku* (Mother Earth) and *Ranginui* (Father Sky). Participatory approaches and spatial multi-criteria analysis are some tools used in geodesign of place-based agroecology practices. GIS supported the preparation of maps in agroecological planning of Henga and Te Kaio farms, to define zoning and areas for locations of *rongoa* for people and livestock, to integrate tourism activities into agroecosystems, exploring the integration among farming systems and livelihood systems. The Aotearoa He Ahuwhenua Taketake project, involving Maori and Moriori link farms, see in agroecology a first step of a long path for the integration of indigenous rights and food sovereignty facilitated by the use of participatory mapping with GIS and geodesign.

Agroecology and smart cities are analyzed in Chapter 9 prepared by Francesca Peroni, John Choptiany and Samuel Ledermann. The authors start with a critical review of globalized universalizing narrative of a smart city, using the generative question on whether smart cities are creating a real inclusive environment for citizens. On the other side, literature and practices on smart cities do not deal with food production and the right to food in the cities, and at the same time, there is a growing research area focusing on UA (urban agriculture). So, the chapter intends to open innovative paths integrating the debate on smart cities, urban agriculture, and ICT through the lens of agroecology. There are different ways of growing food in cities; however, in a debate on smart cities, it is important to avoid any capture of technological dimension subsuming the paradigm of precision farming and Agriculture 4.0. Urban agriculture is a key challenge of agroecological transition and sustainable food systems, asking for a redefinition of the spaces of urban food production and social inclusion. IC technologies can facilitate the spread of agroecological approaches in urban agriculture. The authors present some promising applications (partially in test phases), which may facilitate dialogue, co-creation, and sharing of knowledge among people interested in growing food by adopting agroecological approaches. Urban agroecology can represent a meeting point to overcome the reductive approach of smart cities, to improve the ability of urban ecosystems in providing multiple ecosystem services, and at the same time promoting food sovereignty and inclusion in urban planning and management.

Daniele Codato, Guido Ceccherini and Hugh D. Eva deal with free and open satellite imageries for land rights and climate justice in Chapter 10. The global importance of agroforestry systems of the Amazon region is widely recognized as a casket of biodiversity and cultural diversity of indigenous nations, and its role in the provision of ecosystem services, and to increase the resilience to climate change. The chapter offers an outlook on remote sensing principles and operations, presenting the different typologies of satellite sensors and platforms. The outlook integrates information on the availability of free satellite imageries on the web and on the tools and platforms available to access and process satellite imageries. Some remote sensing techniques are presented with a summary of band combinations and indexes useful for forestry and agriculture. The authors prepared a sort of 'travel guide' to easily navigate the new commons of free geographical information coming from satellite imageries available with a weekly (Sentinel-2) or fortnightly (Landsat 8) update. Despite its global importance, the Amazon territories are under pressure, driven by land-use changes that destroy (agro) forestry ecosystems and violate indigenous land rights. Neo-colonial policies, based on the extraction of commodities (fossil fuel, mineral resources, wood, agricultural products), are devastating this cultural forest, which for millennia was managed by indigenous people who elaborated the agroecological and polycultural systems combined with nomadism, hunting, fishing, and gathering. The final part of the chapter focuses on the use of remote sensing data in analyzing the hardly accessible area of Amazon rain-forest to implement human rights, environmental and climate justice of indigenous people and peasants.

Chapter 11 deals with the role of agrobiodiversity in connecting farms and landscapes. Ingrid Quintero, Yesica Xiomara Daza-Cruz and Tomás Enrique León-Sicard present the MAS (Main Agroecological Structure). The index, developed by León-Sicard, integrates agroecology and landscape ecology, exploring bioecological and socio-cultural dimensions. MAS is based on 10 criteria and 27 indicators measuring and mapping the internal farm agrobiodiversity, the connections with landscapes, and the agroecological practices implemented by farmers. MAS is an evaluation tool, useful to compare farms using different approaches (conventional and agroecological) or to design the agroecological transition, monitoring the change of MAS in a defined period. The methodology to evaluate MAS combines different types of spatial and non-spatial information and tools: satellite or aerial images, interview with farmers, fieldwork for floristic analysis, participatory mapping, and GIS, field survey, and use of drones to collect qualitative and quantitative variables. The use of participatory and desktop GIS provides the calculation of some indicators of MAS. In this procedure, the cartography is useful in the visualization of the internal condition of the farm and the connections with the surrounding ecosystems and agroecosystems. The MAS evaluation facilitates the dialogue among the different dimensions of agroecology, especially between academic research and farm practices in implementing agroecological transition and scaling up of agroecology, starting from a landscape and farm network approach.

The last chapter of the book collects the debate (co-ordinated by Massimo De Marchi) of the conference held at the University of Padova, on 22 September, 2020, in the context of the annual kick-off seminar of the International Joint

Master Degree on Sustainable Territorial Development, Climate Change Diversity Cooperation (STeDe-CCD). In the challenge of finding territorial alternatives to development in the context of climate change, agroecological transition, and food sovereignty represent the key elements to navigate the uncertainty of the pandemic era. Miguel Angel Altieri highlights the role of agroecology either in the rural and urban context, to overcome the social and environmental impacts of conventional farming through the integrated and multi-scale approach among social and natural systems based on rights of farms and citizens connected in sustainable and sovereign food networks. Salvatore Eugenio Pappalardo and Alberto Diantini intervened as discussants to focus on the role of the technological appropriation of the new commons of geographical information and technology in an emancipatory process which is ongoing in many parts of the world, from the Amazon rain-forest supporting the struggle of indigenous groups for safe territories to urban peripheries and conventional farming areas of the global north.

The technologies presented in this book should be handled in the framework of the Nyeleni Declaration of the International Forum of Agroecology (2015), to support livelihood systems in agroecology and the empowerment of the most affected actors in the world food systems: women and youth, herders and pastoralists, fisher-folk, peasant and small-scale farmers, indigenous people, workers, landless, urban communities, and conscious consumers.

This book starts a dialogue between agroecology multi-scale approach from farm to landscape level, and the potential of geographical information and technologies in promoting alliances between farmers and citizens connecting food webs, both in proximity to urban farming and in the quest for land rights in remote areas in the spirit of 2030 SDG.

Dialogue should continue, focusing on the four entry points for agroecological transition (Wezel *et al.*, 2020): responsible governance involving multi-level and multi-actor commitments facilitated by the combination of agroecology and geography experience; circular and solidarity economy being inclusive, technologies and innovation deconstructing the linear accumulation by dispossession; diversity, with all combinations among cultural rights and rights of nature, including the connections among humans and non-humans; the co-creation and sharing of knowledge: the everyday life of farmers and citizens is creative and challenges the unique flow of the disempowering innovation.

Bibliography

Almeida, D.A.O. and A.R. de Biazo (2017). Urban agroecology: For the city, in the city and from the city, *Urban Agriculture*, 33: 13-14.

Alonso-Fradejas, A., L.F. Forero, D. Ortega-Espès, M. Drago and K. Chandrasekaran (2020). 'Junk Agroecology': The corporate capture of agroecology for a partial ecological transition without social justice, ATI, TNI, Crocevia; Retrieved from:

https://www.tni.org/files/publication-downloads/38_foei_junk_agroecology_full_report_eng_lr_0.pdf; accessed on 20 April, 2021.

Altieri, M.A. (2002). Agroecology: The science of natural resource management for poor farmers in marginal environments, *Agriculture, Ecosystems and Environment*, 93: 1-24.

Altieri, M.A. (1989). Agroecology: A new research and development paradigm for world agriculture, *Agriculture, Ecosystems and Environment*, 27: 37-46.

Altieri, M.A., F.R. Funes-Monzote and P. Petersen (2012). Agroecologically efficient agricultural systems for smallholder farmers: Contributions to food sovereignty, *Agron. Sustain. Dev.*, 32: 1-13.

Altieri, M.A., C.I. Nicholls and R. Montalba (2017). Technological Approaches to Sustainable Agriculture at a Crossroads: An agroecological perspective, *Sustainability*, 9: 349.

Altieri, M.A. and C.I. Nicholls (2019). Urban agroecology, *AgroSur.*, 46: 46-60.

Anderson, C.A., C. Maughan and M.P. Pimbert (2019). Transformative agroecology learning in Europe: Building consciousness, skills and collective capacity for food sovereignty, *Agriculture and Human Values*, 36: 531-547.

Anderson, C.R. and C. Maughan (2021). 'The Innovation Imperative': The struggle over agroecology in the international food policy arena, *Frontier Sustainable Food System*, 5: 619185.

Bellon Maurel, V. and C. Huyghe (2017). Putting agricultural equipment and digital technologies at the cutting edge of agroecology, *OCL*, **24**(3): 1-7.

Bellon, S. and G. Ollivier (2018). Institutionalizing agroecology in France: Social circulation changes the meaning of an idea, *Sustainability*, **10**(5):1380.

Brugger, S.O., E. Gobet, J.F.N. van Leeuwen, M. Ledru, D. Colombaroli, W.O. van der Knaap, U. Lombardo, K. Escobar-Torrez, W. Finsinger, L. Rodrigues, A. Giesche, M. Zarate, H. Veit and W. Tinner (2016). Long-term man-environment interactions in the Bolivian Amazon: 8000 years of vegetation dynamics, *Quat. Sci. Rev.*, 132: 114-128.

Campbell, B.M., D.J. Beare, E.M. Bennett, J.M. Hall-Spencer, J.S.I. Ingram, F. Jaramillo, R. Ortiz, N. Ramankutty, J.A. Sayer and D. Shindell (2017). Agriculture production as a major driver of the Earth system exceeding planetary boundaries, *Ecology and Society*, **22**(4).

Conway, G.R. (1987). The properties of agroecosystems, *Agricultural Systems*, 24: 95-117.

Côte, F-X., E. Poirier-Magona, S.B. Perret, P. Roudier, B. Rapidel and M.C. Thirion (Eds.). (2019). *Transition agro-écologique des agricultures du Sud*, Versailles, *Éditions Quae, l'AFD et le Cirad*, Paris, Fr.

Dalgaard, T., N.J. Hutchings and J.R. Porter (2013). Agroecology, scaling and interdisciplinarity, *Agriculture, Ecosystems and Environment*, 100: 39-51.

Daly, A., K. Devitt and M. Mann (Eds.). (2019). *Good Data (Theory on Demand #29)*, Institute of Network Cultures, Amsterdam, Nl.

Declaration of the International Forum for Agroecology, Nyéléni, Mali (27 February 2015), *Development*, 58: 163-168.

de Oliveira Côrtesa, L.E., C. Antunes Zappes and A.P. Madeira Di Beneditto (2019). Sustainability of mangrove crab (*Ucides cordatus*) gathering in the southeast Brazil: A MESMIS-based assessment, *Ocean and Coastal Management*, 179: 104862.

De Schutter, O. (2011). Agroecology and the right to food, *Report of the Special Rapporteur on the right to food. United Nations*; Retrieved from: http://www.srfood.org/images/stories/pdf/officialreports/20110308_a-hrc-16-49_agroecology_en.pdf; accessed on 21 April, 2021.

Díaz León, M.A. and A. Cruz León (Eds.). (1998). *Nueve mil años de agricultura en México: Homenaje a Efraím Hernández Xolocotzi, Grupo de Estudios Ambientales, Universidad Autonoma de Chapingo*, MX.

European Parliament (2016). Precision Agriculture and the Future of Farming in Europe, *Technical Horizon Scan, EPRS, European Parliamentary Research Service Scientific Foresight Unit (STOA)*; Retrieved from: https://www.europarl.europa.eu/RegData/etudes/STUD/2016/581892/EPRS_STU(2016)581892_EN.pdf; accessed on 21 April, 2021.

European Parliament (2017). Precision agriculture in Europe, *Legal, Social and Ethical Considerations – Technical Horizon Scan, EPRS, European Parliamentary Research Service Scientific Foresight Unit (STOA)*; Retrieved from: https://www.europarl.europa.eu/thinktank/en/document.html?reference=EPRS_STU(2017)603207; accessed on 21 April, 2021.

Falkenberg, K. (2015). Sustainability Now! *A European Vision for Sustainability, EPSC Strategic Notes, Issue 18*; Retrieved from: https://sdgtoolkit.org/tool/sustainability-now-a-european-vision-for-sustainability/; accessed on 21 April, 2021.

FAO (Food and Agriculture Organization). (2015). *Final Report for the International Symposium on Agroecology for Food Security and Nutrition*, 18-19 September 2014, Rome; Retrieved from: http://www.fao.org/3/a-i4327e.pdf; accessed on 21 April, 2021.

FAO (Food and Agriculture Organization). (2016a). *Outcomes of the International Symposium and Regional Meetings on Agroecology for Food Security and Nutrition*, COAG 25th Session, 26-30 September, 2016, COAG 2016/INF/4, Rome; Retrieved from: http://www.fao.org/3/amr319e.pdf; accessed on 21 April, 2021.

FAO (Food and Agriculture Organization). (2016b) *Report of the Regional Meeting on Agroecology in Sub-Saharan Africa, Dakar, Senegal*, 5-6 November, 2015, Rome; Retrieved from: http://www.fao.org/3/i6364e/i6364e.pdf; accessed on 21 April, 2021.

FAO (Food and Agriculture Organization). (2018a). *FAO's Work on Agroecology. A Pathway to Achieving the SDGs*, Rome; Retrieved from http://www.fao.org/3/i9021en.pdf; accessed on 21 April, 2021.

FAO (Food and Agriculture Organization). (2018b). *The 10 Elements of Agroecology: Guiding the Transition to Sustainable Food and Agricultural Systems*, Rome; Retrieved from: http://www.fao.org/3/i9037en/i9037en.pdf; accessed on 21 April, 2021.

FAO (Food and Agriculture Organization). (2018c). *International Symposium on Agricultural Innovation for Family Farmers: Unlocking the Potential of Agricultural Innovation to Achieve the Sustainable Development Goals*, 21–23 November 2018, Rome; Retrieved from: http://www.fao.org/about/meetings/agricultural-innovation-family-farmers-symposium/en/; accessed on 21 April, 2021.

FAO (Food and Agriculture Organization). (2018d). *Participatory Guarantee Systems (PGS) for Sustainable Local Food Systems*, Rome; Retrieved from: http://www.fao.org/3/I8288EN/i8288en.pdf; accessed on 21 April, 2021.

Figueroa-Helland, L., C. Thomas and A. Perez Aguilera (2018). Decolonizing food systems: Food sovereignty, indigenous revitalization and agroecology as counter-hegemonic movements, *Perspectives on Global Development and Technology*, **17**(1-2): 173-201.

Gadgil, M. (1995). Prudence and profligacy: A human ecological perspective. *In:* T.M. Swanson (Ed.). *The Economics and Ecology of Biodiversity Decline*, Cambridge University Press, pp. 99-110.

Gebbers, R. and V.I. Adamchuk (2010). Precision agriculture and food security, *Science*, 327: 828-831.

Francis, C., G. Lieblein, S. Gliessman, T.A. Breland, N. Creamer, R. Harwood, L. Salomonsson, J. Helenius, D. Rickerl, R. Salvador, M. Wiedenhoeft, S. Simmons, P. Allen, M. Altieri, C. Flora and R. Poincelot (2003). Agroecology: The Ecology of Food Systems, *Journal of Sustainable Agriculture*, **22**(3): 99-118.

Gkisakis, V. and D. Damianakis (2020). Digital innovations for the agroecological transition: A user innovation and commons-based approach, *J. Sustainable Organic Agric. Syst.*, **70**(2): 1-4.

Gkisakis, V., M. Lazzaro, L. Ortolani and N. Sinoir (2017). Digital revolution in agriculture: Fitting in the agroecological approach? Retrieved from: https://www.agroecology.gr/ictagroecologyEN.html; accessed on 21 April, 2021.

Gliessman, S.R. (2014). Introduction. Agroecology: A global movement for food security and sovereignty. *In:* FAO (Ed.). *Proceedings of the FAO International Symposium*, 19-19 September 2014, Agroecology for Food Security and Nutrition, Rome, It.

Gliessman, S.R. (2016). Transforming food systems with agroecology, *Agroecology and Sustainable Food Systems*, **40**(3): 187-189.

Gliessman, S.R. (2007). *Agroecology: The Ecology of Sustainable Food Systems*, second edition, CRC Press, Boca Raton, USA.

Grey, S. and R. Patel (2014). Food sovereignty as decolonization: Some contributions from indigenous movements to food system and development politics, *Agric. Hum. Values*, **32**(3): 431-444.

Guthman, J. (2004). Agrarian Dreams: The Paradox of Organic Farming in California, *Geographical Review of Japan, Series A*, University of California Press, Oakland, USA.

Hernandez Xolocotzi, E. (Ed.). (1977). *Agroecosistemas de Mexico: Contribuciones a la Ensenanza, Investigación, y Divulgación Agricola, Colegio de Postgraduados*, Chapingo, Mx.

HLPE (High Level Panel of Experts on Food Security and Nutrition). (2014). Food losses and waste in the context of sustainable food systems, *A Report by the High Level Panel of Experts on Food Security and Nutrition of the Committee on World Food Security*, Rome; Retrieved from: http://www.fao.org/3/i3901e/i3901e.pdf; accessed on 21 April, 2021.

HLPE (High Level Panel of Experts on Food Security and Nutrition). (2017a). Second note on critical and emerging issues for food security and nutrition, *A Note by the High Level Panel of Experts on Food Security and Nutrition of the Committee on World Food Security*, Rome; Retrieved from: http://www.fao.org/cfs/cfs-hlpe/critical-and-emerging-issues/en/; accessed on 21 April, 2021.

HLPE (High Level Panel of Experts on Food Security and Nutrition). (2017b). Nutrition and food systems, *A Report by the High Level Panel of Experts on Food Security and Nutrition of the Committee on World Food Security*, Rome; retrieved from: http://www.fao.org/3/a-i7846e.pdf; accessed on 21 April, 2021.

HLPE (High Level Panel of Experts on Food Security and Nutrition). (2019). Agroecological and other innovative approaches for sustainable agriculture and food systems that enhance food security and nutrition, *A Report by the High Level Panel of Experts on Food Security and Nutrition of the Committee on World Food Security*, Rome; Retrieved from: http://www.fao.org/3/ca5602en/ca5602en.pdf; accessed on 21 April, 2021.

Home, R., H. Bouagnimbeck, R. Ugas, M. Arbenz and M. Stolze (2017). Participatory guarantee systems: Organic certification to empower farmers and strengthen communities, *Agroecology and Sustainable Food Systems*, **41**(5): 526-545.

IPBES (Intergovernmental Science-Policy Platform on Biodiversity and Ecosystem Services). (2018a). *Summary for Policy-makers of the Assessment Report on Land Degradation and Restoration of the Intergovernmental Science Policy Platform on Biodiversity and Ecosystem Services*, IPBES secretariat, Bonn, De; Retrieved from: https://www.ipbes.net/system/tdf/spm_3bi_ldr_digital.pdf?file=1&type=node&id=28335; accessed on 21 April, 2021.

IPES-Food (International Panel of Experts on Sustainable Food Systems). (2016). From uniformity to diversity: A paradigm shift from industrial agriculture to diversified agroecological systems; Retrieved from: http://www.ipes-food.org/images/Reports/UniformityToDiversity_FullReport.pdf; accessed on 21 April, 2021.

IPES-Food (International Panel of Experts on Sustainable Food Systems). (2017b). Unravelling the food-health nexus: Addressing practices, political economy, and power relations to build healthier food systems; Retrieved from: http://www.ipes-food.org/_img/upload/files/Health_ExecSummary(1).pdf; accessed on 21 April, 2021.

IPES-Food (International Panel of Experts on Sustainable Food Systems). (2018). Breaking away from industrial food and farming systems: Seven case studies of agroecological transition; Retrieved from: http://www.ipes-food.org/_img/upload/files/CS2_web.pdf; accessed on 21 April, 2021.

Jansen, K. (2015). The debate on food sovereignty theory: Agrarian capitalism, dispossession and agroecology, *The Journal of Peasant Studies*, **42**(1): 213-232.

Klerkxa, L. and D. Rose (2020). Dealing with the game-changing technologies of Agriculture 4.0: How do we manage diversity and responsibility in food system transition pathways? *Global Food Security*, 24: 100347.

López-Ridaura, S., O. Masera and M. Astier (2002). Evaluating the sustainability of complex socio-environmental systems: The MESMIS framework, *Ecological Indicators*, 2: 135-148.

Maezumi, Y.S., D. Alves, M. Robinson, J.G. de Souza, C. Levis, R.L. Barnett, E.A. de Oliveira, D. Urrego, D. Schaan and J. Iriarte (2018). The legacy of 4,500-years of polyculture agroforestry in the eastern Amazon, *Nature Plants*, (4): 540-547.

Malézieux, E. (2012). Designing cropping systems from nature, *Agron. Sustain. Dev.*, 32: 15-29.

Mendez, V.E., C.M. Bacon and R. Cohen (2013). Agroecology as a transdisciplinary, participatory, and action-oriented approach, *Agroecology and Sustainable Food Systems*, **37**(1): 3-18.

Migliorini, P. and A. Wezel (2017). Converging and diverging principles and practices of organic agriculture regulations and agroecology: A review, *Agron. Sustain. Dev.*, **37**(63): 1-18.

Montefrio, M.J.F. and A.T. Johnson (2019). Politics in participatory guarantee systems for organic food production, *Journal of Rural Studies*, 65: 1-11.

Montgomery, D.R. (2007). Soil erosion and agricultural sustainability, *PNAS*, **104**(33): 13268-13272.

Neves, E.G, and M.J. Heckenberger (2019). The call of the wild: Rethinking food production in ancient Amazonia, *Annual Review of Anthropology*, 48: 371-388.

Niggli, U., H. Willer and B.P. Baker (2016). *A Global Vision and Strategy for Organic Farming Research*, TIPI Technology Innovation Platform of IFOAM, Organics International, c/o Research Institute of Organic Agriculture (FiBL), Frick, CH; Retrieved from: https://orgprints.org/id/eprint/31340/1/niggli-etal-2017-TIPI-GlobalVision-Strategy-CondensedVersion.pdf; accessed on 21 April, 2021.

Paracchini, M.L., E. Justes, A. Wezel, P.C. Zingari, R. Kahane, S. Madsen, E. Scopel, A. Héraut, P. Bhérer-Breton, R. Buckley, E. Colbert, D. Kapalla, M. Sorge, G. Adu Asieduwaa, R. Bezner Kerr, O. Maes and T. Negre (2020). *Agroecological Practices Supporting Food Production and Reducing Food Insecurity in Developing Countries.* A Study on Scientific Literature in 17 Countries, Publications Office of the European Union, Luxembourg; Retrieved from: https://op.europa.eu/it/publication-detail/-/publication/cc7852e1-f987-11ea-b44f-01aa75ed71a1/language-lv; accessed on 21 April, 2021.

Patel, R. (2009). Food sovereignty, *The Journal of Peasant Studies*, **36**(3): 663-706.

Perfecto, I., J.H. Vandermeer and A.L. Wright (2009). Nature's matrix: Linking agriculture, conservation and food sovereignty, *Earthscan*, London, UK.

Pimbert, M.P. (Ed.). (2018). Food sovereignty, agroecology and biocultural diversity, *Constructing and Contesting Knowledge*, Routledge, Abingdon, UK.

Raghavan, B., B. Nardi, S.T. Lovell, J. Norton, B. Tomlinson and D.J. Patterson (2016). Computational agroecology. *In:* J. Kaye and A. Druin (Eds.). *Proceedings of the 2016 CHI Conference, Extended Abstracts on Human Factors in Computing Systems*, ACM Press, San Jose, USA.

Rahman, G., M. Reza Ardakani, P. Bàrberi, H. Boehm, S. Canali, M. Chander, W. David, L. Dengel, J.W. Erisman, A.C. Galvis-Martinez, U. Hamm, J. Kahl, U. Köpke, S. Kühne, D.B. Lee, A.K. Løes, J.H. Moos, D. Neuhof, J.T. Nuutila, V. Olowe, R. Oppermann, E. Rembiałkowska, J. Riddle, I.A. Rasmussen, J. Shade, S.M. Sohn, M. Tadesse, S. Tashi, A. Thatcher, N. Uddin, P. von Fragstein und Niemsdorff, A. Wibe, M. Wivstad, W. Wenliang and R. Zanoli (2017). Organic agriculture 3.0 is innovation with research, *Organic Agriculture*, **7**(3): 169-197.

Rentig, H. (2017). Exploring urban agroecology as a framework for transitions to sustainable and equitable regional food systems, *Urban Agriculture*, 33: 11-12.

Ricciardi, V., N. Ramankutty, Z.L. MehrabiJarvis and B. Chookolingo (2018). How much of the world's food do smallholders produce? *Global Food Security*, 17: 64-72.

Ripoll-Bosch, R., B. Díez-Unquera, R. Ruiz, D. Villalba, E. Molina, M. Joy, A. Olaizola and A. Bernués (2012). An integrated sustainability assessment of Mediterranean sheep farms with different degrees of intensification, *Agricultural Systems*, 105: 46-56.

Sánchez-Bayoa, F. and K.A.G. Wyckhuys (2019). Worldwide decline of the entomofauna: A review of its drivers, *Biological Conservation*, 232: 8-27.

Tomich, T.P., S. Brodt, H. Ferris, R. Galt, W.R. Horwarth, E. Kebreab and L. Yang (2011). Agroecology: A review from a global-change perspective, *The Annual Review of Environment and Resources*, 36: 193-222.

Valdez-Vazqueza, I., C. el Rosario Sánchez Gastelum and A. Escalante (2017). Proposal for a sustainability evaluation framework for bioenergy production systems using the MESMIS methodology, *Renewable and Sustainable Energy Reviews*, 68: 360-369.

Wezel, A., S. Bellon, T. Dore, C. Francis, D. Vallod and C. David (2009). Agroecology as a science, a movement and a practice: A review, *Agron. Sustain. Dev.*, **29**(4): 503-515.

Wezel, A., M. Casagrande, F. Celette, J. Vian, A. Ferrer and J. Peigné (2014). Agroecological practices for sustainable agriculture: A review, *Agron. Sustain. Dev.*, 34: 1-20.

Wezel, A., B. Gemmill Herren, R. Bezner Kerr, E. Barrios, A.L. Rodrigues Gonçalves and F. Sinclair (2020). Agroecological principles and elements and their implications for transitioning to sustainable food systems: A review, *Agronomy for Sustainable Development*, **40**(6): 1-13.

Wezel, A., J. Goette, E. Lagneaux, G. Passuello, E. Reisman, C. Rodier and G. Turpin (2018). Agroecology in Europe: Research, education, collective action networks and alternative food systems, *Sustainability*, 10: 1214.

Wolf, S.A. and H.F. Buttel (1996). The political economy of precision farming, *Amer. J. Agr. Econ.*, 78: 1269-1274.

Zhang, N., M. Wang and N. Wang (2002). Precision agriculture – A worldwide overview, *Computers and Electronics in Agriculture*, 36: 113-132.

Part I

Technologies and Geographic Information: Combining Sovereignties in Agroecology

CHAPTER

2

Participatory Geographic Information Science: Disclosing the Power of Geographical Tools and Knowledge in Agroecological Transition

Massimo De Marchi[1]* and Alberto Diantini[2]

[1] Director of the Advanced Master on 'GIScience and Unmanned System for the Integrated Management of the Territory and the Natural Resources - with Majors', Responsible of International Master Degree on Sustainable Territorial Development, Climate Change Diversity Cooperation (STeDe-CCD), Department of Civil Environmental Architectural Engineering, University of Padova

[2] Research Programme Climate Change, Territory, Diversity – Department of Civil Environmental Architectural Engineering – Postdoc Researcher at the Department of Historical and Geographic Sciences and the Ancient World, University of Padova

2.1. Introduction

Can we use the map to change the world? And how the act of mapping can promote awareness and empowerment? This chapter explores the reflections within geography and cartography sciences with a consolidated epistemological and empirical habit about the key role of maps in changing the world, starting from the pre-digital era. With the consolidation of Geographic Information Systems and the emergence of GIScience in the 1990, participatory GIS and critical GIS reinterpreted the 'mapping for change' in the light of inclusive liberation technologies in the empowerment of the weak and marginalized authors in cities and rural contexts. The chapter offers a theoretical compass to orient among the different practices: from 'material' cartography to 'immaterial' participatory GIS, Volunteered Geography, critical geodesign and neogeography. Geographical technologies are a sort of two-faced Janus as they not only unfold a world of possibilities and freedom, but also are part of a world of injustice. Despite the

*Corresponding author: massimo.de-marchi@unipd.it

low interactions, mostly informal, in the last decades among the science of geographical information and agroecology, there are many areas of common interests and mutual interaction and co-operation for technological sovereignty.

2.2. Agroecosystem: Place, Territory, Scale

Place matters (De Blji, 2009); the placed-based approach of agroecology is identifiable in the concept of agroecosystems with the key formalization of Conway (1987). Agroecosystems are ecological systems modified by societies to produce food, fibers and other agricultural products. The structural and dynamic complexity of agroecosystems arises mainly from the complexity of the interactions among socio-economic and ecological processes. Agroecosystems are the form of territories in many contexts where societies co-evolve with ecosystems, basing social reproduction on farming, forestry, and animal husbandry or fishing. Territories (agroecosystems) are bi-modular systems (society-ecosystem) in co-evolution (Nir, 1990; Vallega, 1995). Every system represents the environment of the other and the relations between the two systems are not pure instructions, but interactions (Maturana and Varela, 1987): each of the two systems falls within the fields of possibilities of the other. Systems lie in the quantitative dimension of the parts (and the relationships between them), in the quality of the same, but also in the eye and mind of the observer. Therefore, complexity is not necessarily a property of reality, but can be a characteristic of description: different generations of system thinking generated different views on the agroecosystems (Vallega, 1995; Checkland, 1984). The agroecosystem can be analyzed by focusing on four components: space, time, flows, and decisions (Conway, 1987). The unicity of place and the specificity of time make the difference in observations and actions, either in the seasonal changes or in the short or long time changes. Place and time influence and are influenced by relations (flows of materials, energy, and the immateriality of decisions), creating the complexity of agroecosystem boundaries, more influenced by socio-economic relations than by the physical limits of ecosystems (Conway, 1987). If the physical limits of a rice pond can be easily determined in terms of spatial occupation or water flows, the social and economic relations are more undefined: Where is rice sold? Where are inputs acquired? How is extra agriculture working time invested? Agroecosystems are complex territorial systems, livelihood systems, combining farming and other types of activities, with flexible boundaries and many scales combined by a multiplicity of interacting levels. The co-evolution of agroecosystems is based on some properties (Conway, 1987; Lopez-Ridaura *et al.*, 2002), like productivity, stability, sustainability, equity, and self-reliance. Productivity in agroecology goes beyond the yield – it is the output related to the applied inputs (working time, energy, products). Outputs can be work opportunities, cash, food security, aesthetic values, and a complex combination of personal, collective, social, psychological, economic, and spiritual well-being. For Conway (1987, p. 101-103), stability is the ability of the agroecosystem to grant productivity despite the short time disturbances of the socio-ecological context and then sustainability deals with the ability of the systems to maintain long-time productivity, adapting to important changes. Considering social and ecological interaction in the agroecosystem, the other two properties make the difference in agroecological approach to agroecosystems. Equity is about the distribution of costs and benefits of systems among the different actors; there are no externalities as in conventional farming. Equity is about the distribution of products and ecosystem services not just at field or farm level, but at village, landscape, nation, and world scales. Self-reliance or self-empowerment (Lopez-Ridaura *et al.*, 2002) deals with the ability to govern changes, maintaining identities and values of the system and finding appropriate local alternatives to control and answer to external and global pressures.

An agroecosystem is the point of interactions of different scales (Dalgaard, 2003): on one side the scale of natural systems: cell, organism, population, community, ecosystem, and landscape; on the other, the combination of scales of the farming systems (soil, field, farm, region, nation, and world) and, at the same time, the biological scale of plants or animals managed in the system (cell, body, species, population of animal or plants, etc.). The management of agroecosystems asks for the complex management of nested hierarchies of scales in specific places. This unicity of a place (Francis *et al.*, 2003) and the nested multi-scale contextual approach of agroecology, based on sophisticated local knowledge, have to face the scale gap of standardized technical solutions driven by agricultural policies based on other scales and system approaches (Sinclair, 2019). Scaling up of agroecology needs a reversal approach: breaking the ceiling of the universalizing policies and allowing the local to emerge and consolidate. This is not just an approach to study agroecosystems but to evaluate and design agroecological transition.

2.2.1. Mapping for Change: Critical Cartography, Counter-mapping and Beyond

The challenges for sustainable food systems require the humanization of agricultural extension (Cook *et al.*, 2021) to render in a different way the local logic of relations among place, power, and people and the transformative contribution of agroecology. The collection of the journal, *PLA Participatory Learning and Actions*, offers a vibrant report of the paradigm shift in rural development to overcome the Green Revolution and implement sustainable and inclusive local-based initiatives. IIED (International Institute for Environment and Development) and IDS (Institute of Development Studies, University of Sussex) published 66 issues of the journal between 1988 and 2013. Started as *RRA Notes* in 1988, the journal from No. 22 (1995) was named *PLA Notes*, adopting the name *Participatory Learning and Action* in 2004 (number 50) till the last number 66, of 2013. These steps marked the evolutionary vitality of local practices of rural change outside the universalizing paradigm of technology transfer.

At the turn of 1980 and 1990, PRA (Participatory rural Appraisal) emerged as 'family of approaches and methods to enable local rural people to share, enhance and analyze their knowledge of life and conditions, to plan and to act' (Chambers, 1994a, p. 953). PRA originated in five streams: participatory research and community development (the reference is to the work of Paulo Freire, 1984), agroecosystem analysis, applied anthropology, field research on farming systems, and RRA (Rapid Rural Appraisal).

South-South routes facilitated the spread of PRA, creating a meeting point among local actors, NGOs, and place-based governmental organizations in the context of decentralization. PRA diffused a different approach to development, based on local expertise, participatory behaviors to support empowering processes, consolidating local actions and sustainable local institutions. PRA promotes changes through some key reversal dimensions: from extraction to empowerment (reversal of dominance), and reversal of methods from closed to open, from individual to community, from verbal to visual, and from counting to comparing (Chambers, 1994a, 1994b, 1994c).

In the reversal methodologies, one of the key elements is the transit from verbal to visual, which is very interesting for the connection with visualization in mapping. Chambers (1994b) highlights, how in participatory processes, insiders working with visual tools (maps, diagrams, sequences, etc.) can be presenters and analysts, keeping them far from suppliers of data and information to outsiders and playing the role of researchers or experts. Visual approaches avoid the probing trap of collecting true or false answers, and information is owned and shared by insiders co-creating and circulating knowledge.

The enthusiastic approach to vizuality of PLA/PRA literature and practices can be integrated with the critical cartography point-of-view. Not so much the visual material, *in se*, is the driver of empowerment: the appropriation of visual production by marginalized actors opens the door for change of power relations. In other words, it is counter-mapping to process the opportunity of empowerment. Insiders experience the appropriation of the representation of the space, the enhancement of own knowledge, and the self-reliance in taking decisions on their lives, communities, and places.

There is an enormous value in using participatory mapping and counter-mapping practices with citizens and peasants to collectively shape the existing context and propose changes in a participatory way (Dalton and Mason-Deese, 2012; Monmonnier, 2007; Peluso, 1995).

The special issue of PLA 54, *Mapping for Change: Practice, Technologies and Communication* marked an important moment of interaction among critical cartography, counter-mapping, and alternative rural participatory development approaches (Corbett *et al.*, 2006).

Before exercising an action of manipulation of the territory (physical or mental), humans need a representation of the place that can be a text or an image. The territories are rich in ready-made representations to be used – speeches produced by politicians, companies, the communication market, common sense, and also maps produced by actors who have a more sophisticated technical capacity. Speeches, maps, photos, videos and infographics are different ways of producing territorial images. It is important to understand who are the producers of territorial images and for what reasons do they produce certain types of representations. If the physical and geometric space is univocal and can be represented by a set of co-ordinates, territories existing on the same physical space can be many, because many actors have different projects on the same geometric space (Vallega, 1995). Agroecology, for example, challenges the universalizing approach of industrial conventional farming with place-based specific alternatives. All these conflicts happen somewhere in a place and place matter with different meanings, either for global agribusiness (as commanded place by globalized interests) or for local agroecological practices, as unicity.

The different territorial representations have their own combinations of forces, authorities, influences, and persistence. Among the images, cartography has a unique peculiarity to combine strength, authority, influence, and persistence. The map has extraordinary power to become a theoretical or doctrinal tool (Boulding, 1956). It can be a proposal for discussion, the search for a shared representation of territorial complexity, or the projection on to the ground of an individual project of a strong actor, with a more or less explicit power. People have a universal attitude in locating themselves and representing the territory with mental maps or drawings of personal places in sand or on cloth of a bar, but the maps hung on the walls and which we learned to look at primary schools are constructed with government functions, by the State or strong territorial actors to communicate a territorial project through a specific form of representation. Whoever produces maps knows what is the social effect and the common perception about this sophisticated product. Maps are accepted within a conception of scientism and neutral technicality. It is a graphic instrument capable of displaying a real and non-debatable representation of the territory.

'Maps have an extraordinary authority' (Boulding, 1956, p. 65), which is not found in other images; it is a greater authority 'than the sacred books of all religions' (Boulding, 1956, p. 70. Harley, 1987, p. 2) added how the authority of the map 'can also resist the errors of the map itself' (Bracket, 1987, p. 2).

The map is not the neutral mirror of the world; it is an embedded representation of the culture, social relations, and power of a specific territorial context (MacEachren, 1995; Dorling and Fairbairn, 1997). For these reasons, the map cannot be separated from the cultural environment that makes up the territory (Harley, 1987, 2001). This extraordinary power of maps can be used in different ways to know and reveal that part of the power of the strong actors of the territory which is guaranteed by the ability to produce this sophisticated type of territorial representation. This is the starting point of critical cartography and social-mapping approaches. De-constructing the communication system of the maps and understanding how it works means the ability to use maps also as a tool for citizen geographies, which is an alternative to consolidated geographies. The map is a text that uses a particular form of visual narration (Wood, 1992), combining three basic elements: projection, scale, and symbolization (Monmonier, 2005). The map usually produces the conviction that it is a photograph of the existing reality. However, photos are not selective, except through the resolution. Maps are graphic representations of territories, which by their nature are selective and symbolic, that is, generalized. The maps do not show all the available information: displaying information that is not relevant to the subject would obscure the message; the symbols replace the images of the objects (Tyner, 2010, p. 9).

Different from a picture, the cartographer preparing maps 'lie with maps' (Monmonier, 2005), visualizing and concealing elements through processes of cartographic generalization based on symbolization, simplification, omission, combination, enhancement, and displacement (Tyner, 2010). If a map behaves like a text (Wood, 1992, 2002; Wood *et al*., 2010), counter-mapping can visualize alternative narrations of the agroecosystems, selecting what element to give priority and handling different power relations to promote social justice (Krupar, 2015; Ascselrad, 2010).

Maps can act and actors can act with maps: 'maps are active: they actively construct knowledge, they exercise power, and they can be a powerful means of promoting social change' (Crampton and Krygier, 2006, p. 15). The agency of mapping (Corner, 1999, p. 213) can challenge the 'authoritarian, simplistic, erroneous, and coercive acts of mapping with reductive effects upon both individuals and environments. I focus . . . upon more optimistic revisions of mapping practices . . . situating mapping as a collective enabling enterprise – a project that both reveals and realizes hidden potential'. Mapping can become a creative practice, remaking territories going beyond tracing and 'participate in future unfoldings', challenging the imposed scheme of territorial representation and planning; mapping precedes maps (Corner, 1999).

Adopting the processual approach to mapping, going beyond the absolutism of map object, the critical cartography opens arenas of shifting power and emancipatory inclusive practices – 'maps are of the moment, brought into being through practices (embodied, social, technical), always remade every time they are engaged with; mapping is a process of constant reterritorialization' (Kitchin and Dodge, 2007, p. 335). Critical cartography challenges the practices of mapmaking, revealing the actions behind the object – from craft to performance, from securization to challenge (Kent and Vijakovic, 2018).

One interesting area of mapping is done by indigenous people defending their land rights: these counter-mapping practices offer concrete actions for change and at the same time challenge the embedded colonial vision in mapping, territorial management, participation, and knowledge sharing. Maps used to implement colonial rules can be weaponized by indigenous nations (Bryan and Wood, 2015) not just in the transformation of the map into a weapon, but appropriating the mapping process as highlighted by post-representational cartography (Rossetto, 2019).

Counter-mapping can act as militant research, creating co-operation among researchers and local actors to handle real problems (Dalton and Mason Deese, 2012); at the same time, it offers a theoretical framework to manage grassroots data science in emancipatory processes (Dalton and Stallman, 2017), challenging the ongoing data accumulation for profit or securization (McCalla and Michael, 2011).

2.3. PGIS, Critical GIScience and Voluntary Geography

Star and Estes (1990) define GIS as a 'map of higher order'; this inspiring definition traces a sort of long-lasting connection among pre-digital and digital maps and mapmaking, which is very useful to 'map' continuities and discontinuities between critical cartography and critical GIS.

GIS as geographic information system in six decades witnessed the crossing of five generations and an important paradigm shift. The first generation of GIS started at the beginning of 1960 with the implementation of the geographic information system of land use in Canada by Roger Tomlinson. This first reflection on the use of computers in the electronic processing of geographic information is called the 'generation of pioneers' (Yuan, 2015). In the decade of 1970, GIS entered the second generation driven by the State (the emblematic case is the contribution of United States Census Bureau) and in 1980, with the third generation, GIS spread, driven by software houses diffusing the new GIS packages in firms, public administrations, and universities. The turning point arrived in the decade of 1990 with the fourth generation, 'the GIS of users', facilitating on one hand the diffusion of the personal computers (Yuan, 2015) and on the other, the role of universities implementing research, education, and also the dialogue with civil society.

In October 1993, in Friday Harbor, GIS practitioners and critical human geographers convened the meeting, 'GIS and Society' on the social implications of geographic information systems. John Pickles (1995) with 'ground truth' collected the debate started in 1993 on the emergence of a critical GIS, deconstructing the narrative of neutral technology of GIS and focusing on positionality and value-laden GIS products. Liverman *et al.* (1998) with 'people and pixels' consolidated the connection between geographical information and social sciences, especially regarding the use of satellite imageries.

Two special issues of *Cartography and Geographic Information Systems'* (*GIS and Society* in 1995 and *Public Participation GIS* in 1998) continued the important research area of *GIS and Society*, both as a theoretical reflection on GIS and social implications, and as an applied science in process of territorial changes. Some key research topics of the 1990 agenda are still relevant: epistemologies, technologies and indigenous views, ethical issues, rights and responsibilities, empowerment and marginalization favored by GIS, barriers to effective inclusion, role of GIS in resistance, and advocacy (Goodchild, 2015; Yuan, 2021; Brown and Kyttä, 2018).

In 1996, the NCGIA (National Centre for Geographic Information and Analysis) organized two workshops to reflect on the role of PPGIS (Public

Participation GIS) to facilitate wider public involvement in planning and decision-making processes, considering the increasing applications and the potentiality in PPGIS in urban planning, nature conservation, and rural development (Goodchild, 2015). In this period the debate arises between PPGIS and PGIS (participatory GIS) – the first related to participatory processes using GIS by a public authority to implement top-down decisions, the latter as appropriation of GIS tools by marginalized groups to challenge the status quo.

Michael Goodchild in 1992 with the article 'Geographic Information Science' triggered the second big change of GIS during the fourth generation: the paradigm shift from system to science. The acronym GIS used for 30 years to summarize Geographic Information System was reloaded in different declinations: 'Geospatial Information Science', 'Geospatial Information Studies', 'Geospatial Information Services' consolidating the new research paradigm, and label of 'Geographic Information Science' as the science behind the system. The reflections of Goodchild started from recognizing, as in other sciences, the new tools opening paradigmatic leaps: for example, the microscope in biology or the telescope in astronomy. The availability of the geographic information system (the new tool or paradigmatic artefact) opens new fundamental questions and areas of research for the GIScience, like theories of geographical representations, continuity and discontinuity with the pre-digital cartography, and how to use GIScience in the contested and uncertain representations of the world. The use of GIS tools facilitates visual thinking in exploring the earth and the world and creates different paths on defining fundamental research questions on the tools, the way of knowing, the topic to explore, the approach to scientific research, and the social and ethical implications.

From 2000, with the diffusion of personal portable devices (smartphones), the web and the social network, GIS entered the fifth generation of *produsers*, the portable and the web generation of GIS, driven by the neogeographers. The panorama of geographical data flow, until then characterized by public or private centralized data supply, is transformed by the big amount of data supplied by people (the new geographers) doing different activities, ranging from sharing GPS tracks after trekking, to mapping impacts of pollutants into rivers, to expressing preference on shops.

Crowd-sourced geographic information is the umbrella definition of a large variety of behaviors and processes of data circulation, sharing or accumulation (See *et al.*, 2016; Capineri *et al.*, 2016). To orient in this multifaceted context, it is important to analyze how people contribute to geographic information by looking at how people are involved: from active participation in data collecting, sharing, and analyzing to the passive supply of data to private or public storages. Presented below is a summary of the principal label used to describe different approaches in crowd-sourced geographic information.

VGI (Volunteered Geographic Information) is the name of citizen science in the context of geography and cartography. For Goodchild (2007, p. 2) VGI is 'the harnessing of tools to create, assemble, and disseminate geographic data provided voluntarily by individuals'; Elwood *et al.* (2012, p. 572) define VGI as 'spatial information voluntarily made available, with the aim to provide information about the world'.

Citizen science, according to the white paper on Citizen Science for Europe (*Socientize*, 2014) is 'the involvement of citizens in scientific research activities to which they actively contribute with their intellectual commitment, through widespread knowledge or with their own tools and resources' (*Socientize*, p. 8). However, citizen science and VGI are big containers with different levels of participation. Haklay (2013a) distinguishes four levels of citizen participation and engagement in citizen science projects: level one is the crowdsourcing in which citizens are sensors supplying data and eventually volunteering computing data; at level two, there is the emergence of 'distributed intelligence' when citizens become basic interpreters and apply volunteered thinking; the 'participatory science' arrives at level three, where citizens can participate in problem definition and data collection. Level four of 'extreme citizen science' implements true collaborative science, where citizens define problems, collect, and analyze data.

On the other hand, we can find iVGI (inVoluntary Geographic Information) when 'georeferenced data are not provided voluntarily by individuals for use for many purposes including mapping, but especially for commercial applications, such as geodemographic profiling' (See *et al.*, 2016). 'Contributed geographic information' is defined in opposed to the VGI as 'geographic information collected without the awareness and explicit consent of a user of mobile devices that record the position' (See *et al.*, 2016).

2.4. Technological Sovereignty: Disclosing the Power of Transformative GIScience

As presented in the previous paragraphs, the 1990s marked a turning point (or the meeting point) for GIS and critical approaches. After decades of conventional GIS based on automated cartographic production, data storage management, quantitative computing, definition, and standardization of geoprocessing, in the 1990s, with the encountering among GIS practitioners and critical cartographers, new paths were opened. Participatory, feminist, qualitative, postcolonial, and indigenous GIS (Sui, 2015; Yuan, 2021) and many other GIS themes dealing with inclusion, empowerment, new epistemologies, critical and transformative approaches emerged in the interaction among GIScience and society (Corbett *et al.*, 2016; Schlosseberg and Shuford, 2005; Sieber, 2006).

Sui (2015) names all these emerging practices with the umbrella term of alternative GIS (alt.gis) asking a key question: Is GIS becoming a liberation technology? The interesting question brings the author, through an analysis of the way of thinking behind doing GIS and critical GIS, to the discovery that liberation technology relies upon a different mind. Sui (2015) during the period 1960-1990, sees the first stream of more technical and positivist GIS consolidated expression of the left side of the human brain: slow, sequential, literal, textual, analytical and logical. Meanwhile, the second stream of GIS (since the 1990s) is more narrative, qualitative, systematic, and oriented to empowerment and social justice. This stream would be associated with the right side of the human brain: fast, simultaneous, contextual, metaphorical, aesthetic, and affective. Adopting Pink's framework (2006) on the 'whole new mind' for the contextual age of the 21st century, Sui (2015) explores the relations among the six senses of the new mind (design, story, symphony, empathy, play, and meaning) and the emerging GIS themes. The first sense, design, could be connected with the emergence of geodesign as a participatory way of changing places by leaving the descriptive perspective (what is) to adopt the prescriptive one (what could/should be). Story, the second sense of the new mind, would be connected with the discovering of geographic lore (interesting is the affinity with the reflections on PRA/PLA, Chambers 1994a, b, c) and the roles of geonarratives, story maps, and qualitative GIS. Symphony (new mind) and synthesis (emergent theme in GIS) would be linked (Sui, 2015) in a new framework of consilience in the combination of analysis and synthesis facilitated by VGI of neogeographers. Critical GIS dealing with disenfranchised and powerless actors is still a disruptive and emerging GIS theme considering the challenges of political ecologies, environmental conflicts, climate justice, exclusion, and neo-authoritarian powers. So, critical GIS could be associated with the fourth sense of the new mind – empathy in the struggle, proximity and partnership, not only efficiency and aims. Sui (2015) associates play (fifth sense of the new mind) with the emergence of gaming as the overcoming of geoinformatics. Behind the issue of play, there is an interesting deconstruction of the way of thinking (or applying visual thinking) in GIS – from the God-eye, the vertical top-down view of the world for domination, to a visual stroll of places to enjoy the pleasure of curiosity. Kingsbury and Jones (2009) speak on Dionysian adventures on Google Earth. Meaning, the sixth sense of the new mind would be connected with the emergent GIS theme of place, the paradigm shift from space to place, and the need to deal with emotional and affective relations among people and places.

GIS, to become a liberation technology, should be deconstructed as in critical GIS to highlight the enframing nature of geospatial technology. A first enframing dimension is related to the technical issue – the need to adopt the open GIS paradigm outside the fences of proprietary software, proprietary data, patented technologies, embracing a fully open-source philosophy. Then another line of liberation is related to the theoretical dimension to adopt an alternative way of knowing beyond the Cartesian paradigm and the interactions with indigenous practices and knowledge being fundamental. The third dimension of liberation technology is to increase and diffuse the practices of GIS on human rights and environmental justice e, challenging the monopoly of technology by the military-industrial complex (Sui, 2015). Geographical technologies are a sort of two-faced Janus as they do not only unfold a world of possibilities and freedom but also are a part of a world of accumulation by dispossession, starting from data grabbing.

Klikemberg, in a vibrant article of 2007, reflects on geographies of hope and fear as open possibilities, which are not a future already done, but a future that humanity can shape (Freire, 1994). So, the creation of dangerous agglomerations of power are not the defined destiny; critical GIS and citizens handling new technologies can create points of resistance to power, decentralized global networks and multicultural co-operation to frame collective decisions (Poster, 2004).

The term 'neogeography' describes a way of producing and using geographical information, mainly online mapping through webGIS, by non-professional geographers facilitated by the availability of new technologies. It expresses a process of democratization of geographic data and the production of maps online, including new actors in a sector dominated until a few years ago by the military, companies, administrations, and research centres. It is an ongoing open process, not closed, where it is possible to experience delusions (Haklay, 2013b) and possibilities for 'another politics' (Elwood and Mitchell, 2013).

Politics of neogeography deals with two dimensions – one is the site of citizen's engagement for a change playing the dialectic of conforming the spaces of participation offered by institutions (politics from within) or in alternative transforming the context implementing politics from below or outside adopting geo-visual tactics (Elwood and Mitchell, 2013). The second deals with implementing neogeography politics to learn how to do; so neogeography is framed as a site of personal or community political formation.

This double-site political awareness starts from the deconstruction of the narrative of technological neutrality, recognizing how technology is value-laden and human-controlled, and especially how modern forms of social control are based on technology (Haklay, 2013b). Critical GIS becomes a tool of social transformation, constructing geographies of care and hope and space of critical pedagogy on politics of GIS technologies (Pavlovskaya, 2018) and recognizing technology as a result of political negotiation (Haklay, 2016).

2.5. Redesign in Agroecology: Critical Geodesign in Planning and Evaluation

The transition from efficiency/substitution-based agriculture toward socio-ecological diversity-based agroecology requires integrated management of four domains (Duru *et al.*, 2014): the farming system, the socio-ecological systems, the socio-technical system, and the actor systems. Actors involved in transition should be able to manage different categories of resources – natural resources, the farms, technological complexities of food systems, and knowledge. A participatory design methodology is required to reach a territorial biodiversity-based agriculture (another name for agroecology).

Geodesign, especially critical geodesign, can offer valid support to this inclusive process, having in mind scenarios of change, design processes, pathway definition, and the management of power relations in participatory and inclusive decision making.

Geodesign, defined as a method to change the geography by design (Goodchild, 2010; Steinitz, 2012), is based on the interactions among people living in a place, planners, experts on geographic information facilitating inclusive iterative processes of creating scenarios, making simulations (what if), sharing feedbacks in real time to reach holistic planning and intelligent decisions (Foster, 2016). Geodesign, considered either as a verb and a substantive (Steinitz, 2012, pp. 19-21), is a contextual approach where geography matters: people and place are linked to a specific territorial system; scale matters: it is important to define the scale of transition, from the farm to global food system; the size matters: change on ecological networks can be smaller inside a farm or larger involving bioregion. For assessment of the place and the intervention, geodesign defines a framework to manage data, to integrate the dialogue of knowledge, and to share common values.

Geodesign, as many other participatory GIS approaches, lives the ambiguity of being captured as depoliticizing tool operating inside the structures and generating inequalities (Radil and Anderson, 2018). The challenge of a critical geodesign (Wilson, 2015) is to be engaged in real transformative actions by fusing 'progressive geographic imaginations with concrete and tangible maps' (Pavlovskaya, 2018, p. 40).

The context is not easy; on one hand, we should live in trouble with Anthropocene dealing with the three main treats: climate change, biodiversity loss, and food security; on the other, we face a post-political world (Radil and Anderson, 2018) with a shift to weak democracy or authoritarian populism.

Beyond the reflections on the technology of geographical information, the other key issue on agroecology transition is the management of geographical data. Louikissas (2019) highlights how 'all data are local' and on the need to move from 'data sets to data setting' because data are not neutral things, but result from social located work, giving meaning to data and operating a selection on relevance. Data are stored not only somewhere in the cloud, but are embeded at a higher level of human work to clean, process, standardize, and check the quality to transform local data grabbing (voluntary or involuntary) in the central commodity of datafication economy. So critical thinking is needed to deconstruct the data-driven society and to move to the co-creation of knowledge. Louikissas goes beyond deconstruction, tracing six principles for wide technological sovereignty. The first principle, declared also in the title of the book, is that 'all data are local as produced within human interpretative acts' (Louikissas, 2019, p. 17) in specific places and into a specific local knowledge system. The dialectic between local to global is central to understanding the commodification of data driven by networks granting the flow, aggregation, concentration, and circulation. 'Data have complex attachments to place, which invisibly structure their form', is the second principle. Attachment and invisibility can be directly managed by local actors, giving meaning to data, while the operations of 'detaching' and 'making visible' have to be investigated to understand who is gaining and losing. The third principle of Loukissas (2015, p. 30) is 'data are collected from heterogeneous sources" and heterogeneity transformed into homogeneity needs human work, but also is influenced by the vision and cultural contexts. The fourth principle is 'data and algorithms are entangled'. Algorithms to process data are the results of choice of the data analyst; algorithms allow data to reveal or conceal something, but at the same time data and algorithms conceal human work and human decisions. Data do not 'speaks by themselves' but 'platform recontextualize data' (fifth principle). Data visualization is an important process of giving meaning to data. Geovisualization can be either a process of visual thinking in the private realm of experts and scientists or a public performance of visual communication (Di Biase, 1990), sharing knowledge in a debate driven by experts synthetizing and presenting 'results' or the appropriation by local actors challenging common interpretations. Geographic information and technology deal with three variables: the continuity (or discontinuity) among private and public interaction among actors and data, the level of interaction between people and maps (or digital platforms), and the polarities of presenting the well-known world or discovering and unveil the unknowns (MacEachren, 1995). Finally, in the last principle, which is 'data are indexes of local knowledge' (Loukissas, 2015), there is a sort of circle closure. Data interpretation is again locally, culturally, and historically determined; data can speak, but some cultures are not able to listen. Can culture of yield and conventional farming read the knowledge of territorial biodiversity agroecology? And in the case of reading, what is the result?

The ecosystem of geographic information, from cartography to the new tools and data, the combination of geoinformation and geomedia, desktop GIS, GNSS, Digital Darth, Virtual Geographic Environment and Infrastructures Information Systems, webGIS and geographical CMS, portable GIS on smartphone, drones, wearable, Internet of Things, big data (especially Big Earth data), can be observed in the framework of critical GIS, critical geodesign, to avoid superficial enthusiastic positivism for a transformative technological sovereignty.

About technology of geographic information, we are applying different actions to unveil official soporific speeches, thus opening conversations for possibilities.

Whether it is used as pre-digital tool (paper maps), or as new technology (drones, geographic information systems), there is a critical use of cartography, which is an empowerment of technology, an appropriation of codes for description and transformation becoming practices of citizenship, daily production of new territories of food, and technological sovereignty (Willow, 2013) into a horizon of change (Santos, 2000).

The agroecological transition needs the geovisualization of the present and the future through an empowerment of critical cartography tools. There is data and information available, accessible technologies, engaged farmers, prepared citizens, committed researcher, but we need more awareness to get out of the consumption from the screens to embrace the production of spatial knowledge to act transformative changes.

The challenge of scaling up of agroecology requires (Lopez-Garcia *et al.*, 2021) a different approach in planning: inclusive, participative, flexible, multi-scale, and based on nature matrix beyond the paradigms of land sharing/land sparing.

Vision and inclusive design processes can be positively supported by data, technology, critical geodesign; however, the context to implement the new land planning systems is not only technological driven. In a seminal work of 2001, Jankowsky and Nyerges reflected on criteria to plan and evaluate inclusive participatory GIS processes. They developed a framework called EAST 2 (Enhanced Adaptive Structuration Theory) having as starting point the Antony Giddens' theory of structuration (1984), according to which individuals and society act in dynamics of mutual constitution, detectable in the analysis of structures, continuously produced and reproduced through situated practices (Jones and Karsten, 2008). Among the elements of structuration, Jankowsky and Nyerges added the role of technology (De Sanctis and Poole, 1994) in structuring social processes in mutual interaction. So, the framework of Jankowsky and Nyerges is based on a network of eight constructs, grouped into the three areas well known in the processes of participatory planning: convocation, process, and results (Sclavi and Susskind, 2011). The three constructs of the convocation of EAST 2 are: the social-institutional influences; the influence of each participant; and the influence of participatory GIS. So, the technological dimension in starting a participatory process using technologies of geographic information (PGIS, geodesing) cannot be separated from the context created by institutions and the role of actors engaged in change. From the beginning, we can decide if we really want the transformation or if we are opting for conformation to existing unequal structures. The authors highlight how 'neither technological nor social constructs predominate: they work together to structure and rebuild each other: adaptive structuring' (Jankowski and Nyerges, 2001, p. 352).

After convocation, on entering the step of the process we find three constructs: appropriation, group processes, and emergent influence. The 'appropriation' deals not only with the appropriation of GIS technology, but with the appropriation of the topic at stake (for example, the green infrastructures, the food systems, the regional land-use planning) and the feeling to be part of a group able to decide. The second construct concerns the dynamics within the group of actors in terms of activities, co-operative relationships, conflict management embodied in a creative combination of the working climate, and the task to be carried out. The third construct of the process examines the emergence of information structuration during group processes from the combination of three elements: GIS technology, group of participants, and social-institutional set-up. For the constructs of the result, EAST 2 recalls, as in the tradition of the consensus building, how results have two dimensions: one related to the task (to prepare the regional participatory land-use plan for agroecology) and to the social and institutional context (consolidating trust and creating just and inclusive institutions). Jankowski (2011, p. 358) points out that 'participatory GIS requires reliable, inexpensive, scalable, easy to apply and maintain communication and geographic information technologies, in order to be adopted by planners, local governments, agencies, groups of citizens in local and regional decision-making processes'. The author also emphasizes how use of participatory GIS is not just a question of technologies and settings, but 'requires the activation of a social process in which participants interact with each other and with technology' (Jankowski, 2011, p. 358).

There is enormous value in using participatory mapping and counter-mapping practices with citizens and peasants to draw the existing world and to propose changes in a participatory way (Dalton and Mason-Deese, 2012; Monmonnier, 2007; Peluso, 1995; Verplanke *et al.*, 2016). For decades, counter-mapping, critical cartography, participatory GIS, voluntary geography, crowd-sourcing of geographic information have represented different declinations in the use of geographic information by actors who, in various corners of the earth, challenge the extractivist logic of accumulation by dispossession.

Accumulation by dispossession runs from the Amazon river, crossing the Arctic Shield, looking for minerals and oil and gas, but crossing the human body by looking for patenting genome or by citizens as sensors, to grab data and local knowledge.

The cartographic extra-activism (Kidd, 2019) follows several routes between militant research and the social protagonizm of citizenship based on the issues that geographic information is a common good and that geographic information technology and cartographic representations should be appropriate and shared in their active and emancipatory dimensions (Dalton and Mason-Dees, 2012; Monmonier, 2007; Peluso, 1995). It is not only about theories to be debated in academic contexts, but about inclusive social practices that have been built in the counter-mapping of indigenous lands, the challenges of urban socio-spatial justice, the multiplication of representations of nature and natural resources in the perspectives of eco-citizenship and agro-ecological transition.

These cartographic practices are plurally occurring in different situations, animated by people and resources that act out the extra-activism of possibilities. Visions of change, professional and volunteer time, knowledge and technical capacities are confronted with the official cartographies produced by the institutions of the State, the business intelligence market and the offices of trans-national corporations.

2.6. Participating in the Agroecological Transition

Basso Isonzo is a neighborhood of Padova (the city of our 800 year-old university). Despite the high concentration of buildings and inhabitants, it maintains an important agricultural area with debates and proposals for an urban agro-landscape park. Two farms, were started in 2015 in proximity agroecological production (Terre del Fiume and Terre Prossime) and they have lived, directly in their skin, the challenge of transforming conventionally-farmed land into high diversified agroecosystems. They have dug closed ditches, planted trees and living fences, reclaimed wetlands, and maintained a small forest. But this process has not only occurred as peasant consciousness; it has also promoted a social interaction involving citizens of the city's neighborhoods in participating in the creation of the territorial biodiverse agroecosystem. 'Plant the fence and stop the concrete' (7 May, 2017) and 'A forest is born' (27 March, 2018) are just a couple of initiatives supporting the creation of forests and living fences in the two farms with the collaboration of citizens of all ages.

This is a small example among the thousands existing in different parts of the world, between urban peripheries and agroecological systems in tropical forests, where the agroecological transition of levels three, four and five occurs, accompanied by the creation of an inclusive and participated nature matrix.

Bibliography

Acselrad, H. (2010). *Cartografia social e dinâmicas territoriais: marcos para o debate, Universidade Federal do Rio de Janeiro, Instituto de Pesquisa e Planejamento Urbano e Regional*, Rio de Janeiro, Brazil.

Boulding, K.E. (1956). *The Image: Knowledge in Life and Society*, pp. 65-71, University of Michigan Press, Ann Arbor, USA.

Brown, G. and M. Kyttä (2018). Key issues and priorities in participatory mapping: Toward integration or increased specialization? *Applied Geography*, 95: 1-8.

Bryan, J. and D. Wood (2015). *Weaponizing Maps: Indigenous Peoples and Counterinsurgency in the Americas*, Guilford Press, New York, USA.

Capineri, C., M. Haklay, H. Huang, V. Antoniou, J. Kettunen, F. Ostermann and R. Purves (2016). Introduction. *In:* Capineri, C., Haklay, M., Huang, H., Antoniou, V., Kettunen, J., Ostermann, F. and R. Purves (Eds.). *European Handbook of Crowd-sourced Geographic Information*, pp. 1-11, Ubiquity Press, London, UK.

Chambers, R. (1994a). The origins and practice of participatory rural appraisal, *World Development*, **22**(7): 953-969.

Chambers, R. (1994b). Participatory rural appraisal (PRA): Analysis of experience, *World Development*, **22**(9): 1253-1268.

Chambers, R. (1994c). Participatory rural appraisal (PRA): Challenges, potentials and paradigm, *World Development*, **22**(10): 1437-1454.

Checkland, P.B. (1984). System thinking in management: The development of soft systems methodology and its implications for social sciences. *In:* Ulrich, H. and G. Probst (Eds.). *Self-organization and Management of Social Systems: Insights, Promises, Doubts and Questions*, pp. 94-104, Springer-Verlag, Berlin, Germany. Conway, G.R. (1987). The properties of agroecosystems, *Agricultural Systems*, 24: 95-117.

Cook, B., P. Satizábal and J. Curnow (2021). Humanizing agricultural extension: A review, *World Development*, 140: 105337.

Corbett, J.M., L. Cochrane and M. Gill (2016). Powering up: Revisiting participatory GIS and empowerment. *The Cartographic Journal*, **53**(4): 335-340.

Corbett, J.M., G. Rambaldi, P. Kyem, D. Weiner, R. Olson, J. Muchemi, M. McCall and R. Chambers (2006). Overview: Mapping for change – The emergence of a new practice, *Participatory Learning and Action*, 54: 13-19.

Corner, J. (1999). The agency of mapping: Speculation, critique and invention. *In:* Cosgrove, D. (Ed.). *Mappings*, pp. 213-252, Reaktion, London, UK.

Crampton, J. and John Krygier (2006). An introduction to critical cartography, *ACME: An International E-Journal for Critical Geographies*, **4**(1): 11-33.

Dalgaard, T., N.J. Hutchings and J.R. Porter (2003). Agroecology, scaling and interdisciplinarity, *Agriculture, Ecosystems and Environment*, **100**(1): 39-51.

Dalton, C. and T. Stallmann (2018). Counter-mapping data science, *The Canadian Geographer/Le Geographe Canadien*, **62**(1): 93-101.

Dalton, C. and L. Mason-Deese (2012). Counter (mapping) actions: Mapping as militant research, *ACME: An International E-Journal for Critical Geographies*, **11**(3): 439-466.

De Blji, H. (2009). *The Power of Place, Geography Destiny and Globalization's Rough Landscape*, Oxford University Press, New York, USA.

DeSanctis, G. and M. Poole (1994). Capturing the complexity in advanced technology use: Adaptive structuration theory, *Organization Science*, **5**(2): 121-147.

Di Biase, D. (1990). Visualization in the earth sciences, *Earth and Mineral Sciences*, **59**(2): 13-18.

Dorling, D. and D. Fairbairn (1997). *Mapping Ways of Presenting the World*, Longman, London, UK.

Duru, M., O. Therond and M. Fares (2015). Designing agroecological transitions: A review, *Agron. Sustain. Dev.*, 35: 1237-1257.

Elwood, S. and K. Mitchell (2013). Another politics is possible: Neogeographies, visual spatial tactics, and political formation, *Cartographica: The International Journal for Geographic Information and Geovisualization*, **48**(4): 275-292.

Elwood, S., Goodchild, M.F. and D.Z. Sui (2012). Researching volunteered geographic information: Spatial data, geographic research, and new social practice, *Annals of the Association of American Geographers*, 102: 571-590.

Foster, K. (2016). Geodesign parsed: Placing it within the rubric of recognized design theories, *Landscape and Urban Planning*, 156: 92-100.

Francis, C., G. Lieblein, S. Gliessman, T.A. Breland, N. Creamer, R. Harwood, L. Salomonsson, J. Helenius, D. Rickerl, R. Salvador, M. Wiedenhoeft, S. Simmons, P. Allen, M. Altieri, C. Flora and R. Poincelot (2003). Agroecology: The ecology of food systems, *Journal of Sustainable Agriculture*, **22**(3): 99-118.

Freire, P. (1986). *Pedagogia do oprimido*, *Paz e Terra*, Rio de Janeiro, Brazil.

Freire, P. (1994). *Cartas à Cristina*, *Paz e Terra*, Rio de Janeiro, Brazil.

Giddens, A. (1984). *The Constitution of Society: Outline of the Theory of Structuration.* Cambridge, Polity Press.

Goodchild, M. (2010). Towards geodesign: Repurposing cartography and GIS? *Cartographic Perspectives*, 66: 7-22.

Goodchild , M. (2015). Two decades on: Critical GIScience since 1993, *The Canadian Geographer/Le Géographe Canadien*, **59**(1): 3-11.

Goodchild, M.F. (1992). Geographical information science, *International Journal of Geographical Information Systems*, **6**(1): 31-45.

Goodchild, M.F. (2007). Citizens as sensors: The world of volunteered geography, *GeoJournal*, 69: 211-221.

Haklay, M. (2016). Why is participation inequality important? *In:* Capineri, C., Haklay, M., Huang, H., Antoniou, V., Kettunen, J., Ostermann, F. and Purves, R. (Eds.). *European Handbook of Crowdsourced Geographic Information*, pp. 35-44, Ubiquity Press, London, UK.

Haklay, M. (2013a). Citizen Science and volunteered geographic information: Overview and typology of participation. *In:* Sui, D., Elwood, S. and M. Goodchild (Eds.). *Crowd-sourcing Geographic Knowledge, Volunteered Geographic Information (VGI) in Theory and Practice*, Springer, Dordrecht, The Netherlands.

Haklay, M. (2013b). Neogeography and the delusion of democratisation, *Environment and Planning A: Economy and Space*, **45**(1): 55-69.

Harley, B. (1987). The map and the development of the history of cartography. *In:* Harley, J.B. and D. Woodward (Eds.). *The History of Cartography. Vol. I: Cartography in Prehistoric, Ancient, and Medieval Europe and the Mediterranean*, pp. 1-42, University of Chicago Press, Chicago, USA.

Harley, B. (2001). *The New Nature of Maps*. The Johns Hopkins University Press, London, UK.

Jankowski, P. (2011). Designing public participation geographic information systems. *In:* Nierges, T., Couclelis, H. and R. McMaster (Eds.). *The Sage Handbook of GIS and Society*, pp. 347-360, Sage, London, UK.

Jankowski, P. and T. Nyerges (2001). Geographic Information Systems for Group Decision Making, *Towards a Participatory, Geographic Information Science*, Taylor & Francis, London, UK.

Jones, M.R. and H. Karsten (2008). Giddens's structuration theory and information systems research, *MIS Quarterly*, **32**(1): 127-157.

Kent, A.J. and P. Vujakovic (2018). *The Routledge Handbook of Mapping and Cartography*, Routledge, London, UK.

Kidd, D. (2019). Extraactivism: Counter-mapping and data justice, *Information, Communication & Society*, **22**(7): 954-970.

Kingsbury, P. and J.P. Jones (2009). Walter Benjamin's Dionysian Adventures on Google Earth, *Geoforum*, **40**(4): 502-513.

Kitchin, R. and M. Dodge (2007). Rethinking Maps, *Progress in Human Geography*, **31**(3): 331-344.

Klikemberg, B. (2007). Geospatial technologies and the geographies of hope and fear, *Annals of the Association of American Geographers*, **97**(2): 350-360.

Krupar, S. (2015). Map power and map methodologies for social justice, *Georgetown Journal of International Affairs* **16**(2): 91-101.

Liverman, D., E.F. Moran, R.R. Rindfuss and P.C. Stern (1998). *People and Pixels: Linking Remote Sensing and Social Science*, National Academy Press, Washington, USA.

López-García, D., M. Cuéllar, A. Azevedo-Olival, N.P. Laranjeira, V.E. Méndez, S. Peredo, C.A. Barbosa, C. Barrera-Salas, M. Caswell, R. Cohen, A. Correo-Humanes, V. García-García, S.R. Gliessman, A. Pomar-León, A. Sastre-Morató and G. Tendero-Acín (2021). Building agroecology with people. Challenges of participatory methods to deepen on the agroecological transition in different contexts, *Journal of Rural Studies*, 83: 257-267.

López-Ridaura, S., O. Masera and M. Astier (2002). Evaluating the sustainability of complex socio-environmental systems. The MESMIS framework, *Ecological Indicators*, 2: 135-148.

Loukissas, Y.A. (2019). *All Data are Local: Thinking Critically in a Data-driven Society*, The MIT Press, London, UK.

MacEachren, A.M. (1995). *How Maps Work: Representation, Visualization and Design*, Guilford Press, New York, USA.

Maturana, H.R. and F.J. Varela (1987). *L'albero della conoscenza*, *Garzanti*, Milan, Italy.

McCall and K. Michael (2011). *Mapeando el territorio: Paisaje local, conocimiento local, poder local*, pp. 221-246. *In:* G. Bocco, P.S. Urquijo and A. Vieyra (Eds.). *Geografía y Ambiente en América Latina, Proceedings CIGA Symposium, Geografia y Ambiente en Latina America*, Morelia, Mexico.

Monmonier, M. (2005). Lying with maps, *Statistical Science*, 20: 215-222.

Monmonier, M. (2007). Cartography: The multi-disciplinary pluralism of cartographic art, geospatial technology, and empirical scholarship, *Progress in Human Geography*, **31**(3): 371-379.

Nir, D. (1990). Region as socio-environmental system, *An Introduction to a Systemic Regional Geography*, Kluwer, Dordrecht, The Netherlands.

Pavlovskaya, M. (2018). Critical GIS as a tool for social transformation, *The Canadian Geographer/Le Geographe Canadien*, **62**(1): 40-54.

Peluso, N.L. (1995). Whose woods are these? Counter-mapping forest territories in Kalimantan, Indonesia, *Antipode*, 27: 383-406.

Pickles, J. (1995). Ground Truth, *The Social Implications of Geographic Information Systems*, Guilford Press, New York, USA.

Pink, D.H. (2006). *A Whole New Mind: Why Right-Brainers Will Rule the Future. Riverhead Trade*. New York.

Poster, M. (2004). The information empire, *Comparative Literature Studies*, **41**(3): 317-334.

Radil, S.M. and M.B. Anderson (2018). Rethinking PGIS: Participatory or (Post)-Political GIS? *Progress in Human Geography*, **43**(2): 195-213.

Rossetto, T. (2019). *Object-oriented Cartography, Maps as Things*, Routledge, London, UK.

Santos, M. (2000). *La naturaleza del espacio: técnica y tiempo: Razón y emoción*, Ariel, Barcelona, Spain.

Schlosseberg, M. and E. Shuford (2005). Delineating 'public' and 'participation' in PPGIS, *URISA Journal*, **16**(2): 15-26.

Sclavi, M. and L.E. Susskind (2011). *Confronto creativo, Dal diritto di parola al diritto di essere ascoltati et al.*, Edizioni, Milan, Italy.

See, L., P. Mooney, G. Foody, L. Bastin, A. Comber, J. Estima, S. Fritz, N. Kerle, B. Jiang, M. Laakso, H.Y. Liu, G. Milcinski, M. Nikšic, M. Painho, A. Podör, A.M. Olteanu-Raimond and M. Rutzinger (2016). Crowd-sourcing, Citizen Science or Volunteered Geographic Information? The Current State of Crowd-sourced Geographic Information, *ISPRS International Journal of Geo-Information*, **5**(5): 55-77.

Sieber, R. (2006). Public participation geographic information systems: A literature review and framework, *Annals of the American Association of Geography*, **96**(3): 491-507.

Sinclair, F. and R. Coe (2019). The options by context approach: A paradigm shift in agronomy, *Experimental Agriculture*, **55**(S1): 1-13.

Socientize (2014). White Paper on Citizen Science for Europe; Retrieved from https://ec.europa.eu/futurium/en/system/files/ged/socientize_white_paper_on_citizen_science.pdf; accessed on 21 April, 2021.

Star, J. and J.E. Estes (1990). *Geographic Information Systems: An Introduction*, Prentice-Hall, Englewood Cliffs, USA.

Steinitz, C. (2012). *A Framework for Geodesign: Changing Geography by Design*, ESRI Press, Redlands, USA.

Sui, D. (2015). Emerging GIS themes and the six senses of the new mind: Is GIS becoming a liberation technology? *Annals of GIS*, **21**(1): 1-13.

Tyner, J.A. (2010). *Principles of Map Design*, The Guilford Press, New York, USA.
Vallega, A. (1995). *La regione sistema territoriale sostenibile, compendio di geografia regionale sistematica*, *Mursia*, Milan, Italy.
Verplanke, J., M.K. McCall, C. Uberhuaga, G. Rambaldi and M. Haklay (2016). A shared perspective for PGIS and VGI, *The Cartographic Journal*, **53**(4): 308-317.
Willow, A.J. (2013). Doing sovereignty in Native North America: Anishinaabe counter-mapping and the struggle for land-based self-determination, *Human Ecology*, **41**(6): 871-884.
Wilson, M. (2015). On the criticality of mapping practices: Geodesign as critical GIS? *Landscape and Urban Planning*, 142: 226-234.
Wood, D. (1992). *The Power of Maps*, Guilford, New York, USA.
Wood, D. (2002). The map as a kind of talk: Brian Harley and the confabulation of the inner and the outer voice, *Visual Communication*, 1: 139-161.
Wood, D., J. Fels and J. Krygier (2010). *Rethinking the Power of Maps*, Guilford Press, New York, USA.
Yuan, M. (2015). Frontiers of GIScience: Evolution, state-of-art, and future pathways. *In:* P. Thenkabail (Ed.). *Remote-sensing Handbook*, CRC Press, Boca Raton, USA.
Yuan, M. (2021). GIS research to address tensions in geography, *Singapore Journal of Tropical Geography*, 42: 13-30.

CHAPTER
3

Sustainable Agricultural Development to Achieve SDGs: The Role of Livestock and the Contribution of GIS in Policy-making Process

Alice Morandi

Advanced Master on GIScience and Unmanned Systems for Integrated Management of Territory and Natural Resources,University of Padova

3.1. Introduction

The transformation of the planet over the past century due to intense human activity has affected ecosystems more rapidly and decisively than in any other comparable period of history; the emerging environmental change is making the production of 'good' decisions a challenging issue for the future. The current ways in which our food is produced and distributed are now recognized as a leading cause of soil degradation, depletion of water resources and greenhouse gas (GHG) emissions that contribute to climate change (Zhang *et al.*, 2018; IPCC, 2019).

It is important to consider that the specific characteristics of the agro-food sector are closely linked to natural resources and the surrounding environment. According to climate change projections, in the most probable future scenario, the absence of radical interventions will lead to a decrease in global agricultural yield (Koellner *et al.*, 2013; IPCC, 2019). The effects of climate change could negatively affect specific geographical areas and their ability to guarantee current production levels, exacerbating an unsolved problem: hunger in the world. The paradox is that although more and more food is being produced, the number of undernourished people has started to increase again in recent years (Pérez de Armiño, 1998; Breton, 2009; Godfray *et al.*, 2010; FAO, 2018b).

The publication of data on the persistence of hunger in the world, as well as an understanding of the interlinkages between environmental change and the globally

*Email: alice.morandi78@gmail.com

dominant agro-food system, has moved experts to research new approaches to the international food supply chain that promote the shared objective of sustainability. World leaders at the 2015 UN General Assembly reaffirmed the urgent need to 'end hunger, achieve food security and improved nutrition, and promote sustainable agriculture' (UN General Assembly, 2015, p. 14), recognizing the links between supporting sustainable agriculture, empowering small farmers, promoting gender equality, ending rural poverty, ensuring healthy lifestyles, tackling climate change, and other issues addressed within the set of 17 Sustainable Development Goals in the 2030 Agenda.

According to the FAO, the challenge of feeding an expanding world population will peak in 2050, when the number of inhabitants of our planet will reach 9 billion (FAO, 2018b). As a result of demographic trends and changes in dietary habits related to increased urbanization, demand for food is expected to increase by 60 per cent as compared to 2006 (FAO, 2016). In this respect, the increase in the production and consumption of meat from intensive livestock farming in developing countries is of particular concern, both in terms of environmental impact and food security. Livestock farming activities require extensive use of food resources: one-third of the world's cereal production is consumed as feed by farmed animals, diverting important resources that could be used to feed people in the world's poorest regions (McMichael *et al.*, 2007; HLPE, 2016; FAO, 2018b). At the global level, meat production is also a major contributor to the production of GHG, which is responsible for the increase in average temperatures on earth (FAO, 2013).

The FAO's report (2013), *Tackling Climate Change through Livestock: A Global Assessment of Emissions and Mitigation Opportunities*, is today considered the most comprehensive report to date on how livestock farming contributes to global warming, and of the sector's capacity to respond to the problem. The FAO report sets the percentage of anthropogenic emissions generated by the livestock sector at 14.5 per cent..

The most widely accepted definition of sustainability requires society 'to meet the needs of the current population without compromising the ability to meet the needs of future generations' (WCED, 1987, p. 34). To meet the needs of future generations, it is necessary to develop a more sustainable and equitable food system. A sustainable agro-food system is achievable if decision-makers develop and utilize tools that support the implementation of targeted, robust, and long-term mitigation policies in which the livestock sector plays a key role. Reversing unsustainable global trends requires great effort and integration across policy areas, as well as data analysts and information systems (IS) (Sugumaran *et al.*, 2010).

It is in this context that decision support system (DSS) comes into focus, offering solutions, supporting policymakers and practitioners in decision-making processes. DSS helps users to access, interpret, and understand information from data, to perform analyses and to create models to identify possible actions. In short, a DSS is a software system that provides the user (the decision-maker) with a series of interactive and user-friendly data analysis and modeling functionalities to increase the efficiency and effectiveness of the decision-making process (Sharda *et al.*, 2015). DSSs are increasingly being combined with geographic information systems (GIS) to form a hybrid decision support tool, known as a spatial decision support system (SDSS). Such systems combine the data and logic of a DSS with the powerful spatial referencing and spatial analytic capabilities of GIS to shape a new system that is even more valuable than the sum of its parts (Basil, 2010).

This article describes a spatial decision support system (SDSS) that analyzes sustainable livestock production systems. It includes a brief overview of the SDSS and its various components and an example application of an interactive tool, GLEAM-I, developed by FAO, which is applied to the poultry meat value chain in Colombia.

3.2. Sustainability Challenges for Livestock

In the second half of the last century, global meat consumption increased exponentially as compared to global population growth. Indeed, between 1950 and 2000, the inhabitants of our planet doubled from 2.7 to over 6 billion people, while the total meat consumption increased fivefold, from 45 million tonnes per year in 1950 to 233 million tonnes per year in 2000 (Steinfeld *et al.*, 2006). Global meat consumption is set to continue to grow rapidly and the FAO has estimated that annual global meat production will reach 465 million tonnes by 2050 (FAO, 2018b).

The consumption of animal products, though still consumed at a higher rate in developed countries, is growing particularly rapidly in developing countries, in part because meat consumption may be equated with status and wealth. As illustrated in Fig. 1, according to estimates published in the *Agricultural Outlook* by OECD and FAO (2019), growth in per capita meat consumption between 2016-18 and 2028 will be most evident in Latin American and Caribbean countries.

The global livestock population, according to the 2016 FAO study, is at the highest level of all time, with 28 billion animals, 82 per cent of which are chickens. Over the past 20 years, the farm chicken population has increased from 14 to 23 billion animals. In terms of tonnes of meat produced, pork is currently the most produced meat in the world, followed by poultry, beef, and sheep (FAO, 2016).

Chicken meat production has experienced the highest growth rates compared to other meat production, nearly reaching the levels of pork production. According to the FAO's database, chicken meat production in 2017 was 109 million tonnes, just below pork production (119 million tonnes) (FAOSTAT[1]).

Accelerated growth in poultry production leads to the expectation that it might soon overtake pork production, becoming the world's most produced meat

[1] FAOSTAT provides free access to food and agriculture data for over 245 countries and territories and covers all FAO regional groupings from 1961 to the most recent year available

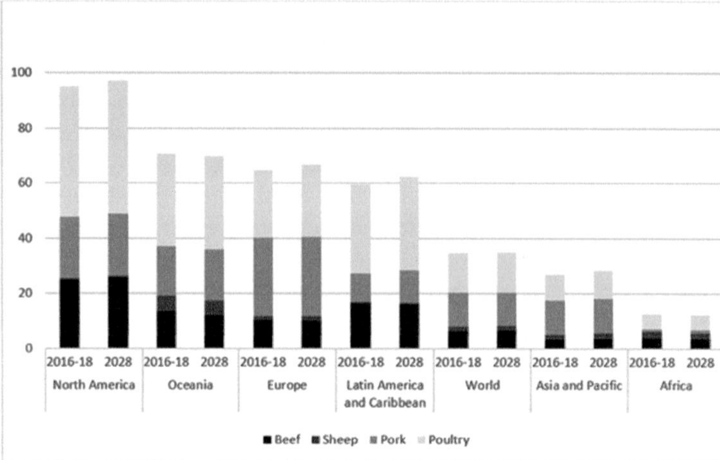

Fig. 1: Per capita meat consumption by region, 2016-18 vs 2028 kg/person/year
(*Source*: Author's elaboration on OECD and FAO, 2019)

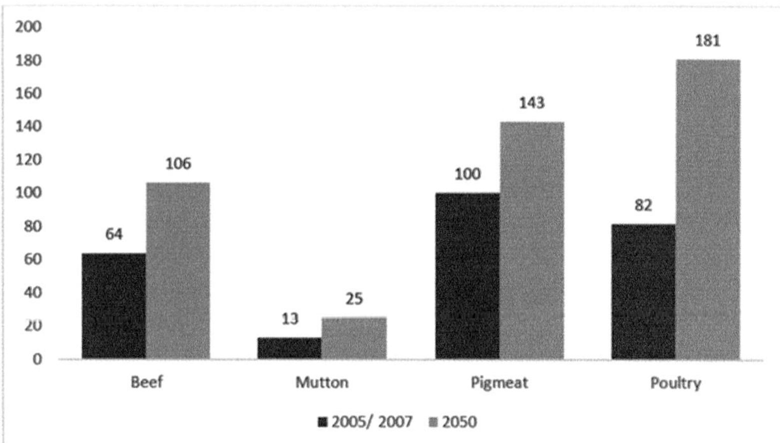

Fig. 2: Global demand for meat 2005/2007 vs 2050 projection (in million tonnes)
(*Source*: Author's elaboration on FAO, 2012)'s data)

(OCDE and FAO, 2019). The rapid increase in market demand for poultry meat (Fig. 2) is, on the one hand, attributed to the lower production costs required, resulting in lower prices compared to other meats, and on the other, reflects the public perception that it is a healthier product with fewer calories and higher nutritional value. However, production standards, environmental outputs, quality of meat, animal welfare and food security are critical research and ethical issues that must be considered (Soler and Fonseca, 2011).

In general, the expected increase in world demand for meat and the resulting increase in production could have serious repercussions on the competition for soil, water, and other food commodities, and further complicate efforts at sustainable production. Widespread meat-based diets contribute significantly to the increase in the demand for agricultural land: of all land not covered by ice, 26 per cent is used for grazing, while 12 per cent is used for agricultural purposes. However, because one-third of agricultural production is used as animal feed, the total percentage of land dedicated to meat production is 30 per cent while the percentage of land used to grow crops that feed humans is 8 per cent (Smith *et al.*, 2010; FAO, 2019).

3.2.1. Impact on Food Resources and Food Security

Farmed animals need to be fed to live, grow, and produce meat. However, the food resources consumed by these animals are greater than what they produce in the form of meat, milk, and eggs for the market. The amount of food ingested by an animal organism does not directly produce a similar amount of body mass: in fact, only part of the food ingested is used by the organism for the growth of its body structure, while the rest is burned as energy to maintain the body's vital functions (McMichael *et al.*, 2007; Gómez *et al.*, 2011; HLPE, 2016).

Traditionally, animal feed production was based on locally available food resources, such as crop residues or tree leaves, which were of no value for human consumption. However, as livestock production grows and intensifies, dependence on local resources is increasingly reduced, while dependence on concentrated feed, which is traded locally or internationally, increases (Steinfeld *et al.*, 2006; HLPE, 2016). How can we explain the increased use of cereals to feed livestock? The root cause was the decline in cereal prices beginning in the 1950s, due to overproduction in the United States. Even after the United States' grain surplus was exhausted, supply has kept up with growth in demand: low-cost cereal production has been sustained by the expansion of cultivated land and intensification of crops (McMichael *et al.*, 2007).

The rate between ingested food and growth of the organism is known as the food conversion index (FCI), which measures the amount of feed, expressed in kilograms, needed to grow one kilogram of live weight of the animal. To produce 1 kg of chicken meat requires about 2 kg of cereals, for 1 kg of pork about 4 kg of cereals and for 1 kg of beef, about 7 to 8 kg of cereals. Although there are substantial differences between species and how they are farmed, in general, the food conversion index always remains unfavorable, leading to a net loss of food (Steinfeld *et al.*, 2006; Mottet *et al.*, 2017).

The production of animal-origin food, and in particular meat, requires extensive use of food resources. One-third of world cereal production (745 million tonnes in 2007) is consumed by farmed animals (McMichael *et al.*, 2007). Maize is the main cereal used on farms: about 60 per cent of global production is used as feed. In 2007, 463 million tonnes of maize out of a total of 787 million tonnes were used as feed, while only 110 million tonnes were used for direct human consumption. For soybeans, over 70 per cent of world production is used on farms, as it is another main component of modern feed (FAO, 2018b).

Although ruminants (cattle, sheep and goats) have a low conversion rate compared to non-ruminants (chickens and pigs), the latter group consumes a greater share of concentrated feed, which is a complex industrial product. Chickens consume the largest share of feed, with 30 per cent, followed by pigs, with a slightly lower share (29 per cent), dairy cows (25 per cent) and beef steers (14 per cent) (FAO, 2018b).

In a world where the number of food-insecure people continue to grow, meat production exacerbates problems with distribution and access to food, absorbing important resources that could be used to feed the world's poorest people (HLPE, 2016). However, abandoning livestock farming for vegan or vegetarian diets worldwide does not appear to be a feasible or desirable way forward in the short term. In fact, the people in the world's poorest regions depend heavily on meat protein intake, even though meat represents a small percentage of their diets when compared to their counterparts in developed countries. In addition, livestock plays a crucial economic role for about 60 per cent of rural households in developing countries, contributing to the livelihoods of about 1.7 billion people living in poverty (FAO, 2018a). The growth of the livestock sector has the potential to play a key role in preventing people from falling into poverty and thus improving food security in poorer countries. To continue to support these livelihoods, it is important to rethink the way meat is produced today, building on more traditional forms of production, expanding the use of animal feeds that have little or no value as human foods, and returning to the use of livestock feed derived from local crop residues (Delgado, 2003; ATKearney, 2018; FAO, 2018a). The challenge is how to achieve this result without compromising current production levels that must meet the expected growth in global demand for meat.

3.2.2. Impact on the Environment

The relationship between the consumption of food of animal origin, intensive livestock farming, and environmental impact has long been ignored by the scientific community, the media, and even the green community. Awareness has increased, thanks to growing interest within the scientific community and the dissemination of many scientific studies on the relationship between animal husbandry and environmental impact. One of the most important of these studies was published by the FAO, *Livestock's Long Shadow* (2006), a 390-page scientific report that accurately assesses the global impact of the livestock sector on the environment.

According to this report, the livestock sector is responsible for the majority of anthropogenic land use. In fact, directly and indirectly, modern meat production methods use 30 per cent of all land area not covered by ice and 70 per cent of all agricultural land (Steinfeld *et al.*, 2006). The search for land for intensive livestock farming is one of the main causes of deforestation, especially in Latin

America, where the most intense deforestation is taking place: 70 per cent of land in the Amazon, once forest, has now been turned into grassland, and fodder crops cover a large part of the remaining area (McMichael *et al.*, 2007). To make way for pastures, forest land is razed to the ground with the use of huge bulldozers or the land is set on fire. However, deforested land in the Amazon is not suitable for grazing in the long term; after a few years, the soil becomes barren and farmers must cut down another section of forest to move their herds, leaving vast expanses of wasteland behind (Finer and Mamani, 2019). Several reports show a critical overlap between deforestation in the Amazon rain-forest, linked to the production of animal feed and fire alerts (Greenpeace, 2009; Finer and Mamani, 2019). Fires in the Amazon had increased by 30 per cent in 2019 (INPE, 2020).

Intensive livestock farming also has a critical impact on soil degradation, mainly due to overgrazing. Soil compaction results in loss of flora and plant roots that play an essential role in water absorption and nutrient recycling. Degraded soil is much more susceptible to wind and water erosion, and less likely to be sufficiently fertile for agriculture (Steinfeld *et al.*, 2006).

According to the FAO, 20 per cent of the world's pastures are considered degraded, and the arid and semi-arid environments of Africa and Asia, as well as the semi-wetlands of Latin America, are particularly affected. At the same time, the expansion of feed cultivation in natural ecosystems also contributes to land degradation (FAO, 2016).

In recent years, public attention has increasingly turned to livestock farming as a primary contributor to climate change. Different global studies give a wide range of estimations on the contribution of livestock to GHG emissions. This variability generates uncertainty among policy makers, suggesting that there is a lack of consensus among scientists about the contribution of livestock to global GHG emissions (Herrero *et al.*, 2011).

In 2006, the FAO estimated that the processes involved in animal farming generate a GHG production equivalent to 18 per cent of all global emissions from human activities, more than the entire transport sector (Steinfeld *et al.*, 2006). Subsequently, in 2009, the Worldwatch Institute published a critical analysis of the FAO's report, with a modified methodology that considered all emissions produced by farms and all activities related to them, concluding that total GHG emissions attributable to the livestock sector account for 51 per cent or more of total anthropic emissions, much higher than the FAO estimate of 18 per cent (Goodland and Anhang, 2009). However, later the FAO in its 2013 report (considered the most comprehensive global estimate of the effects of intensive livestock) estimated total GHG emissions from livestock supply chains as 14.5 per cent of all human-induced emissions, or 8.1 gigatonnes of CO_2-eq per annum for the 2010 reference period (FAO, 2013).

The FAO has mapped the distribution of GHG emissions from the livestock sector: Latin American and Caribbean region have higher levels of emissions than any other region in the world (almost 1.9 gigatonnes of CO_2-eq). Although there has been a slight decrease in emissions in recent years, changes in land use, deforestation, and the expansion of intensive feed crops contribute drastically to the high GHG emissions in the region (FAO, 2013).

As explained in the FAO report (2013), emissions from the livestock sector originate in four processes: enteric fermentation, manure management, feed production, and energy consumption, which are described below:

- Enteric fermentation refers to methane generated during the digestion of ruminants and monogastrics, although levels in the latter are much lower. Food quality is closely related to enteric emissions. for example, diets with a high percentage of fiber-rich ingredients are related to higher enteric emissions;
- Manure causes methane (CH_4) and nitrous oxide (NO_2) emissions. Methane is generated during the anaerobic decomposition of organic matter. Nitrous oxide is a product of the decomposition of ammonia contained in manure. Different manure-management systems give rise to different levels of emissions. In general terms, methane emissions are higher when manure is stored and treated in liquid systems. On the other hand, solid storage and treatment systems tend to favor nitrous oxide emissions;
- There are several emissions linked to feed production. Carbon dioxide (CO_2) emissions come from the expansion of pastures and agricultural land used for animal feed in nature areas and forests, the production of fertilizers and pesticides for such crops, and their processing and transport. On the other hand, the use of nitrogen fertilizers and the application of manure cause nitrous oxide emissions;
- Energy consumption occurs throughout the entire production chain. The production of fertilizers, the use of agricultural machinery, and the processing and transport of crops for animal feed generate GHG. These emissions are attributed to feed production. There is also energy consumption on the farms themselves due to ventilation, lighting, air conditioning, etc. Finally, the processing, packaging, and transport of animal products consume energy and generate emissions (FAO, 2013).

Emissions from enteric fermentation account for about 44 per cent of the total sector (slightly below 3.5 gigatonnes of CO_2-eq). Animal feed and diet production is the second most important source with 3.3 gigatons of CO_2-eq, or 41 per cent of the total. Manure management causes about 10 per cent, or 0.8 gigatonnes of CO_2-eq. Energy consumption generates 0.4 gigatonnes of CO_2-eq, practically 5 per cent of the total. Animal feed production therefore emerges as a major contributing factor to emissions produced in the livestock sector, both because of the high environmental impact generated during its production and because it is able to mitigate or exacerbate enteric emissions (FAO, 2013).

Climate change is one of the most worrying risk factors for the future of agri-food systems, affecting productive capacity and threatening to exacerbate uneven distribution of food resources. Because of its high impact on the environment, the livestock sector has greatest mitigating potential for GHG (FAO, 2018a).

3.3. SDG's Strategic Framework

By signing the 2030 Agenda for Sustainable Development, with its 17 Sustainable Development Goals (SDGs) and 169 targets in 2015, governments around the world are committed to address urgent economic, social, and environmental global challenges over the next years. They seek to address, in a sustainable manner, the root causes of poverty and the universal need for development (FAO, 2018a).

The SDGs cover the five dimensions of sustainable development: prosperity (economic growth), people (social dimensions), planet (environmental protection), peace (just and inclusive society), and partnership (multilevel inclusive governance). Governments are expected to take ownership of the SDGs and establish national frameworks for their achievement. Implementation and success will depend on the commitment of individual nations to promote sustainable development policies together with inter-sectoral co-ordination mechanisms, and focused plans and programs (UN General Assembly, 2015).

The five pillars of sustainable development are all interlinked within the Agenda. Indeed, while each goal has a clear starting point in one of the pillars, most goals are in effect, embedded in all dimensions. Traditionally, however, sustainability analysis has used a partial, sectoral approach that gauges the effects of development on a single dimension of sustainability. One pitfall to this approach is that it fails to take account of simultaneous contributions, feedback effects, dynamics, synergies, and trade-offs between different policy goals and targets. This is particularly important because of the complex, non-linear interactions at play in the SDGs, where the achievement of one target can have positive, neutral, or negative effect on one or several others (FAO, 2018a).

3.3.1. Livestock and the Sustainable Development Goals

For decades, the livestock debate has focused on how to sustainably increase production. The UN 2030 Agenda for Sustainable Development has added a new and broader dimension to the debate. It has shifted the emphasis of the conversation from fostering sustainable production per se, to enhancing the contribution of the sector to the SDGs achievement (FAO, 2018a).

The livestock sector can contribute directly or indirectly to each of the SDGs, by addressing many present challenges to development. The sector can provide the world with adequate and reliable supplies of safe, healthy, and nutritious food; create employment opportunities upstream and downstream in the food chain; stimulate smallholder entrepreneurship and close inequality gaps; improve natural resource-use efficiency; promote sustainable consumption and production patterns; and increase the resilience of households to cope with climate shocks. In order to fulfil its potential, the sector must face a new set of intersecting challenges (FAO, 2018a).

The major challenge will be to translate the role of livestock as envisioned by the SDGs into national and local policies and strategies. The SDGs and targets are aspirational and global (FAO, 2018a). Thus, each country will have to decide how the role of livestock in the SDGs should be incorporated into national planning processes, policies, and strategies, and how to set national targets that take into account national and local contexts, while also being guided by the global level of ambition laid out in the SDGs (HLPE, 2016).

3.3.2. Operating Livestock Sector Value Chain

In order to analyze the current fragmented agro-food production system, it is necessary to introduce the concept of global value chain (GVC), which allows the analysis of the relationships between companies that are part of a production chain, the power relationships within them, and the flow of value. The concept of GVC allows stakeholders to identify how transnational production chains mediate the role of companies, workers, and territories in the global economy.

The global food value chain can be defined as the set of activities needed to bring agricultural products to consumers, including agricultural inputs, crops and livestock, processing, preservation, marketing, distribution, and consumption (Gómez *et al.*, 2011). The GVC model applied to the livestock sector makes it possible to assess which steps in the chain produce the most negative externalities and therefore, which areas require priority intervention (Van der Vorst *et al.*, 2007).

According to the analysis recently conducted by the ATKearney group (2018), the global meat value chain consists of seven steps organized into the three sections presented below and in Fig. 3:

Section 1. Feed production:

1.1. *Agricultural input*: Enterprises in this sector provide agricultural inputs and services to farmers. The main products and services concern biotechnology, agro-chemicals and fertilizers, and agricultural machinery.
1.2. *Cultivators*: Agricultural enterprises responsible for the production of maize, soya and other crops needed for the production of animal feed.
1.3. *Transformation*: Agrochemical companies processing farmers' raw products into industrially produced concentrated feed. They also provide animal health services.

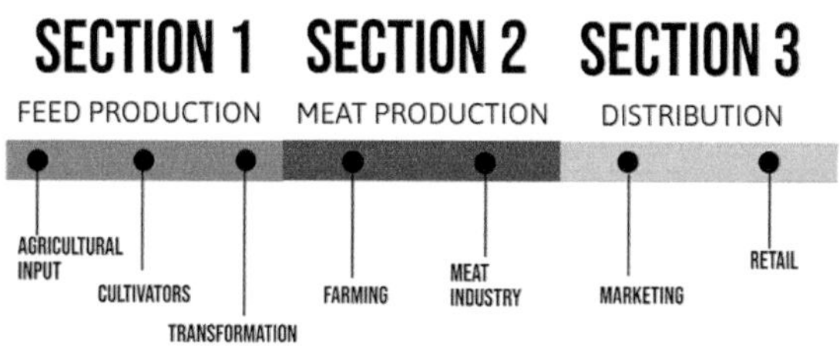

Fig. 3: Global meat value chain
(*Source*: Author's elaboration on ATKearney group, 2018)

Section 2. Meat production:

2.1. *Farmers*: Livestock enterprises involved in the growth, reproduction, and feeding of livestock. The main products are dead animals, whole or in pieces.

2.2. *Meat industry*: Enterprises in this sector convert raw meat into food products that are either processed or simply packaged and ready to be displayed on supermarket shelves.

Section 3. Distribution and marketing:

3.1. *Marketing*: Communication agencies aim to create the image and brand of the companies that distribute the products, creating an added value that consists of identity and reliability of the brand.

3.2. *Distribution*: Companies in this sector distribute and market finished meat products. They include both restaurant services and large-scale organized distribution, developed in a widespread network of sales outlets.

3.4. Spatial Decision Support System (SDSS) for Planning Sustainable Livestock Production

To achieve the SDGs through a sustainable approach to the livestock sector, it is necessary to face a multitude of complex spatial problems with multiple conflicting goals. An acceptable solution must reconcile these conflicting goals. A variety of analytical techniques have been developed to help decision-makers solve problems with multiple criteria. Consequently, decision-makers have turned to analysts and analytical modeling techniques to enhance their decision-making capabilities: firstly, exploring the problem, to increase the level of understanding and to refine the definition; and secondly, planning and evaluating alternative solutions to investigate the possible trade-offs between conflicting objectives, and to identify unanticipated and potentially undesirable characteristics of the solutions (Sugumaran *et al.*, 2010).

Analyzing the huge amount and variety of data characteristic of highly complex systems, such as the meat production chains described above, is not an easy task. It is in this context that decision support system (DSS) comes into focus, offering solutions, and supporting policymakers and practitioners in decision-making processes.

3.4.1. A Short Overview of DSS and SDSS

Decision support systems have become a cultural area of scientific research, thanks to the work of Gorry and Scott Morton in the early 1970s, who highlighted the usefulness and potential of these systems. The first definition provided by the authors describes DSSs as interactive computer-based systems that help decision-makers use data and models to solve unstructured problems (Gorry and Morton, 1971).

Since then, there have been several definitions of DSS. Another classic definition of DSS, provided by Keen and Morton (1978), defines DSS as a system that combines the intellectual resources of individuals with the capabilities of the computer to improve the quality of decisions. It is a computer-based support system for management decision-makers dealing with semi-structured problems.

More precisely, Turban (1995) defines the DSS as 'an interactive, flexible and adaptable computer-based information system developed specifically to support the solution of an unstructured management problem to improve decision-making. It uses data, provides a friendly interface, and stimulates decision makers' ' intuitions'. De facto, the notion of DSS is an 'umbrella concept' with a plurality of meanings, it is a 'conceptual methodology' (Sharda *et al.*, 2015).

According to Andrew (1991), a DSS has three fundamental components:

1. The database management subsystem, which includes a database that contains data relevant to the class of problems for which the DSS was designed;
2. The model management subsystem, which includes a library of models (Model Base) related to financial science, statistics, management and other models that provide the DSS with analytical capabilities;
3. The user interface subsystem, which covers all aspects of communication between the user and the different components of the DSS. As many users are often managers who have no computer training, DSS must be equipped with intuitive and easy-to-use interfaces.

DSSs are increasingly being combined with geographic information systems (GIS) to form a hybrid type of decision-support tool, known as a spatial decision support system (SDSS). Such systems combine the data and logic of a DSS with the powerful spatial referencing and spatial analytic capabilities of a GIS. The idea of an SDSS system evolved in the late 1980s (Armstrong *et al.*, 1986) and since 1995, the concept of SDSS has been firmly established in the literature (Crossland *et al.*, 1995). It is possible to define an SDSS briefly and effectively as a computer-based system that combines conventional data, spatially referenced data, and decision logic as a tool for assisting a human decision-making process (Basil, 2010).

SDSS are tools that can integrate the dimensions of sustainability (society, economy, environment) and offer a systemic approach to problems, identify relationships and feedbacks, explicitly introduce limits or constraints, and demonstrate the importance of 'where' in combination with 'what' and 'how much'. In this sense, the added value provided by SDSS is mainly linked to the explicit consideration of the spatial dimension of decision-making problems, which is intrinsic to issues related to the development, transformation, and management of agro-food systems (Basil, 2010).

SDSS are structured procedures for the generation and comparison of alternatives, able to combine real environmental data with economic and social information, compare them through group work, and represent the final outcome according to specific maps, thus ensuring a relevant support tool for decision-making processes (Marotta, 2010).

Sustainable development of the livestock sector requires extensive and evidence-based diagnostic tools to avoid undesirable environmental effects. With these tools, all stakeholders in the sector can identify problem areas and evaluate options for intervention. In order to understand the application potential of SDSS in achieving shared sustainability goals through sustainable livestock farming, the next paragraph will describe the interactive GLEAM model developed by FAO.

3.5. Global Livestock Environmental Assessment Model (GLEAM) by FAO

GLEAM is a GIS framework for life-cycle analysis (LCA) of natural resource use and environmental impacts associated with the livestock value chain. The objective of GLEAM is to quantify livestock production, natural resource use in the sector, and GHG emissions, in order to evaluate mitigation scenarios that support a more sustainable livestock sector.

The GLEAM model has also been developed in a free web application, GLEAM-i (FAO, 2018c). The web application is the first user-friendly global public tool for the livestock sector that supports governments, planners, producers, industry, and civil society in the estimation of GHG emissions, using Tier 2 methodologies of the intergovernmental panel on climate change (IPCC), as well as in the assessment of mitigation measures in the livestock, diet, or manure management sectors (MacLeod *et al*., 2018).

GLEAM-i distinguishes key production stages, such as feed production, processing, and transport; population and livestock feeding dynamics; manure management; and the processing and transport of products, such as meat and milk. The model captures the specific impacts of each stage, providing a broad, yet detailed picture of production and natural resource use.

GLEAM-i allows direct comparison of baseline conditions and scenarios and incorporates 2010 reference data imported from FAOSTAT and other data archives. For all countries, the user starts by reviewing the reference data proposed by the tool and can replace it with project data or more accurate or recent data. GLEAM-i is a key instrument, both for identification and evaluation of best practices. The model offers, on the one hand, a detailed view of the current state of the sector and, on the other, a tool for evaluating different mitigation options (MacLeod *et al*., 2018).

This flexibility of GLEAM derives from the fact that it is based on a GIS environment consisting of: (a) input data levels; (b) scripts written in Python that perform calculations; and (c) procedures to run the model, check calculations, and extract output. The basic spatial scale used in GIS is the 0.05×0.05 degree cell (which measures approximately 5 km by 5 km at the equator). Emissions and output are calculated for each cell using input data with different spatial resolution levels (FAO, 2018c). The data used in GLEAM can be classified into (a) basic input data, and (b) intermediate data. Basic input data is defined as primary data, such as the number of animals, herd or block parameters, mineral fertilizer application rates, and temperature. This data is derived from sources, such as literature, databases, and surveys. Intermediate data are values generated within GLEAM and then used for subsequent calculations and include values for parameters, such as herd structures and manure application rates (MacLeod *et al.*, 2018).

Data availability, quality, and resolution vary depending on the parameter and country involved. For countries belonging to the organization for economic co-operation and development (OECD), there are often complete national or regional data sets and, in some cases, sub-national data (e.g. for manure management on US dairy farms). In contrast, data is often not available in non-OECD countries, requiring the use of regional default values (e.g. for many physical performance parameters of pigs and poultry).

The size of the livestock population is based on FAOSTAT data and its geographical distribution is based on the gridded livestock of the world (GLW) model. GLW density maps are based on observed densities and explanatory variables, such as climate data, land cover, and population parameters (Fig. 4).

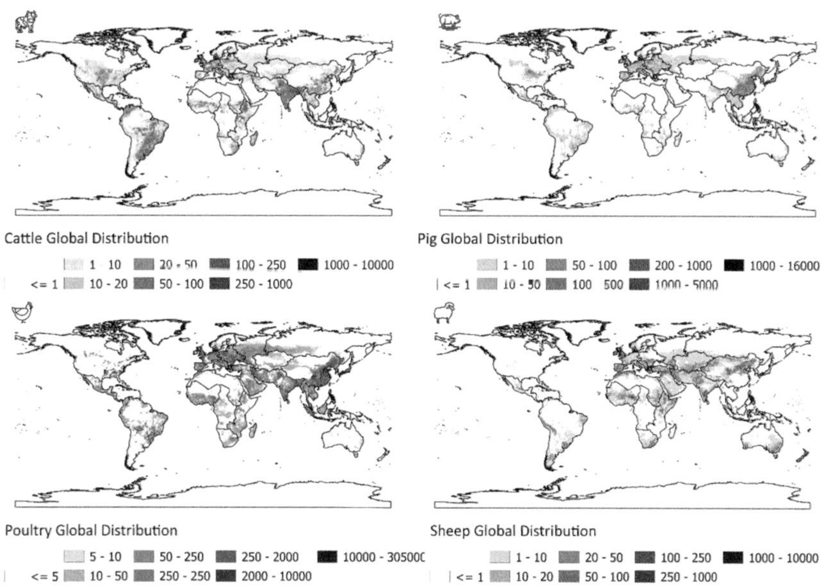

Fig. 4: Gridded livestock of the world (GLW) (*Source*: Author's elaboration on gridded livestock of the world vs 2.01 {GLW} – FAOSTAT dataset)

3.6. Case Study on Climate Change Mitigation in Colombian Poultry Supply Chains

Since chicken meat is projected to become the most produced and widespread meat in the world, this study uses the interactive tool GLEAM-i 2.0 to analyze the sustainability of the chicken meat value chain and its impacts in Colombia, a developing country.

For poultry, GLEAM distinguishes between 'backyard' and 'industrial' systems. Animals in industrial systems have been further divided into Layers and Broilers. The characteristics of each system are detailed in Table 1. In the present study, only the broilers category in Colombia is considered because it is completely oriented to the chicken meat market. According to GLEAM's simulation for 2010, the total number of broilers bred in Colombia was 84.860.385 animals, which generated 1.094.795.160 kg of meat (Table 2).

The total emissions produced was 4.623.418.792 kg CO_2-eq, with a meat emission intensity coefficient of 27.3 kg of CO_2-eq/kg protein. It is particularly interesting to analyze which of the links in the chain contributed most to this result (emission sources detailed in Table 3). Emissions caused by land-use-change (LUC) due to the expansion of intensive soybean crops account for 45 per cent of the emissions produced in the entire chain, the highest of all factors. The second highest contributor is from animal feed production with 30 per cent of

Table 1: Poultry Herd Type in GLEAM (2018)

Backyard	Animals producing meat and eggs for the owner and local market, living freely. Diet consists of swill and scavenging (20 to 40 per cent) while locally produced feed constitutes the rest.	Simple, using local wood, bamboo, clay, leaf material and handmade construction resources for support, plus scarp wire netting walls and scrap iron for roof.
Layers	Fully market-oriented; high capital input requirements; high level of overall flock productivity; purchased non-local feed or on-farm intensively produced feed.	Layers housed in a variety of cages, barn and free-range systems, with automatic feed and water provision.
Broilers	Fully market-oriented; high capital input requirements; high level of overall flock productivity; purchased non-local feed or on-farm intensively produced feed.	Broilers assumed to be primarily loosely housed on litter, with automatic feed and water provision.

(*Source*: Author's Elaboration on FAO, 2018c)

Table 2: Colombian Broiler Value Chain Emissions Estimation

Emss: total GHG emissions	kg CO_2-eq /year	4.623.418.792
Emss: total CO_2	kg CO_2-eq/year	3.832.794.545
Emss: total CO_4	kg CO_2-eq/year	37.449.535
Emss: total N_2O	kg CO_2 -eq/year	753.174.711
Emss: feed – N_2O fertilizer and crop residues	kg CO_2-eq/year	630.329.116
Emss: feed – N_2O manure applied and deposited	kg CO_2-eq/year	2.720.103
Emss: feed – CO_2 feed production	kg CO_2-eq/year	1.381.600.207
Emss: feed – CO_2 luc soy	kg CO_2-eq/year	2.085.476.350
Emss: manure – CH_4 from manure management	kg CO_2-eq/year	37.449.535
Emss: manure – N_2O from manure management	kg CO_2-eq/year	120.125.493
Emss: energy – CO_2 direct energy use	kg CO_2-eq/year	314.753.609
Mss: energy – CO_2 indirect energy use	kg CO_2-eq/year	50.964.380
Prod: meat – carcass weight	kg CW/year	1.094.795.160
Prod: meat – protein amount	kg protein/year	156.008.310
Herd: total number of animals	heads/year	84.860.385
Herd: adult females	heads/year	2.031.720
Herd: adult males	heads/year	203.172
Herd: replacement females	heads/year	668.657
Herd: replacement males	heads/year	65.121
Herd: fattening animals	heads/year	81.891.714
Intake: total intake	kg DM/year	3.253.154.271
Intake: total intake – swill & roughages	kg DM/year	-
Intake: total intake – grains & food crops	kg DM/year	2.309.739.532
Intake: total intake – agro-industrial by-products	kg DM/year	878.351.653
Intake: total intake – additives	kg DM/year	65.063.085
Ei: emission intensity of meat	kg CO_2-eq/kg protein	27.3

(*Source*: Author's calculations by GLEAM 2.0)

emissions, followed by fertilizer residues and crops with 14 per cent of emissions, and energy use with 7 per cent of emissions (Fig. 5).

Table 3: Colombian Broiler Value Chain Emissions by Source

Feed - N_2O fertilizer and crop residues	kg CO_2-eq/year	630.329.116
Feed - N_2O manure applied and deposited	kg CO_2-eq/year	2.720.103
Feed - CO_2 feed production	kg CO_2-eq/year	1.381.600.207
Feed - CO_2 LUC soy	kg CO_2-eq/year	2.085.476.350
Manure - CH_4 from manure management	kg CO_2-eq/year	37.449.535
Manure - N_2O from manure management	kg CO_2-eq/year	120.125.493
Energy - CO_2 direct energy use	kg CO_2-eq/year	314.753.609
Energy - CO_2 indirect energy use	kg CO_2-eq/year	50.964.380

(*Source*: Author's Elaboration of GLEAM 2.0 Results)

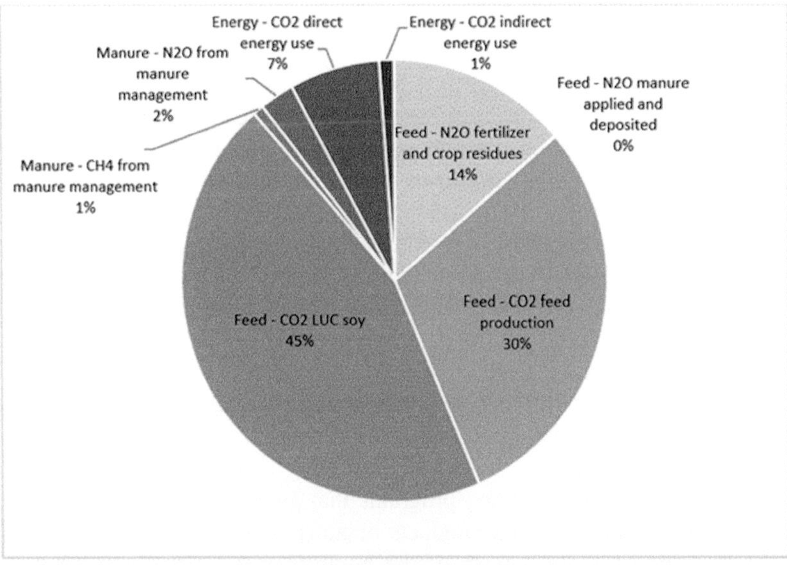

Fig. 5: Colombian broiler value chain emissions by source
(*Source*: Author's elaboration of GLEAM 2.0 results)

The activities related to Section 1 in the Colombian chicken meat value chain (as defined in the previous paragraph) constitute a large environmental burden in terms of GHG, contributing to global warming and at the same time endangering the industry's ability to maintain itself over the long term.

3.7. Possible Improvement Scenarios

As stated by Graziano da Silva, Director of the FAO, it is possible to create a more sustainable and environmentally friendly livestock sector. Da Silva claims that emissions related to meat production could quickly be reduced by 20 to 30 per cent in all production systems by adopting a better selection of animal feed, optimizing manure management, and improving animal health. These changes are also essential for increasing livestock production and preventing further deforestation (FAO ,2018a, Foreword). At the global level, it is observed that many advanced countries, influenced by market demand for healthier and greener consumption patterns, are trying to implement alternative feed practices for the poultry sector, with policies and objectives varying by country (Soler and Fonseca, 2011).

A study conducted by the FAO on alternative animal feeding systems for the poultry sector has shown that in the Caribbean region, when alternative feed (derived from local crops) replaces commercial concentrated feed, production levels are maintained and, at the same time, production costs are reduced (FAO, 2013). In Latin American countries, and in developing countries in general, the animal feeding systems that are often presented as alternatives, are rather traditional practices that resulted from the slow adaptation of humans to their environment (FAO, 2013). Traditional poultry feeding systems are still in widespread use in rural areas, often used by poorer families that practice subsistence farming, and are in fact the most common method for feed production in rural areas within developing countries. The resource base available to feed local poultry includes household organic waste, crop residues, forage and wild plants, and by-products of small local agricultural industries (cereal by-products, sugar cane, etc.) (FAO, 2013).

Between the two extremes represented by traditional and intensive production systems, there are semi-industrial systems, characterized by small- to medium-sized livestock (50 to 500 units) which may include local species, crossbreeds, or selected species. These businesses are only partly dependent on the purchase of food from commercial feed manufacturers, as they use alternative forms of poultry nutrition by mixing concentrated feed with locally available resources from the traditional systems described above (Soler and Fonseca, 2011).

Therefore, interventions aimed at preserving and improving this type of production, through research and programs that guarantee good percentages of productivity and quality of the products obtained, could not only considerably reduce the environmental impact of the poultry sector, but also facilitate the distribution of wealth through the development of small- and medium-sized enterprises within the sector (Soler and Fonseca, 2011).

The GLEAM-i tool will be used to predict the impacts (in terms of production and emissions) of substituting industrial feeding with backyard feeding methods in Colombia's broiler meat production chain. The food rations of broiler chickens for meat production in Colombia is composed of 71 per cent corn, 27 per cent agro-industrial products derived from soya, and 2 per cent additives (Fig. 6). On the other hand, as far as 'backyard' production is concerned, the birds' diet appears much more varied: 23 per cent is based on cereals (including sorghum, soya, and maize), 23.5 per cent on agro-industrial products (including mainly residues from sugar cane production) and 53 per cent on alternative sources deriving from various types of local crops and food and agricultural waste (FAO, 2018c).

Fig. 6: Feed settings screen in GLEAM-i 2.0

Feed ration percentages of broilers are modified on the basis of the suggestions made by the Soler and Fonseca study (2011) proposing alternative feeding systems for poultry in Colombia, and these modifications are adjusted by GLEAM to ensure the right nutritional intake of the animal. The use of maize is increased to 31 per cent, but the use of other cereals is increased to 20 per cent (sorghum, soya), the use of agro-industrial products derived from soya is eliminated, but the use of agricultural residues (mainly sugar cane) is increased to 27 per cent, and the use of alternative sources derived from various types of local crops is increased to 20 per cent (Fig. 7).

As can be seen in Table 4, the change in the composition of the feed system, as shown by the studies previously mentioned, does not change poultry production levels, but the effects in terms of emissions are very incisive: the total reduction of emissions estimated by GLEAM-i is 43.7 per cent, the intensity of chicken meat emissions drops from 27.4 to 14.3 kg of CO_2-eq/kg protein.

However, the change of feed leads to a considerable increase in emissions produced in manure disposal (manure management) from 37.449.535 to 154.151.957 kg of CO_2-eq/year. To mitigate this effect, a further intervention is

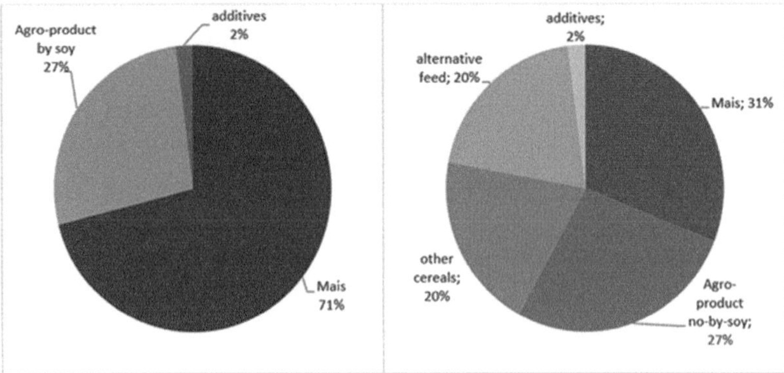

A B

Figs. 7: Baseline feed composition (A); Scenario feed composition (B)
(*Source*: Author's elaboration of GLEAM 2.0 results)

proposed for poultry manure management. Intensive chicken farms usually use the litter system, collecting animal manure, leaving it to dry, and then reselling it to the fertilizer industry.

A second scenario is then defined (Fig. 8), in which the above feed changes are combined with changes in GLEAM-i to the manure disposal data, replacing the 'litter' system with a 'daily spread' system. The latter system requires manure to be removed daily from grassland and applied to crops within 24 hours as direct fertilizer. The 'daily spread' system strongly mitigates the increase in emissions

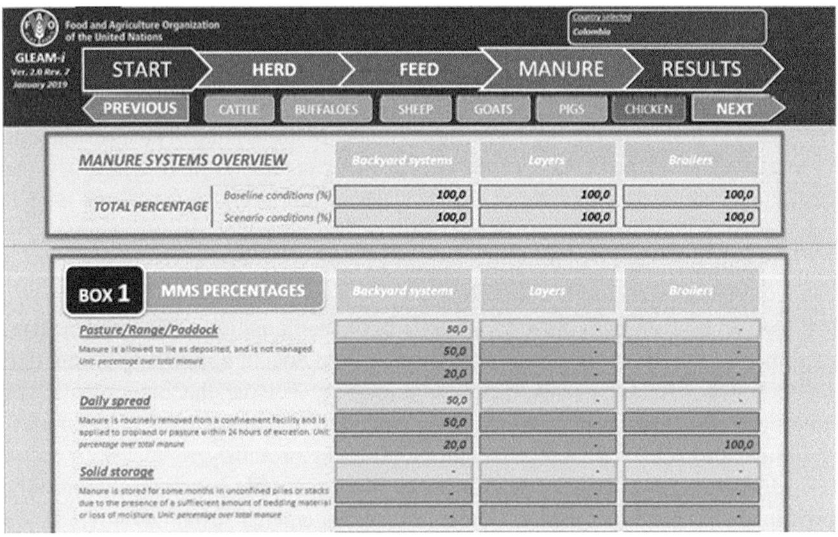

Fig. 8: Manure settings screen in GLEAM-i 2.0

Table 4: Results GLEAM-i Calculation on Broiler Value Chain Colombia, Baseline, Feed Scenario, Feed + Manure Scenario

	Baseline	**Feed Scenario**	**Feed + Manure Scenario**	
EMSS: Total GHG emissions	4.623.418.792	2.602.021.210(-43,7%)	2.422.312.297(-47,6%)	kg CO_2-eq/year
EMSS: Total CO_2	3.832.794.545	1.813.879.567(-52,7%)	1.813.879.567(-52,7%)	kg CO_2-eq/year
EMSS: Total CH_4	37.449.535	154.151.957 (311,6%)	51.383.986 (37,2%)	kg CO_2-eq/year
EMSS: Total N_2O	753.174.711	633.989.687(-15,8%)	557.048.745 (-26,0%)	kg CO_2-eq/year
EMSS: Feed - N_2O **fertilizer and crop residues**	630.329.116	530.916.389(-15,8%)	530.916.389 (-15,8%)	kg CO_2-eq/year
EMSS: Feed - N_2O **manure applied and deposited**	2.720.103	3.249.054 (19,4%)	5.068.524 (86,3%)	kg CO_2-eq/year
EMSS: Feed - CO_2 **feed production**	1.381.600.207	1.448.161.578 (4,8%)	1.448.161.578 (4,8%)	kg CO_2-eq/year
EMSS: Feed - CO_2 **LUC soy**	2.085.476.350	0 (-100,0%)	0 (-100,0%)	kg CO_2-eq/year
EMSS: Manure - CH_4 **from manure management**	37.449.535	154.151.957 (311,6%)	51.383.986 (37,2%)	kg CO_2-eq/year
EMSS: Manure - N_2O **from manure management**	120.125.493	99.824.243 (-16,9%)	21.063.831 (-82,5%)	kg CO_2-eq/year
EMSS: Energy - CO_2 **direct energy use**	314.753.609	314.753.609 (0,0%)	314.753.609 (0,0%)	kg CO_2-eq/year
EMSS: Energy - CO_2 **indirect energy use**	50.964.380	50.964.380 (0,0%)	50.964.380 (0,0%)	kg CO_2-eq/year
EI: Emission intensity of meat	27,3	14,3 (-47,5%)	13,2 (-51,7)	kg CO_2-eq/kg protein
PROD: Meat - carcassweight	1.094.795.160	1.094.795.160 (0,0%)	1.094.795.160 (0,0%)	kg CW/year

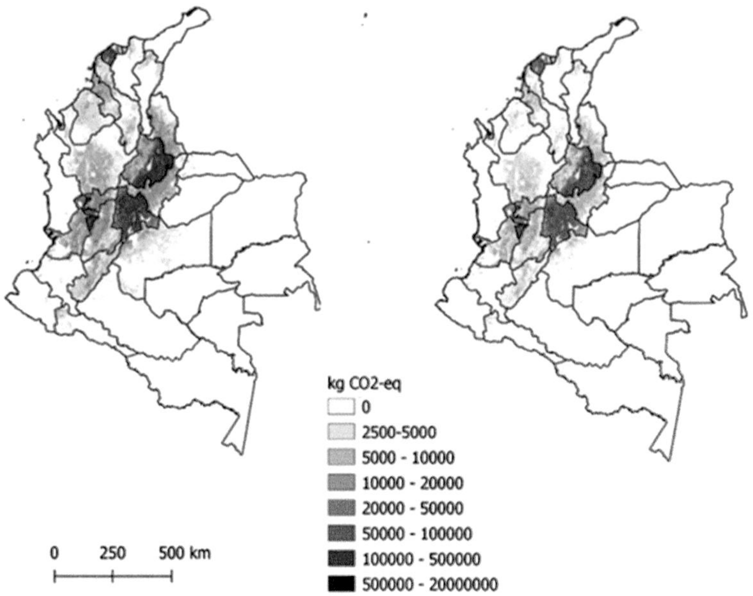

A B

Fig. 9: Colombian broiler value chain GHG emission baseline (A) and feed + manure scenario (B) (*Source*: Author's elaboration of GLEAM-i calculations transposed on distribution of broilers in Colombia by Gridded Livestock of the World vs 2.01 (GLW) – FAOSTAT dataset)

resulting from manure disposal, with emissions of 51,383,986 – a figure much lower than the emission estimate of 154,151,957 for the first scenario. When compared to baseline conditions, 'daily spread' reduces total emissions (Fig. 8) (-47.6 per cent) and the coefficient of intensity of chicken meat emissions drops further to 13.2. Fig. 9 presents a map representation of the emission reduction between baseline conditions and the feed-manure scenario.

In practice, the solution to make the chicken meat value chain in Colombia more sustainable is to promote collaboration between small- and medium-sized agricultural and poultry farmers. Poultry farms will have to diversify their feed by using not only commercial feed, but also waste from local crops and by-products from neighbouring cultivators. Cultivators will have to reduce the use of processed fertilizers by using poultry farm waste directly.

3.8 Conclusions and Recommendations

The intensive production methods of the current agri-food system have a growing environmental impact. The survival and prosperity of the agri-food sector is highly dependent on the conditions of the environment in which it operates and on access to natural resources (mainly water and fertile soil). At the same time, agricultural and agro-industrial activities involve a large expenditure of non-renewable resources and contribute significantly to anthropogenic greenhouse gas (GHG) emissions.

In the analysis conducted in this work, attention has been paid to the intensive livestock sector, recognized as crucial for food security in developing countries, and for its large environmental impact. Indeed, the expected increase in global demand for meat in the near future and the consequent increase in production will have serious repercussions on the availability of food resources, as livestock farming requires extensive use of agricultural products that could instead be used to feed the world's poorest people.

Moreover, the scientific community has raised the alarm about the enormous environmental impact generated by the livestock sector, which is recognized as responsible for most anthropogenic GHG emissions, contributing significantly to global warming. Climate change is one of the most worrying risk factors for the future of agro-food systems, affecting productive capacity and heightening global food insecurity for lack of adequate distribution and access.

Poultry farms are the most widespread in the world and soon the consumption of chicken meat will exceed that of pig meat. Increase in market demand for poultry is, on the one hand, supported by lower production costs and low consumer prices and, on the other, reflective of the perception that chicken is a healthy product.

The global meat industry exists within complex global value chains that, if managed sustainably, can significantly mitigate food insecurity and environmental impact in countries with agri-food sectors. As chicken meat will become the most widely produced and consumed meat, this study has chosen to analyze the sustainability of the chicken meat value chain and its effects in a developing country like Colombia.

The analysis shows that activities related to feed production constitute the largest GHG emission burden of the Colombian chicken meat value chain, contributing to global warming and diverting food resources away from humans. In order to increase the sustainability of the global meat value chain, it is necessary to rethink current production methods, opting for solutions that favor local development and lower environmental impact. As far as the chicken meat value chain in Colombia is concerned, it is suggested in the present work to favor the development of small and medium poultry enterprises, through greater integration and collaboration with neighboring agricultural stakeholders, where waste material of one becomes the productive input for the other, and vice versa. Possible scenarios for improvement of the chain include the diversification of poultry feed by including crops and waste from local agriculture and the direct application of manure collected daily from farms as organic fertilizer for crops.

Technology certainly has the potential to support progress in human development, but sustainable development requires us to look back and draw inspiration from the oldest and most traditional practices. Considering traditional farming methods could help us returning to a more 'human', balanced, and environmentally friendly food system.

Bibliography

Andrew, P.S. (1991). *Decision Support Systems Engineering*, John Wiley & Sons Inc., New York, USA.

Armstrong, M.P., P.J. Densham and G. Rushton (1986). Architecture for a microcomputer-based spatial decision support system, *Proceedings of the Second International Symposium on Spatial Data Handling, IGU Commission on Geographical Data Sensing and Processing*, Williamsville, USA.

ATKearney group (2018). *How will Cultured Meat and Meat Alternatives Disrupt the Agricultural and Food Industry?* Retrieved from https://www.atkearney.com/retail/article/?/a/how-will-cultured-meat-and-meat-alternativesdisrupt-the-agricultural-and-food-industry; accessed on 6 November, 2020.

Basil, M., Paparrizos, K., N. Matsatsinis and J. Papathanasiou (2010). *Decision Support Systems in Agriculture, Food and the Environment: Trends, Applications and Advances.* 10.4018/978-1-61520-881-4. IGI Global.

Breton, V. (2009). *Continuarán muriendo de hambre millones de personas en el siglo* XXI? *Revista Española de Estudios Agrosociales y Pesqueros*, **4**(224): 69-110.

Crossland, M.D., B.E. Wynne and W.C. Perkins (1995). Spatial decision support systems: An overview of technology and a test of efficacy, *Decision Support Systems*, **14**(3): 219-235.

Delgado, C. (2003). *The Livestock Revolution: A Pathway from Poverty*? Australia: Crawford Fund.

FAO (Food and Agriculture Organization). (2012). World Agriculture Towards 2030/2050. The 2012 revision, ESA Working Paper No. 12-03.

FAO (Food and Agriculture Organization). (2013). Tackling Climate Change through Livestock: A Global Assessment of Emissions and Mitigation Opportunities; Retrieved from http://www.fao.org/docrep/018/i3437e/i3437e.pdf; accessed on 8 November, 2020.

FAO (Food and Agriculture Organization). (2015). Global Livestock Environmental Assessment Model; online tool; Retrieved from http://www.fao.org/gleam/en/; accessed on 18 April, 2019. (not in text)

FAO (Food and Agriculture Organization). (2016). The State of Food and Agriculture, 2016; Retrieved from http://www.fao.org/3/i6030e/i6030e.pdf; accessed on 18 December, 2020.

FAO (Food and Agriculture Organization). (2018a). World Livestock: Transforming the Livestock Sector through the Sustainable Development Goals; Retrieved from http://www.fao.org/3/CA1201EN/ca1201en.pdf; accessed on 22 April, 2020.

FAO (Food and Agriculture Organization). (2018b). The State of Food Security and Nutrition in the World (SOFI): Building Climate Resilience for Food Security and Nutrition; Retrieved from http://www.fao.org/state-of-food-security-nutrition/2018/en; accessed on 18 December, 2020.

FAO (Food and Agriculture Organization). (2018c). Global Livestock Environmental Assessment Model. Model Description Version 2.0; Retrieved from http://www.

fao.org/fileadmin/user_upload/gleam/docs/GLEAM_2.0_Model_description.pdf; accessed on 21 March, 2021.

FAO (Food and Agriculture Organization). (2019). The State of Food and Agriculture, 2019. Moving Forward on Food Loss and Waste Reduction; Retrieved from http://www.fao.org/3/ca6030en/ca6030en.pdf; accessed on 22 April, 2020.

Finer, M. and N. Mamani (2019). Fires and Deforestation in the Brazilian Amazon; Retrieved from https://maaproject.org/2019/fires-deforestation-brazil-2019/; accessed on 2 December, 2020.

GLEAM (2018).

Godfray, H.C.J., J.R. Beddington, I.R. Crute, L. Haddad, D. Lawrence, J.F. Muir, J. Pretty, S. Robinson, S.M. Thomas and C. Toulmin (2010). Food security: The challenge of feeding 9 billion people, *Science*, **327**(5967): 812-818.

Gómez, M., C. Barrett, L. Buck, H. De Groote, S. Ferris, O. Gao, E. Mccullough, D.D. Miller, H.Outhred, A.N. Pell, T. Reardon, M. Retnanestri, R. Ruben, P. Struebi, J. Swinnen, M.A.Touesnard, K. Weinberger, J.D.H. Keatinge, M.B. Milstein and R.Y. Yang (2011). Food value chains, sustainability indicators and poverty alleviation, *Science*, **332**(6034): 1154-1155.

Goodland, R. and J. Anhang (2009). Livestock and climate change, *World Watch*, 22: 10-19.

Gorry, A.G. and S.S. Morton (1971). Framework for Management Information Systems, *Sloan Management Review*, **13**(1).

Greenpeace (2009). Slaughtering the Amazon – Executive Summary; Retrieved from https://www.greenpeace.org/usa/wpcontent/uploads/legacy/Global/usa/planet3/PDFs/slaughtering-the-amazon.pdf; accessed on 8 November, 2020.

Herrero, M, P. Gerber, T. Vellinga, T. Garnett, A. Leip, C. Opio, H. Westhoek, P. Thornton, J. Olesen, N. Hutchings, H. Montgomery, J. Soussana, H. Steinfeld and T. Mcallister (2011). Livestock and greenhouse gas emissions: The importance of getting the numbers right, *Animal Feed Science and Technology*, 166-167; 779-782.

HLPE (High Level Panel of Experts on Food Security and Nutrition of the Committee on World Food Security). (2016). *Sustainable Agricultural Development for Food Security and Nutrition: What Roles for Livestock?* Retrieved from http://www.fao.org/3/mq860e/mq860e.pdf; accessed on 8 November, 2020.

INPE (Instituto Nacional de Pesquisas Espaciais) (2020). Monitoramento do Desmatamento da Floresta Amazônica Brasileira por Satélite; Retrieved from http://www.obt.inpe.br/OBT/assuntos/programas/amazonia/prodes; accessed on 2 August, 2020.

IPCC (Intergovernmental Panel on Climate Change). (2013). Fifth Assessment Report of the Intergovernmental Panel on Climate Change; Retrieved from https://www.ipcc.ch/report/ar5/syr/; accessed on 13 December, 2020.

IPCC (Intergovernmental Panel on Climate Change). (2019). Climate Change and Land. An IPCC Special Report on Climate Change, Desertification, Land Degradation, Sustainable Land Management, Food Security, and Greenhouse Gas Fluxes in Terrestrial Ecosystems; Retrieved from https://www.ipcc.ch/report/srccl/; accessed on 6 December, 2020.

Keen, G.P. and M.S. Morton (1978). *Decision Support Systems: An Organizational Perspective*, Addison Wesley, Massachusetts, USA.

Koellner, T., L. de Baan, T. Beck, M. Brandão, B. Civit, M. Margini, L. Milái Canals, R. Saad, D. Maia de Souza and R. Müller-Wenk (2013). UNEP-SETAC guideline on global land use impact assessment on biodiversity and ecosystem services in LCA, *The International Journal of Cycle Assess*, **18**(6): 1188-1202.

MacLeod, M.J., T. Vellinga, C. Opio, A. Falcucci, G. Tempio, B. Henderson, H. Makkar, A. Mottet, T. Robinson, H. Steinfeld and P.J. Gerber (2018). Invited review: A position on the global livestock environmental assessment model (GLEAM), *Animal*, **12**(2): 383-397.

Marotta, L. (2010). ICZM technologies for integrating data and support decision making, *Instrumentation Viewpoint*, 8: 93.

McMichael, A.J., J.W. Powles, C.D. Butler and R. Uauy (2007). Food, livestock production, energy, climate change, and health, *Lancet*, 370: 1253-1263.

Mottet, A., C. de Haan, A. Falcucci, G. Tempio, C. Opio and P. Gerber (2017). Livestock: On our plates or eating at our table: A new analysis of the feed/food debate, *Global Food Security*, 14: 1-8.

OECD (Organization for Economic Co-operation and Development), World Bank and WTO (World Trade Organization). (2014). Global Value Chains: Challenges, Opportunities, and Implications for Policy; Retrieved from https://globalvaluechains.org/publication/global-value-chains-challenges-opportunities-and-implications-policy; accessed on 30 March, 2021.

OECD (Organization for Economic Co-operation and Development), FAO (Food and Agriculture Organization). (2019). OECD-FAO Agricultural Outlook 2019-2028; Retrieved from https://doi.org/10.1787/agr_outlook-2019-en; acessed on 6 December, 2020.

Pérez de Armiño, K. (1998). El futuro del hambre, Población, alimentación y pobreza en las primeras décadas del siglo XXI, *Cuadernos de trabajo de Hegoa*, 22: 1-52.

Sharda, R., E. Dursun and E. Turban (2015). *Business Intelligence and Analytics: Systems for Decision Support*, Pearson Education Inc., New Jersey, USA.

Smith, P., P.J. Gregory, D. van Vuuren, M. Obersteiner, P. Havlík, M. Rounsevell, J. Woods, E. Stehfest and J. Bellarby (2010). Competition for land, *Philosophical Transactions of the Royal Society B*, **365**(1554): 2941-2957.

Soler, D.M. and J.A. Fonseca (2011). Producción sostenible de pollo de engorde y gallina ponedora campesina: Revisión bibliográfica y propuesta de un modelo para pequeños productores, *Revista de Investigación Agraria y Ambiental RIAA*, **2**(1): 29-43.

Steinfeld, H., P. Gerber, T. Wassenaar, V. Castel, M. Rosales and C. de Haan (2006). Livestock's long Shadow: FAO and Livestock, Environment and Development (LEAD); Retrieved from http://www.fao.org/3/a0701e/a0701e00.htm; accessed on 12 December, 2020.

Sugumaran, R. and J. Degroote (2010). *Spatial Decision Support Systems: Principles and Practices*, CRC Press, Boca Raton, USA.

Turban, E. (1995). *Decision Support and Expert Systems Management Support Systems*. Prentice-Hall, Inc.

UN General Assembly (2015). Transforming Our World: The 2030 Agenda for Sustainable Development, 21 October 2015, A/RES/70/1; Retrieved from https://www.refworld.org/docid/57b6e3e44.html; accessed on 18 November, 2020.

Van der Vorst, J.G.A.J., C.A. da Silva and J.H. Trienekens (2007). Agro-industrial supply chain Management: Concepts and applications. FAO; Retrieved from http://www.fao.org/3/a1369e/a1369e.pdf; accessed on 8 April, 2020.

WCED (World Commission on Environment and Development) (1987). Our Common Future; Retrieved from https://sustainabledevelopment.un.org/content/documents/5987our-common-future.pdf; accessed on 6 April, 2020.

Zhang, W., J. Dearing, M.D.S. Hossain, J. Dyke and A.M. Augustyn (2018). Systems thinking: An approach for understanding 'eco-agri-food systems', pp. 17-55. *In:* W. Zhang, J. Gowdy, A.M. Bassi, M. Santamaria, F. deClerck, A. Adegboyega, G.K.S. Andersson, A.M. Augustyn, R. Bawden, A. Bell, I. Darknhofer, J. Dearing, J. Dyke, J. Failler, P. Galetto, C.C. Hernandez, P. Johnson, P. Kleppel, P. Komarek, A. Latawiec, R. Mateus, A. McVittie, E. Ortega, D. Phelps, C. Ringler, K.K. Sangha, M. Schaafsma, S. Scherr, M.S.A. Hossain, J.P.R. Thorn, N. Tyack, T. Vaessen, E. Viglizzo, E. Walker, L. Willemen and S.L.R. Wood. The Economics of Ecosystems and Biodiversity (TEEB), TEEB for Agriculture & Food: Scientific and Economic Foundations, Geneva, Switzerland.

Part II

Agroecology at Farm Level: Contribution of New Basket of Growing Geographical Technologies

CHAPTER
4

Revolution in Precision of Positioning Systems: Diffusing Practice in Agroecology and Organic Farming

Angela Gatti[1]* and Alessio Zanoli[2]

[1] Advanced Master on GIScience and Unmanned Systems for Integrated Management of Territory and Natural Resources, University of Padova

[2] Master's degree in Sciences and Technologies for the Environment and the Territory, University of Udine and University of Trieste

4.1. Introduction

Positioning systems are the most significant technology developed in the last 20 years. Their applications in agriculture have revolutionized the working method. The possibility of knowing soil and crops variability at centimetre-level precision permits managing the activity in a very precise way. Even livestock management has improved by means of positioning systems. These technologies, combined with a sustainable agriculture management system, can contribute to mitigating the environmental impacts of agroecosystems. The FAO (Food and Agriculture Organization) have qualified organic farming and agroecology practices as two kinds of environmentally-friendly agriculture that respect the balance of the natural cycle and drastically reduce environmental impacts (FAO, 2018b). These types of agriculture management appear well suited to be substantially improved by position system technologies. For this reason, this chapter intends to present an overview of positioning system technologies with a focus on their applications in sustainable agriculture management. In the first paragraph, an essential overview of GNSS technology is given with particular reference to the different

*Corresponding author: angi.gatti94@gmail.com

types of positioning (absolute and differential). Next, the applications of GNSS technology in agriculture are explained. Efforts have been made to pay attention to the methods applied in sustainable agriculture: farm machinery guidance, soil sampling, harvest yield monitors, biomass monitoring, and livestock tracking. Some innovative examples of the application of GNSS in agroecology and organic agriculture are also noted. Finally, the characteristics of low-cost GNSS and their potential diffusion in sustainable agriculture are given.

4.2. Global Navigation Satellite Systems (GNSS)

Before the advent of man-made satellites, navigation and positioning mainly depended on ground-based radio navigation systems that were developed during the World War II (Shi and Wei, 2020). In the 1950s and 1960s, the USSR and the USA managed to realize three satellite navigation and positioning systems based on the Doppler shift of a radio signal. These 'Doppler shift-based' positioning systems needed long-term observations to realize navigation and positioning and the positioning accuracy was also unsatisfactory. To overcome these limitations, the joint development in the early 1970s of a new US satellite navigation system, the GPS (Global Positioning System), opened a new chapter for the development of satellite navigation systems (Shi and Wei, 2020; Kaplan and Hegarty, 2006; El Rabbany, 2002). GPS is fully operational nowadays and provides accurate, continuous, worldwide, three-dimensional position, and velocity information to users with the appropriate receiving equipment (Kaplan and Hegarty, 2006). Originally limited to use by the United States military, civilian use was allowed from the 1980s. Since then, many countries in the world have begun to develop independent GNSS (Global Navigation Satellite Systems). GNSSs have evolved from a single GPS constellation to multiple GNSS constellations and in the coming decades, the number of navigation satellites in orbit may increase to several hundred (Shi and Wei, 2020).

Although GPS is only one of the GNSS constellations, it is commonly used to refer to the general satellite positioning service.

GPS constellation will be explained in detail in the next paragraphs. Aside from GPS, the most notable GNSS constellations are:

- *GLONASS (GLObal Navigation Satellite System)*: Recent operating Russian constellation, consisting of about 24 satellites. Its reference system is PZ-90, while GPS refers to WGS84 (*see* paragraph 'GNSS basic idea'). Due to the low number of satellites, GLONASS does not constitute an independent system of positioning. However, its combination with GPS increases the overall number of visible satellites
- *GALILEO*: UE constellation. It is currently in the final phase of experimentation (fully operative since 2020) with the launch of the first satellite in 2006. In March 2019, 26 satellites existed

- *Quasi-Zenith Satellite System (QZSS)*: It is a four-satellite augmentation system developed by the Japanese government to enhance the GPS in the Asia-Oceania regions with a focus on Japan
- *IRNSS (India, seven satellites at the end of 2018)*: It is an autonomous regional satellite navigation system. It covers India and an area extending 1,500 kms around it with plans for further extension. The system currently consists of a constellation of seven satellites and with two additional satellites on ground as stand-by
- *BeiDou (BDS)*: It is a Chinese satellite navigation system. In addition to the PNT (Positioning, navigation, and Timing) service provided by all GNSSs, BDS-3 also provides regional message communication and global short message communication, global search, and SAR (rescue service), regional PPP (precise point positioning) service, BDSBAS (embedded satellite-based augmentation service), and space environment monitoring function. By the end of 2019, 28 BDS satellites had been successfully launched (Yang *et al.*, 2020)

All of the GNSS satellite systems provide an accurate, continuous, and worldwide assessment of the three geographic coordinates (latitude, longitude, and altitude) of every GNSS receiver in real or deferred time (El Rabbany, 2002; Czajewski and Michalski, 2004; Kaplan and Hegarty, 2006; Shi and Wei, 2020).

Any GNSS system is composed of three segments:

- Space segment
- Operational ground control segment
- User segment

In the next few paragraphs, the segments and functioning of the GPS constellation are explained, given that the main functions and structures of the various satellite navigation systems are similar (Shi and Wei, 2020).

4.2.1. The Space Segment

The GPS space segment consists of a constellation of satellites transmitting radio signals to users. GPS satellite orbits are nearly circular with a semi-major axis of about 26,500 km, with 60° longitude, and an inclination of 55° to the equator plane. Their revolution period is 11 hrs and 58 min. Each satellite circles the Earth twice a day (US Government, 2016).

The United States is committed to maintaining the availability of at least 24 operational GPS satellites for 95 per cent of the time. Overall, there are about 30 satellites positioned in six Earth-centered orbital planes with four satellites in each plane. This satellite constellation is built to allow any GPS receiver to view at least four to 10 satellites from virtually any point at any time on the planet. This is the necessary requirement for always providing the positioning information and its fulfilment makes GPS a fully operational capability system (El Rabbany, 2002).

4.2.2. The Operational Ground Control Segment

The CS (Control Segment) is responsible for maintaining the satellites and their proper functioning. This includes maintaining the satellites in their proper orbital positions (called station keeping), tracking satellite position, and monitoring satellite subsystem health and status. The CS also monitors the satellite solar arrays, battery power levels, and propellant levels used for manoeuvers. Furthermore, the control segment activates spare satellites (if available) to maintain system availability (Kaplan and Hegarty, 2006). It is composed of an MCS (Master Control Station), a worldwide network of monitor stations, and ground control stations. There are five monitor stations arranged along the equatorial line and located in Colorado Springs (with the MCS), Hawaii, Kwajalein, Diego Garcia, and Ascension Island. Their goal is to monitor, track, and predict the ephemerides which are the co-ordinates of satellites along their orbit. Ephemerides, the plural form of ephemeris, is a tabulation of computed positions and velocities (and/or various derived quantities, such as right ascension and declination) of an orbiting body at specific times (NASA, 2020).

The predicted ephemeris (or broadcast ephemeris) is the reference ephemeris recalculated and corrected by means of the collected satellite position data of the previous 12-24 hours. The precise ephemeris is computed in post-processed mode and made available with a delay of two to four weeks (Jia *et al.*, 2014).

The positions of the monitor stations are known very precisely. Each monitor station is equipped with high-quality GPS receivers and a caesium oscillator for continuous tracking of the ephemerides of all GPS satellites in view. Three of the monitor stations (Kwajalein, Diego Garcia, and Ascension Island) are also equipped with ground antennas for uploading the information to the GPS satellites (El Rabbany, 2002).

Every kind of GPS observation collected at the monitor stations is transmitted to the MCS for processing. The outcome of the processing is predicted satellite navigation data that includes, along with other information, the satellite positions as a function of time (ephemeris), the satellite clock parameters, atmospheric data, satellite almanac, and others. This fresh navigation data is sent to one of the ground controlstations to upload it to the GPS satellite (El Rabbany, 2002).

4.2.3. The User Segment

It consists of users equipped with a GPS receiver to obtain real- or deferred-time three-dimensional positioning. A GPS receiver is composed of:

- *An antenna*: Receives the incoming satellite signal and then converts its energy into an electric current which can be handled by the GPS receiver (Langley, 1991a; Langley, 2000)
- *A controller*: To control the receiver through a keyboard and a display
- *A software*: Present on the ROM memory which manages the acquisition and storage processes

- *A microprocessor*: To execute software operations, filter raw data, and calculate or convert geographic coordinates
- *A precision quartz-clock-oscillator*: To determine GPS time reference
- *A data recording system*: To store GPS acquisition data and receiver software
- *A power supply*

There are different kinds of GPS receivers, some are light and easy-portable tools (e.g. smartphones) (Fig. 1A) and can achieve precision within metres, while others are heavier, larger, more sophisticated (Fig. 1B), and are used for centimeter precision applications.

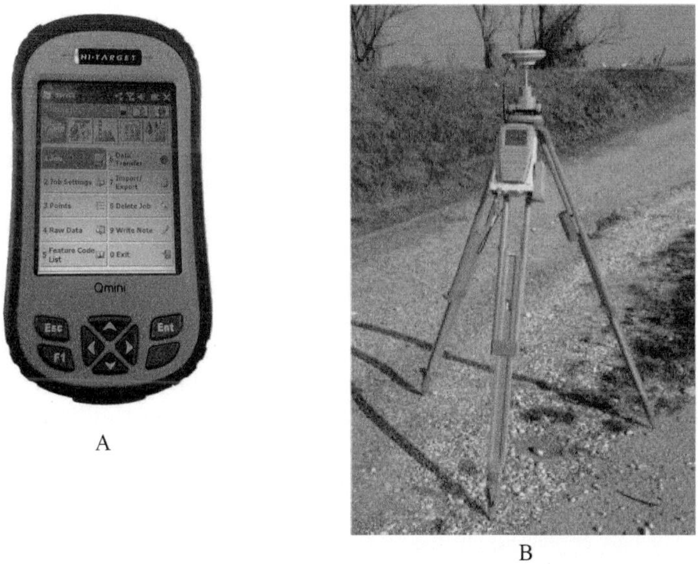

Fig. 1: Easy, portable GPS receiver (A) and sophisticated GPS receiver (B) (*Source*: Authors' elaboration)

4.2.4. GPS Co-ordinates System References

The standard physical model of the Earth used for GPS positioning applications is the WGS 84 (World Geodetic System 1984) (NIMA, 2000). WGS 84 is an ellipsoidal model of the Earth's shape. Such information is necessary to derive accurate satellite ephemeris information (Kaplan and Hegarty, 2006). Its characteristics are:

- *Center*: In the center of mass of the Earth
- *Z axis*: Passing through the North Pole
- *X axis*: Chosen so that the Greenwich meridian lies on the XZ plane
- *Y axis*: Chosen to give a right-handed triad, i.e. such that an observer placed along the Z axis sees the X axis overlapping Y with counter clockwise motion

- *Semi-major axis*: A = 6 378,137.000000 m
- *Semi-minor axis*: C = 6356, 752.314 245 m
- *Ellipticity*: F = 1 / 298.257223563
- *Geocentric gravitational constant*: U = 3986,005 × 108 m^3/s^2

4.2.5. The GNSS Basic Idea

The basic principle of GNSS work is rather simple. If the distances from a point on Earth (a GNSS receiver) to four GNSS satellites are known alongwith the satellite locations, then the location of the point (or receiver) can be determined simply by figuring out the position that concurrently accounts for the four distances. To understand how the satellite-receiver distance is calculated, it is necessary to know the GNSS signal structure (Langley, 1991b; US Government, 2016).

4.2.5.1. The GNSS Signal Structure

Each GNSS satellite transmits a microwave radio signal to GNSS receivers, travelling at the speed of light (c = 299,792.458 m/s), and is composed of two carrier frequencies modulated by two digital codes and a navigation message (El Rabbany, 2002). The two carrier frequencies are generated at 1,575.42 MHz (referred to as the L1 carrier), 1,227.60 MHz (referred to as the L2 carrier), which are obtained by multiplying the fundamental frequency of 10.23 MHz by 154 and 120 times. The corresponding carrier wavelengths are approximately 19 and 24.4 cm respectively (El Rabbany, 2002). All the GNSS satellites transmit the same L1 and L2 carrier frequencies. However, the code modulation is different for each satellite to significantly minimize the signal interference. The two GNSS codes are called coarse acquisition (or C/A-code) and precision (or P-code). Each code consists of a stream of binary digits, zeros, and ones, known as bits (El Rabbany, 2002). The carrier phase is shifted by 180° when the code value changes from zero to one or from one to zero (Wells *et al.*, 1987) (Fig. 2). Presently, the C/A-code is modulated on to the L1 carrier only, while the P-code is modulated onto both the L1 and the L2 carriers so that instruments capable of receiving the P code will have greater accuracy. The GNSS navigation message is a data stream added to both the L1 and the L2 carriers as a binary biphase modulation, at a low rate of 50 kbps. It consists of 25 frames of 1,500 bits each, or 37,500 bits in total (Hoffman *et al.*, 1994). The navigation message contains the information previously transmitted to the satellite from the master ground station (the co-ordinates of the GNSS satellites as a function of time, the satellite health status, the satellite clock correction, the satellite almanac, and atmospheric data). Each satellite transmits its own navigation message with information on the other satellites, such as the approximate location and health status (Hoffman *et al.*, 1994).

Actually, two more civil carrier frequencies at 1,176 MHz (referred to as the L5 carrier) and 1,575 MHz (referred to as the L1C carrier) are currently under development and will become fully operational in the next few years (US Government, 2016). L5 was developed to meet the demands of navigation

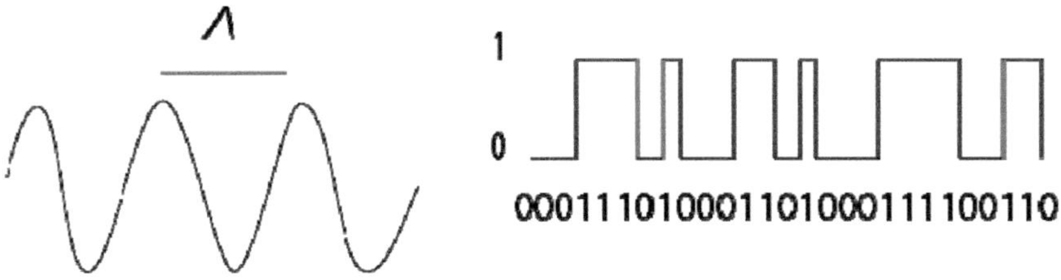

A B

Fig. 2: A sinusoidal wave (A) and a digital code (B) (*Source*: Authors' elaboration)

users in the field of safety-of-life-related transportation and other high-precision applications while L1C was designed for compatibility and interoperability between GPS and other GNSSs (Shi and Wei, 2020; US Government, 2016).

4.2.5.2. Pseudo Range Measurement

The pseudo range is a measure of the range, or distance, between the GNSS receiver and the satellite. As previously stated, the ranges from the receiver to the satellites are needed for the position GNSS details computation. The procedure of the GNSS range determination, or pseudo ranging, can be described as follows. The code produced by the satellite arrives at the receiver on the ground with a certain delay due to the distance between them. Then, the same code is produced inside the receiver: the phase shift between these two codes will therefore be a function of the satellite-receiver distance (Fig. 3). The time delay between the incoming signal and the replica produced by the receiver is obtained from subsequent comparisons between the signals until the correlation becomes maximum. Such delay is then multiplied by the speed of the radio signal (c) to detect the satellite-receiver distance. Unfortunately, the synchronization between receiver and satellite clocks is not perfect. For this reason, the measured range is contaminated and therefore, this distance measure is referred to as the pseudo range and not the range (Langley, 1993). Either the P-code or the C/A-code can be used for measuring the pseudo range.

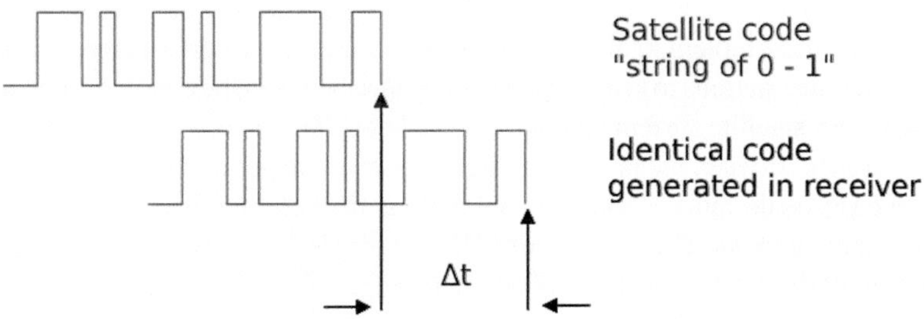

Fig. 3: Pseudorange measurements (*Source*: Authors' elaboration)

4.2.5.3. Carrier-phase Measurement

Another way of measuring the receiver-satellites ranges can be obtained through the carrier phases purified from the code signal modulated to them (superimposed). The range would simply be the sum of the total number of full carrier cycles plus fractional cycles at the receiver and the satellite, multiplied by the carrier wavelength (El Rabbany, 2002). The ranges determined with the carriers are far more accurate than those obtained with the codes (i.e. the pseudorings) (Langley, 1993). This is because the wavelength (or resolution) of the carrier phase, 19 cm in the case of L1 frequency, is much smaller than those of the codes. However, there is one problem. The carriers are just pure sinusoidal waves, which means that all cycles look the same. Therefore, a GNSS receiver has no means of differentiating one cycle from another (Langley, 1993). In other words, the receiver, when it is switched on, cannot determine the total number of the complete cycles between the satellite and the receiver. It can only measure a fraction of a cycle very accurately (less than 2 mm), while the initial number of complete cycles remains unknown or ambiguous. This is commonly known as the initial cycle ambiguity, or the ambiguity bias. The problem is typically solved by using the differential techniques (that will be described below). However, the problem can also be solved using a single receiver (El Rabbany, 2002).

4.2.6. Types of Positioning (Absolute, Differential)

There are two main methods of carrying out GNSS positioning: the absolute method and the differential method. Both can be achieved using the predicted or the precise ephemeris and with the pseudo range measurement, while, as previously stated, the carrier-phase measurement is applied almost exclusively to the differential method.

The absolute method is the simplest one. It needs only one receiver which determines the absolute position in real time. Its precision ranges from 1 to 50 m. It is a cheap method that does not include expensive equipment. In fact, it is utilized daily by people on their smartphone for such activities as reaching a location (Øvstedal, 2002).

The differential method (DGNSS) improves the positioning accuracy compared to the absolute method, enabling a GNSS accuracy of about 1-3 cm in cases of the best implementations. Each DGNSS uses a network of fixed ground-based reference stations to broadcast the difference between the positions indicated by the GNSS satellite system and known fixed positions. These stations broadcast the difference between the measured satellite pseudo ranges and actual (internally computed) pseudo ranges, and receiver stations may correct their pseudo ranges by the same amount. The digital correction signal is typically broadcast locally over ground-based transmitters of shorter range (Kaplan and Hegarty, 2006) (Fig. 4).

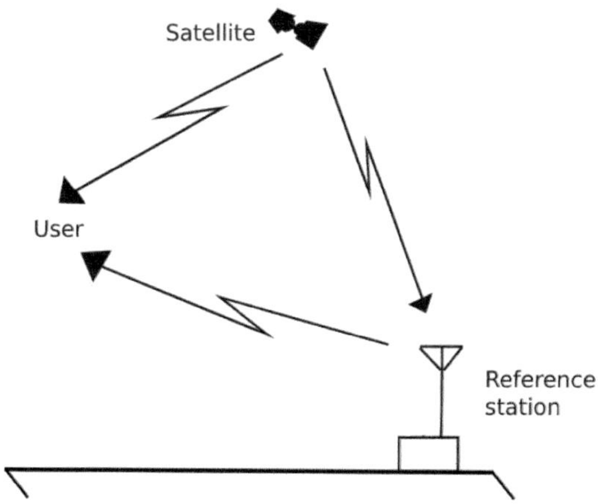

Fig. 4: Local-area DGNSS concept (*Source*: Authors' elaboration)

4.2.6.1. GNSS Positioning Errors

GNSS satellites broadcast their signals in space with a certain accuracy, but there are a series of errors that always influence absolute positioning. However, these are drastically reduced by differential methods. The main types of error are due to:

- *Orbit of the satellites*: The predicted ephemeris transmitted by satellites are characterized by an accuracy of the order of 10 m, while precise ephemeris is characterized by an accuracy of 1-2 cm (El Rabbany, 2002)
- *Clock non-synchronization*: Substantially due to the clock of the receiver which, for economic reasons, has low precision compared to satellites clocks (Kaplan and Hegarty, 2006)
- *Signal propagation*: It is assumed that the signal is propagating in a vacuum (at the speed of light in vacuum), while actually, the signal is influenced by atmospheric conditions (El Rabbany, 2002)
- *Atmospheric conditions*: Atmospheric rarefactions, especially in the ionosphere and troposphere, can cause a rapid variation in the amplitude and phase of the radio signals (Smita *et al*., 2006)
- *Multiple reflections (multipath)*: The whole antenna can receive, in addition to the signal coming directly from the satellite, other signals that have undergone reflections on reflective surfaces like buildings walls, rocky walls, bodies of water, etc. These signals tarnish the data and complicate the reception and analysis of the signal by the receiver (Kos *et al*., 2010)
- *Cycle slip (signal blockage)*: Temporary interruption of satellite visibility due to wood, electromagnetic sources, passage of vehicles near roads, short galleries, paths
- *Receiving station*: The eventual instability of the oscillator which generates the sample frequency (and therefore the signal replicas) (Kaplan and Hegarty,

2006), or of the phase center of the antenna (i.e. the point where the signal is currently received and, in essence, instrumental center) can affect the positioning precision (Hoffman *et al.*, 1994)

- *The operator*: If it makes mistakes setting up and measuring the height of the antenna

4.2.7. The Most Important Kind of GNSS Technologies for Agriculture

Currently, the most three utilized GNSS technologies in agriculture are: PPP (Precise Point Positioning), RTK (Real-Time Kinematic), and SBAS (Satellite-based Augmentation Systems).

PPP (Precise Point Positioning): This employs readily available satellite orbit and clock correction data, generated from a network of global reference stations to perform absolute positioning, using measurements from a single GNSS receiver. The corrections are delivered to the end user via satellite or internet, thus ensuring worldwide coverage. PPP can achieve decimeter-level accuracy, even under 1 cm (ESA, 2020a), without the need for a base station in the proximity. However, this comes at a price since PPP requires a rather long timeframe (15-30 min) to resolve any local biases, such as the atmospheric conditions, multipath environment, and satellite geometry to converge to decimeter-level accuracy (Cosentino *et al.*, 2006).

RTK: This is a carrier signal-based differential GNSS method. It enables highly accurate and repeatable positioning in the vicinity (typically 10-20 km) of a base station receiver placed in a known position. RTK utilizes a real-time communication channel (usually short-range radio) to transmit the corrections from the base station to the rover. The base station broadcasts its well-known location together with the code and carrier measurements at L1 and L2 frequencies for all in-view satellites. This information allows the rover equipment to fix the phase ambiguities and determine its location relative to the base, with an accuracy up to 2 cm. By adding up the location of the base, the rover is positioned in a global co-ordinate framework (Cosentino *et al.*, 2006).

SBAS: Solutions based on SBAS are becoming increasingly available in precision agriculture applications, frequently being the preferred option for farmers entering the market. SBAS systems provide services for improving the accuracy, integrity, and availability of the basic GNSS signals. This is achieved through a ground infrastructure consisting of reference stations receiving the data from the GNSS satellites and a processing facility center that computes the integrity, corrections, and ranging data forming the SBAS SIS. This is then transmitted or relayed through geostationary satellites back to users on the ground. Apart from integrity assurance, this correction service increases the positioning accuracy to end-users' receivers (getting both the primary and the SBAS signals) to sub-meter level (EGSA, 2019a), less than 1 meter (ESA, 2020b).

4.3. Applications of GNSS Technology in Agriculture

Since 1961, more than 3.2 million km^2 of land, an area equivalent to Australia, have been converted for agricultural use (IPCC, 2019). According to the IPCC's *Climate Change and Land Report* (2019), agriculture, forestry, and other intensive soil uses are responsible for almost a quarter (23 per cent) of all greenhouse gas emissions caused by human activity. In particular, intensive agricultural management reduces fertility, increases the compaction and erosion phenomena which, in the long run, increase the risk of desertification (D'Odorico *et al.*, 2013). The properties and functions of soil, and consequently those of the whole agro-ecosystem, are compromised by intensive agricultural management. Management strategies based on ecology and greater attention to natural balances can increase the sustainability of agricultural production while reducing the environmental impacts (Matson *et al.*, 1997).

According to the objectives of the 2030 Agenda for Sustainable Development, the insurance of sustainable food production systems and the mitigation of climate change are a priority to implement resilient agricultural practices that help maintain ecosystems (FAO, 2017). Migliorini and Wezel (2017) individuated two approaches proposed by different stakeholder groups about the future scenario and development of local, national, and global agriculture systems. On the one hand, there is an approach that increasingly relies on technology such as precision farming; on the other, there are the more ecologically-based traditional farming systems, like agroecology and organic farming (Migliorini and Wezel, 2017). Precision farming has strongly expanded in the last decade by means of GPS and big-data technology and is famous for its motto 'The right thing, at the right place, at the right time' (Shanwad *et al.*, 2002, p. 1). GPS can broadcast real-time signals which allow GPS receivers to calculate their precise locations and position. This information is provided on a real-time basis, implying that the position data are continually provided while still in motion (DGPS positioning) (Shanwad *et al.*, 2002). GPS technology allows specific site management of agricultural production, like soil preparation, sowing, and precision harvesting. Farmers can apply fertilizers and pesticides only where and when crops need it, reducing the use of fuel and consequently reducing environmental loading (EGSA, 2019a). Precision farming is a low sustainable system (Bongiovanni and Lowenberg-DeBoer, 2004) based on an industrial approach that influences socio-economic and farmer perception limiting its diffusion (Tey and Brindal, 2012).

More ecologically-based farming systems are agroecology and organic farming: two examples of sustainable agriculture that improve natural resources, provide ecosystem services, and produce lower greenhouse gas emissions than conventional agriculture. These two agriculture systems are in many parts quite similar in principles and practices (Migliorini and Wezel, 2017). Organic farming regulations mainly focus on the restriction of external inputs and limitation of chemical inputs, but include a holistic vision of sustainability (Migliorini and Wezel, 2017). Agroecology has a defined set of principles for ecological management of agri-food systems and also includes some socio-economic principles (Wezel *et al.*, 2020). Both agroecology and organic farming offer promising contributions to the future development of sustainable agricultural production and food systems, especially if their principles and practices converge to a transformative approach that impedes the conventionalization of agro-food systems (Migliorini and Wezel, 2017). The IAASTD (International Assessment of Agricultural Knowledge, Science and Technology for Development) in 2009 stated that agroecological methods were already available and used, and will allow smallholder farmers (farm area < 2 ha) in the world[1] to double their food production within 100 years in food-insecure areas of the planet (Migliorini and Wezel, 2017). Globally, 1.4 per cent of farmland is organic, and from 1999 to 2017, organic agriculture land increased by 533 per cent (Willer and Lernoud, 2019).

The revolution of sustainable agriculture is the application of precision technologies in conjunction with geolocalization, which provide the farmers with useful information for understanding crop needs and adapting their practices for the benefit of environmental conservation (Maurel and Huyghe, 2017).

Maurel and Huyghe (2017) have demonstrated that farm machinery and digital technologies are compatible with agroecology, contrary to the myth of agroecology being based on natural processes only, and therefore, not suitable for digital technologies.

In organic farming, the slurry application is fundamental and GNSS technology can permit applying the right dose of manure according to particular location conditions (Jacobsen *et al.*, 2005). By means of precision technologies, like GNSS, human intervention can be limited to ensure that plants' or animals' needs are met, making it possible to reduce chemical inputs (Maurel and Huyghe, 2017). Many socio-economic and environmental advantages will be reached with diffusion of the GNSS technology application in organic farming and agroecology.

To understand the potential of positioning systems' diffusion in agriculture, it is important to know their main applications. In literature, there is not a complete list of GNSS applications in agriculture and most information derives from precision farming reports. In the next paragraphs, the main applications are summarized on agriculture user needs and requirements (EGSA, 2019a).

4.4. Farm Machinery Guidance

Farm machinery guidance was the first activity where GNSS technology was applied in the mid-1990s and this system has significantly enhanced farm-field operations, such as spraying, fertilizing, planting, and harvesting. Corrected GNSS signals are utilized for the precise determination of the tractor deviations

[1] These make up 80 per cent of the total farm numbers and produce over 50 per cent of the world's food on 20 per cent of agricultural land (Migliorini and Wezel, 2017)

from a reference line, thus aiding farmers in following the desired path (EGSA, 2019a).

The two main GNSS-based guidance techniques are the lightbar (or manual) guidance system and the automatic steering system. Lightbar, being the simplest and least expensive solution, requires an operator to manually drive the vehicle. Its systems include a GNSS (typically RTK or SBAS-enabled) receiver and antenna, a computer/microprocessor for the computation of cross-track errors relative to a guidance line, an interface that allows user inputs and LED light bars. Automatic steering is a more advanced version of guidance that follows the same principles as lightbar guidance, but instead of prompting the driver to make slight corrective manoeuvers, it enables the vehicle to steer itself. Automatic steering systems consist of a GNSS receiver and antenna, controller, user interface module, attitude, and steering feedback sensors, and a steering actuator. Automatic steering systems using RTK can provide year-to-year repeatable accuracy to the level of 2.5 cm. RTK-based auto-steer guidance is used for applications, such as planting, harvesting, installing drip irrigation, and controlled traffic patterns (EGSA, 2019a).

4.4.1. Soil Sampling

When collecting soil samples, GNSS is used to precisely locate the sample points from a predefined grid. After testing the soil samples, information, such as nitrogen, organic material content, and moisture content, can be obtained. This type of information is mapped and used as a reference to guide farmers to efficiently and economically treat soil problems (Banu, 2015). The locational information can save money and time by allowing variable rate applications and treating only those areas with a documented need. There are two primary methods of precision soil sampling: grid sampling and zone management. In grid soil sampling, the field is divided into square grid sections (typically 0.5-1 ha). In zone management, areas within a field with similar soil or yield properties are grouped together and managed accordingly. The accuracy required for precision soil sampling is at meter/sub-meter level and can be satisfied by SBAS (EGSA, 2019a).

To test nitrogen, organic material, and moisture content of soil, it is possible to carry out a direct measure with specific sensors mounted on the tractor. For example, to measure organic matter content optical sensors are used based on the reflection of a visible (red) or infrared (1700-2600 nm) light beam. To measure moisture content, TDR (Time Domain Reflectometer) sensors, FDR (Frequency Domain Reflectometry), electromagnetic induction sensors, or infrared or microwave reflection sensors could be used (*Basso and Sartori*, 2013).

4.4.2. Harvest Yield Monitoring

During the 1990s, combine harvesters were equipped with yield monitors based on GNSS location. Harvest or yield monitoring systems enable the collection of accurate yield/crop data at a specific location and time. They are installed on a combine harvester and typically consist of a DGPS receiver, a computer, a user interface, and dedicated sensors that measure the amount and specific characteristics of crops harvested at the exact point where the combine harvester is located (EGSA 2019a). Apart from accumulated grain weight, a readout of the harvested area, and the corresponding yield rates, yield monitors may provide information such as soil moisture content and field elevation. The various yield data are stored and can be then plotted on maps, allowing post-analyses and identification of crop performance trends. This is very beneficial for year-to-year farm management decisions regarding the application of inputs in different areas of a field (EGSA, 2019a).

To apply the appropriate amounts of inputs at a precise time and/or location, farmers utilize the VRA (variable rate application) system. This is achieved through the utilization of a variable-rate control system that is linked to the application equipment. Following the accurate mapping and measurement of characteristics, such as acidity levels, and phosphorous, nitrogen, or potassium content, farmers use VRA to match the quantities of fertilizers, seeds, and herbicide to the need (EGSA, 2019a). Low fertilizer prices and high technology costs initially limited the adoption of variable rate technology, but then this system was applied in all phases of crop growth: soil tillage, sowing, mineral and organic fertilizers, crop protection, and irrigation treatments (Basso and Sartori, 2013). There are two methodologies to apply the VRA:

- *VRA based on maps*: Modifies the quantity of the products to be distributed on the basis of information in prescription maps
- *VRA based on sensors*: Utilizes sensors that detect important data in real-time like state of crops or physical-chemical states of soil, which are utilized as indicators to adjust the distribution of chemical products or other

4.4.3. Biomass Monitoring

The optical properties of the canopy are affected by changes in health, density, vigor, and productivity. Generally, to monitor the crop development and growth across an agricultural area, remote sensing is used. GNSS is used for on-site inspections of crop health and validations of the maps produced via other means. The information about crop health and stress is detected with SBAS GPS, thanks to its sub-meter level accuracy, and then translated into farming strategies to apply inputs by means of VRA method. The main sensors mounted on the tractor for monitoring crop health are radiometers, infrared image analyzers that can discriminate crops area from bare soil and any crop stress conditions. Infrared thermometers are used to measure the temperature of plants for irrigation purposes. Three different types of sensors can be used to quantify the biomass (Basso and Sartori, 2015)

- mechanical, to detect the resistance opposed by the crop correlating it with its mass

- photoelectric or electromechanical, to measure the distance between plants on the wire or their number
- mechanical or infrared probes for height determination

In addition, novel applications of GNSS-R (GNSS-Reflectometry) are being introduced for agriculture. This approach is based on the reflected L-band signal of the GNSS, which is sensitive both to the dielectric constant (humidity) and to the structure (vegetation), allowing effective monitoring of vegetation and biomass.

4.4.4. GNSS in Greenhouse Farming: The Case of GreenPatrol

An innovative project for sustainable agriculture which employs GNSS technology is the EU-funded H2020 GreenPatrol project. It makes direct use of the Galileo European GNSS to develop an innovative robotic solution for integrated pest management in greenhouses (GreenPatrol, 2017).

Nowadays, greenhouses are acquiring an important role in agriculture with their total surface in the world reaching 489,000 ha in 2017, 22 per cent more than 2011. Greenhouses protect crops from adverse weather conditions and allow farmers to control temperature, water, and nutrients of plants. Pests can cause losses of up to 15 per cent of production in a greenhouse. The goal of H2020 GreenPatrol project is to develop a prototype scouting robot to detect pests in the early stages for minimizing (or avoiding) the use of pesticides and to highly improve pest treatment success. The GreenPatrol robot has the capability to navigate inside a greenhouse and to perform pest detection autonomously by inspecting the leaves through its vision system, which combines image processing techniques with machine learning algorithms. Furthermore, it provides decision support for pest treatment.

To navigate inside the greenhouse, the GreenPatrol robot relies on the increased accuracy provided by Galileo GNSS system, thanks to its signal strength inside the greenhouse and the availability of multiple frequencies. GNSS is also used to localize the position of the ill plant detected inside the greenhouse (GreenPatrol, 2017).

4.4.5. Tracking Livestock

The location of valuable animals on a large farm can be monitored by GNSS transmitters embedded in the animals' collars. About 20 years ago, GNSS collars became commercially available and began to be used in livestock grazing research that was often a challenge (Bailey *et al.*, 2018) since finding all animals in a pasture is not easy. The first commercially available GNSS collars on the market involved the use of GNSS receivers embedded into the collars of individual animals (the most common cows), which were used to track their position and behavior in relation to grazing habits, thus allowing optimal use of grasslands and food resources (EGSA, 2019a). In the past, livestock motion and location data were stored on the sensor's device (collar or ear tag) and could not be accessed until the device was removed from the animal. Nowadays, real-time or near real-time monitoring of location and animal motion data is commercially available and permits readily finding livestock in extensive and rugged rangeland pastures (Bailey *et al.*, 2018). The geo-referenced data can be collected continuously and stored in dedicated databases, thus permitting post-processing and use for the elaboration of farming strategies related to animal feeding, pasture area management, and herd management. GNSS receivers are also used to detect cow fertility and illness along with other tools, as accelerometers and pedometers. The last most innovative use of livestock GNSS is the so-called virtual fencing, whereby animals reaching the boundary of a predefined area receive a sound or electrical stimulus that prevents them from exiting it. The accuracy requirement for livestock tracking and virtual fencing is at the meter level and can be provided by SBAS or even (multi)GNSS receivers. However, it must be noted that livestock tracking has not yet been taken up by end-users primarily due to the high cost of collars. An exception to that is the use of GNSS-enabled collars for scientific monitoring of wildlife (EGSA, 2019a).

4.5. Revolution of GNSS in Organic Farming and Agroecology

The main applications of GNSS in agriculture were summarized in the previous chapter and the importance of the diffusion of sustainable systems, like organic farming and agroecology, was underlined. The main impediment may lie in the lack of demand from farmers since specialized farm machinery is a niche market that is expensive and brings high investment costs (Maurel and Huyghe, 2017). A move towards agroecology could be done through different ways, such as sharing equipment between farmers, outsourcing farm services, extending the useful life of machinery, putting in place tax incentives, or a regulatory framework to support the acquisition of specialized farm machinery. Another way to be encouraged is use of public incentives to cut costs and target wider markets and economies of scale (Maurel and Huyghe, 2017). However, a low-cost GNSS technology is now available and can be adopted with a low investment (*see* paragraph 'Low-cost GNSS'). They can achieve sub-centimeter accuracy in RTK acquisition mode even with a short receiving signal period (Poluzzi *et al.*, 2020). In addition, it is possible to use this technology in an area where neither power line nor internet service are available: an advantage for poor farmers who live in remote rural areas.

There is no database of world farms that practice organic farming and agroecology with the support of positioning systems, but in literature there are some encouraging and interesting examples which are explained in the following paragraphs.

4.5.1. Nitrogen Efficiency in Organic Farming Using a GPS Precision Farming Technique

In their study, Koopmans and Zanen (2005) assessed the effects of a GPS-controlled precision tillage system in an organically-managed arable farm in The Netherlands. From 2003 to 2007, the impact of lowering manure input levels in organic farming was studied in combination with GPS-controlled precision tillage. Effects on soil structure, nutrient use efficiency, and spinach yield were evaluated. Half of the plots were treated, using GPS-controlled tillage and half of the plots with traditional organic tillage, using no specific tracks in the field. Fertilization was applied at two levels: 40 and 14 ton/ha dairy manure sludge (NPK = 4:1.5:5.5) corresponding respectively to farmers' practice (100 per cent) and phosphate equilibrium (35 per cent) in the spinach, based on total rotation. Interestingly, there was no significant difference between yields in plots with GPS-controlled precision tillage with 35 per cent fertilization and traditional tillage with 100 per cent fertilizations. Nutrient use efficiency was significantly higher in GPS-tillage with 35 per cent fertilization treatment as compared to the 100 per cent fertilization treatment. The GPS-controlled tillage showed higher mean nutrient efficiency (71 per cent) compared to the traditional one (59 per cent). The GPS-controlled precision tillage using the same tracks in the field, year after year, offers the opportunity to improve soil structure and nutrient use efficiency. The GPS-controlled precision tillage resulted in a significantly higher yield than traditional tillage.

Organic agriculture should play a leading role and set an example for sustainable soil management, thus implying greater nutrient use efficiency and fewer inputs. The higher nutrient use efficiency at lower fertilization levels stretches the possibilities for reducing inputs in organic agriculture. If fertilizer inputs are reduced in the next few years toward phosphate equilibrium at the crop rotation level, the GPS-controlled precision tillage system could become an important tool for organic farmers to maintain high-level yields. For farmers, the GPS systems may be a solution for improving their soil structure and increasing the sustainability of their practices without substantially lowering yields.

4.5.2. Innovations in Agroecology – A Case Study from The Netherlands

In Noord-Brabant, a region in southwest Netherlands, Govert van Dis and his wife Phily Brooijmans run a smart organic arable farm. It is a family farm of around 100 ha dating back many generations. Crops are cultivated without the use of pesticides or chemical fertilizers. Therefore, crop rotation forms the basis of the farming system. As the farm is pesticide-free, weed management is one of the major challenges. They manage the farm with the support of GPS technology. In particular, they use a GPS tractor, both to sow in very straight rows and remove weeds very close to the crop without damaging the crops themselves. They use a system of 'fixed paths' where the tractors are always driving in the same fixed paths throughout the seasons, so as not to compact the soil of the whole field and to conserve the soil health. Therefore, they started using a recent innovation, the eco-plough. This plough rotates the soil, but at a very shallow level compared to conventional ploughs. Weeds and residues are covered, but soil functions like mineralization are optimised, run-off of nutrients is limited, soil compaction is reduced, and fuel use is reduced. Because of the transition to an agroecological farming system, the soil health has significantly increased. Since adopting this method four years ago, soil organic matter has increased by several decimals. In addition, Govert and Phily have the notion that the crops need less and less nitrogen (N) compared to the calculations on what they should receive. In general, the crops keep up very well in the end of the growing seasons, without any additional manure (FAO, 2018a).

4.5.3. Agricare, Integrated Application of Innovation in Agriculture

Agricare (Furlan *et al.*, 2015, 2018) acronym for 'Introducing innovative precision farming techniques in AGRIculture to decrease CARbon Emissions' is a European LIFE + project (LIFE13 ENV / IT / 000583) which was born with the ambitious goal of combining two agricultural techniques which, like few others, are believed to be capable of facing the current challenges of agriculture in the third millennium: conservative agriculture and precision agriculture.

From a technical-operational point of view, Agricare aims to demonstrate in the field that land management in line with the principles and techniques of conservative agriculture, implemented with operating machines equipped with the most advanced mechatronic innovations as GNSS, has an important potential in terms of reducing greenhouse gase (GHG) emissions and protecting soils from potential threats of degradation of their fertility.

Operationally the project took place in ValleVecchia farm, located between the beach towns of Caorle and Bibione, in the province of Venice (Italy). The cultivation with conventional soil tillage techniques (B1) of four different crops in rotation (common wheat, rapeseed, corn, and soy) was compared with three different soil tillage conservative methods: minimum tillage, strip tillage, and no tillage for a period of two years. These latter methods were carried out applying one or two different GNSS-enabled techniques: automatic guidance with a uniform dosage of production factors (seeds, fertilizers…), and automatic guidance with variable rate application system of production factors based on a specific variability map (*see* paragraph 'Harvest Yield Monitoring').

The results showed that both minimum tillage and no-tillage combined with the full application of GNSS-enabled techniques (variable rate dosage) get closer to the yield and gross income performance of conventional agriculture, while having better energy balance and better economic and environmental potential. Conservative techniques combined with GNSS-enabled techniques (assisted driving and variable dosage) were less energy-intensive than the techniques applied without this aid. For these reasons, they offer good prospects for more sustainable agriculture (Furlan *et al.*, 2015, 2018).

4.5.4. Grape Mundo: App for Grape-Farming in India

'Grape Mundo' is an open-source application freely downloadable that helps farmers to chalk out a schedule for input application, which is based on the best practices followed by growers in the Nashik belt (northwest India). Thanks to GNSS technology, each farmer can geolocalize the position of his fields and monitor the cultivation. Grape Mundo has been developed to help farmers to identify problems: as preventing pre-harvest and post-harvest losses, estimating yields, and calculating and enhancing grape farm productivity. This application guides farmers in performing precision and sustainable grape farming in order to produce high-quality grapes using minimum chemicals and thus lowering costs.

To develop this application, the inventors have travelled more than 40,000 km across various regions of India in order to identify farmers' challenges, directly observing the people who hold the tools to practice agroecology (FAO, 2018b). The innovation of 'Grape Mundo' is to connect family farmers with efficient and low-cost techniques, allowing them to achieve high-quality production along with better yields. The diffusion of this app has social, economic, and environmental impacts:

- Increase income and decrease farming expenses
- Easier market access resulting in better rates for yields
- Minimize pre-post-harvest losses
- Efficient and controlled use of natural resources
- Reduce agrochemical pollution
- Improve farming practices through co-learning

To permit a major diffusion of the app in Nashik belt, the App's basic and main key points are written in the local language (Marathi) to be easily understandable for grape farmers (scientific language is only used when necessary). Grape family farmers have been using traditional methods for farming and thanks to this technology, they can increase grape productivity while at the same time maintaining sustainable agriculture. This quality product is directly sold to end consumers so that farmers can obtain full value for their efforts (FAO, 2018b).

4.6. Low-cost GNSS

As anticipated in Chapter 4, low-cost GNSS technology is currently available and can be adopted with low investment. The *EGSA GNSS Market Report* (2019b), that comprises device revenues, revenues derived from augmentation, and added-value services attributable to GNSS, states that the global installed base of GNSS devices in use is forecast to increase from 6.4 billion in 2019 to 9.6 billion in

2029. In terms of global annual GNSS receiver shipments (number of devices sold in a given year), the market is forecast to increase from 1.8 billion units in 2019 to 2.8 billion units in 2029 with an increase of devices per capita from 0.8 to 1.1 in the world.

It is possible to individuate three price segments of GNSS receivers (EGSA, 2019b):

- less than 5 €, mass-market receivers (the majority of shipments), 90 per cent used for smartphones and wearables
- between 5 € and 150 €, for the rise receivers (estimated annual growth of 6 per cent), mainly used by unpowered assets, as well as in road and drone applications
- more than 150 €, high-end receivers, account for less than 3 per cent of the total GNSS receiver shipments, they are used across all professional market segment.

If we consider the global GNSS downstream market, the growth is mainly due to the revenues from mass-market and mid-end devices (<150 €) and from augmentation services that will grow from 150 € billion in 2019 to 325 € billion in 2029 (EGSA, 2019b).

4.6.1. Smartphone and U-blox Receivers' Performances

Generally high-end receivers are based on dual signal frequency (L1+L2) that guarantee centimeter level of accuracy and reduction of biases, while mass-market receivers use single signal frequency (L1) more sensible to ionospheric residual and so their performances depend on this limitation (Cina and Piras, 2015). For this reason, in different monitoring fields (like landslide monitoring) where it is important to obtain a more precise and reliable solution, dual signal frequency receivers are traditionally preferred (Cina and Piras, 2015).

The modern smartphones and mass-market receivers, like u-blox, a new generation of single-frequency GNSS receivers, are able to reach a very impressive level of quality, both in static and kinematic positioning (Dabove *et al.*, 2020). The improvement is also allowed by the quality of the GNSS signals, the modern infrastructure dedicated to GNSS positioning that permits differential corrections (e.g. CORS, network, NRTK, etc.), and by the increasing interest for user communities and big players in the usage of these technologies for high-quality positioning (Dabove *et al.*, 2020).

Cina and Piras (2015) have demonstrated that the coupling between mass-market receivers and products offered by a network of GNSS permanent stations (e.g. Virtual RINEX) enables monitoring landslides with high accuracy and low cost.

The comparison of positioning performances obtained with a modern smartphone and a u-blox GNSS receiver, both in real-time and post-processing, demonstrated that the precisions and accuracies obtained with the u-blox receiver were about 5 cm and 1 cm, respectively, while those obtained with a smartphone were slightly worse (few meters in some cases), due to the noise of its measurements (Dabove *et al.*, 2020). The quality of the signals collected using these technologies is completely able to reach good positioning and surely, by combining the sensors with a better external antenna, the performances could be better (Dabove *et al.*, 2020).

4.6.2. Potential Diffusion in Organic Farming and Agroecology

Mass market receivers and modern smartphones could be a solution for small organic and agroecological farmers who cannot afford investments in expensive machinery like that utilized in precision farming. An interesting Slovenian study (Osterman *et al.*, 2013) demonstrated the possibility of building a low-cost GNSS navigation system for agriculture, using a single-channel GPS receiver, u-blox, connected to a laptop for data elaboration. The approximate price of the system was around 500 €, thanks to low-cost GNSS components and open-source programs used in it. In future, persons who have a modern smartphone could have a potential low-cost GNSS receiver in their hands (Dabove *et al.*, 2020): a technology that could help farmers, especially in rural areas to resist climate change by doing 'smart' organic farming and agroecology.

4.7. Final Considerations

GNSS currently represents an interesting tool for agriculture and farmers. Its usage is expanding, thanks to the diffusion of precision agriculture. The main GNSS provided advantage is certainly the farm machinery guidance and set of benefits derived from it. Such benefits can be even larger if GNSS technology is combined with other technologies, such as remote sensing, Bluetooth, or other types of sensors allowing sophisticated and useful tools, such as Variable Rate Application, biomass and yield monitoring, or herd monitoring for pasture management.

Basically, GNSS linked technology allows better precision for 'field works", less soil compaction, higher field input efficiency, as well as saving farmers time and fatigue. With this in mind, it certainly represents a useful and promising tool for a more sustainable agriculture, which is a priority for achieving the Sustainable Development Goals 2030 in sustainable agriculture. Thanks to the low costs of GNSS technology, it is possible to expand its use in organic farming and agroecology, which are key agricultural practices for responding to climate change and aiding the protection of natural resources, health food supply, and ending poverty. In future, agriculture will be digital and sustainable; therefore, research and investments in GNSS will be important for allowing the survival of small agroecological farmers who contribute to 50 per cent of the world's food production (Migliorini and Wezel, 2017).

Bibliography

Bailey, D.W., M.G. Trotter, M.G. Knight and C.W. Thomas (2018). Use of GPS tracking collars and accelerometers for rangeland livestock production research. *Translational Animal Science*, **2**(1): 81-88.

Banu, S. (2015). Precision agriculture: Tomorrow's technology for today's farmer, *J. Food Process. Technol.*, **6**(8): 468-473.

Basso, B. and L. Sartori (2013). Agricoltura di precisione per la sostenibilità degli agroecosistemi, pp. 271-298. *In:* M. Pisante (Ed.). *Agricoltura sostenibile, Principi, sistemi e tecnologie applicate all'agricoltura produttiva per la salvaguardia dell'ambiente e la tutela climatica*, Edagricole - New Business Media, Milan, Italy.

Bongiovanni, R. and J. Lowenberg-DeBoer (2004). Precision agriculture and sustainability, *Precision Agriculture*, **5**(4): 359-387.

Cina, A. and M. Piras (2015). Performance of low-cost GNSS receiver for landslides monitoring: Test and results, *Geomatics, Natural Hazards and Risk*, **6**(5-7): 497-514.

Cosentino, R.J., D.W. Diggle, M.U. de Haag, C.J. Hegarty, D. Milbert and J. Nagle (2006). Differential GPS. *In:* E.D. Kaplan and C.J. Hegarty (Eds.). *Understanding GPS: Principles and Applications*, pp. 429-453, Artech House, Norwood, USA.

Czajewski, J. and A. Michalski (2004). The accuracy of the global positioning systems. *IEEE Instrumentation & Measurement Magazine*, **7**(1): 56-60.

Dabove, P., V. Di Pietra and M. Piras (2020). GNSS Positioning Using Mobile Devices with the Android Operating System, *ISPRS International Journal of Geo-Information*, **9**(4): 220.

D'Odorico, P., A. Bhattachan, K.F. Davis, S. Ravi and C.W. Runyan (2013). Global desertification: Drivers and feedbacks, *Advances in Water Resources*, 51: 326-344.

EGSA (European Global Navigation Satellite Systems Agency) (2019a). Outcome of the European GNSS' User Consultation Platform; Retrieved from https://www.gsc-europa.eu/sites/default/files/sites/all/files/Report_on_User_Needs_and_Requirements_Agriculture.pdf; accessed on 15 February, 2020.

EGSA (European Global Navigation Satellite Systems Agency) (2019b). GNSS Market Report. Issue 6; Retrieved from https://www.gsa.europa.eu/market/market-report; accessed on 20 April, 2020.

El Rabbany, A. (2002). *Introduction to GPS: The Global Positioning System*, Artech House, Inc. USA.

ESA (European Space Agency). (2020a). Navipedia: Precise Point Positioning; Retrieved from https://gssc.esa.int/navipedia/index.php/Precise_Point_Positioning; accessed on 3 August, 2020.

ESA (European Space Agency). (2020b). Navipedia: SBAS General Introduction; Retrieved from https://gssc.esa.int/navipedia/index.php/SBAS_General_Introduction; accessed on 1 March, 2020.

FAO (Food and Agriculture Organization of the United Nations). (2017). Climate Smart Agriculture Sourcebook; Retrieved from http://www.fao.org/climate-smart-agriculture-sourcebook/concept/module-a1-introducing-csa/a1-overview/en/; accessed on 16 June, 2020.

FAO (Food and Agriculture Organization of the United Nations). (2018a). Profiles on Agroecology: Innovations in Agroecology – A Case Study from the Netherlands Case Study; Retrieved from http://www.fao.org/agroecology/detail/en/c/882848/; accessed on 6 April, 2020.

FAO (Food and Agriculture Organization of the United Nations). (2018b). Grape Mundo: An Ecosystem for Grape Farming; Retrieved from http://www.fao.org/agroecology/database/detail/en/c/1144153/; accessed on 15 June, 2020.

Furlan, L., G. Crocetta, L. Sartori, A. Pezzuolo, D. Cillis, N. Colonna, S. Canese and E. Bragatto (2015). Agricare, applicazione integrata dell'innovazione in agricoltura, *L'Informatore Agrario S.r.l.*, 27: 36-38.

Furlan, L., S. Barbieri, N. Colonna, F. Colucci, L. Sartori, A. Pezzuolo, D. Cillis, F. Marinello, D. Misturini, F. Gasparini, C.M. Centis and G. Donadon (2018). Lavorazione del terreno sostenibile, *L'Informatore Agrario S.r.l.*, 24/25: 33-42.

GreenPatrol (017). GreenPatrol Robot; Retrieved from https://www.greenpatrol-robot.eu/; accessed on 8 April, 2020.

Hoffmann-Wellenhof, B., H. Lichtenegger and J. Collins (1994). *Global Positioning System: Theory and Practice*, Springer-Verlag, New York, USA.

IAASTD International Assessment of Agricultural Knowledge, Science and Technology for Development (2009). Agriculture at a crossroads, in international assessment of agricultural knowledge, science and technology for development global report. *Island Press*, Washington DC.

IPCC (Intergovernmental Panel on Climate Change). (2019). Climate Change and Land: An IPCC Special Report on Climate Change, Desertification, Land Degradation, Sustainable Land Management, Food Security, and Greenhouse Gas Fluxes in Terrestrial Ecosystems; Retrieved from https://www.ipcc.ch/srccl/; accessed on 20 June, 2020.

Jacobsen, B. H., N. Madsen and J.E. Ørum (2005). Organic Farming at the Farm; Level Scenarios for the Future Development; Retrieved from https://www.sociology.ku.dk/research-projects/research_projects/current-projects/proper-food-under-economic-restraints/?pure=en%2Fpublications%2Forganic-farming-at-the-farm-level(f75eb460-a1be-11dd-b6ae-000ea68e967b).html; accessed on 4 February, 2020.

Jia, R.X., X.Y. Li, C.F. Xia and D.Y. Jin (2014). Broadcast ephemeris accuracy analysis for GPS based on precise ephemeris, *Applied Mechanics and Materials*, 602-605, 3667-3670.

Kaplan, E.D. and C.J. Hegarty (2006). *Understanding GPS: Principles and Applications*, ArtechHouse, INC. USA.

Koopmans, C.J. and M. Zanen (2005). Nitrogen Efficiency in Organic Farming Using a GPS Precision Farming Technique; Retrieved from https://orgprints.org/id/eprint/4504/; accessed on 16 March, 2020.

Kos, T., I. Markezic and J. Pokrajcic (2010). Effects of multipath reception on GPS positioning performance, *Proceedings of ELMAR-2010*, Zadar, Hr.

Langley, R.B. (2000). Smaller and smaller: The evolution of the GPS receiver, *GPS World*, **11**(4): 5458.

Langley, R.B. (1991a). The mathematics of GPS, *GPS World*, **2**(7): 4550.

Langley, R.B. (1991b). The GPS receiver: An introduction, *GPS World*, **2**(1): 5053.

Langley, R.B. (1993). The GPS observables, *GPS World*, **4**(4): 5259.

Matson, P.A., W.J. Parton, A.G. Power and M.J. Swift (1997). Agricultural intensification and ecosystem properties, *Science*, **277**(5325): 504-509.

Maurel, V.B. and C. Huyghe (2017). Putting agricultural equipment and digital technologies at the cutting edge of agroecology, *OCL*, **24**(3): D307.

Migliorini, P. and A. Wezel (2017). Converging and diverging principles and practices of organic agriculture regulations and agroecology: A review, *Agron. Sustain. Dev.*, 37, 63.

NASA (National Aeronautics and Space Administration). (2020). Glossary: Ephemeris; Retrieved from https://cneos.jpl.nasa.gov/glossary/ephemeris.html; accessed on 20 March, 2020.

NIMA (National Imagery and Mapping Agency, Department of Defense). (2000). *World Geodetic System 1984: Its Definition and Relationships with Local Geodetic Systems*, US Government Printing Office, Fairfax, USA.

Osterman, A., T. Godeša and M. Hočevar (2013). Introducing low-cost precision GPS/GNSS to agriculture, *Actual Tasks on Agricultural Engineering*, pp. 229-239. *Proceedings of the 41st International Symposium on Agricultural Engineering*, Opatija, Hr.

Øvstedal, O. (2002). Absolute positioning with single-frequency GPS receivers, *GPS Solutions*, **5**(4): 33.

Poluzzi, L., L. Tavasci, F. Corsini, M. Barbarella and S. Gandolfi (2020). Low-cost GNSS sensors for monitoring applications, *Appl. Geomat.*, 12: 35-44.

Shanwad, U.K., V.C. Patil, G.S. Dasog, C.P. Mansur and K.C. Shashidhar (2002). Global positioning system (GPS) in precision agriculture, *Proceedings of Asian GPS Conference*, Bangkok, Thailand.

Shi, C. and N. Wei. (2020). Satellite navigation for digital earth. *In:* Guo, H., M. Goodchild and A. Annoni (Eds.). *Manual of Digital Earth*, Springer, Singapore.

Smita, D., W. Rashmi and A.K. Gwal (2006). Ionospheric effects on GPS positioning, *Advances in Space Research*, **38**(11): 2478-2484.

Tey, Y.S. and M. Brindal (2012). Factors influencing the adoption of precision agricultural technologies: A review for policy implications, *Precision Agriculture*, **13**(6): 713-730.

US Government (2016). Space Segment; Retrieved from https://www.gps.gov/systems/gps/space/; accessed on 6 April, 2020.

Wells, D., N. Beck, A. Kleusberg, E.J. Krakiwsky, G. Lachapelle, R.B. Langley, K. Schwarz, J.M. Tranquilla, P. Vanicek and D. Delikaraoglou (1987). *Guide to GPS Positioning, Fredericton*, New Brunswick: Canadian GPS Associates.

Wezel, A., B.G. Herren, R.B. Kerr, E. Barrios, A.L. Rodriguez Goncalves and F. Sinclair (2020). Agroecological principles and elements and their implications for transitioning to sustainable food systems: A review, *Agronomy for Sustainable Development*, **40**(40): 4.

Willer, H. and J. Lernoud (2019). The World of Organic Agriculture. Statistics and Emerging Trends, *Research Institute of Organic Agriculture FiBL and IFOAM Organics International*, pp. 1-336.

Yang, Y., Y. Mao and B. Sun (2020). Basic Performance and Future Developments of BeiDou Global Navigation Satellite System, *Satell. Navig.*, 11.

CHAPTER

5

Hyperspectral Remote Sensing and Field Spectroscopy: Applications in Agroecology and Organic Farming

András Jung[1]* and Michael Vohland[2]

[1] Faculty of Informatics, Institute of Cartography and Geoinformatics, Budapest, Hungary

[2] Geoinformatics and Remote Sensing, Institute for Geography, Leipzig University, Germany

5.1. Introduction

Remote sensing performs non-destructive measurements without manipulating the measured material, while providing the possibility of a broad spatial overview and high temporal flexibility of measurements. High-resolution remote sensing applications can consolidate sustainable, prevention- and precision-oriented crop management strategies by decreasing production risks. In this chapter, we present and analyze the main aspects, perspectives and technical foci of hyperspectral remote sensing and field spectroscopy in the context of agroecology and organic farming. Moreover, we provide an overview of currently available measurement techniques and methods, and identify areas of interest for their future development.

When spectral imaging information is requested on a regular temporal basis for large-scale areas, remote sensing is often applied in many areas of agriculture. Nowadays, many of the classical remote sensing tools are also available for small-scale farming, site-specific acquisition of information, and applications in daily practice. There is a great potential for organic and sustainable land-use practices to increase information availability in everyday farming using proximal- and remote-sensing technologies. Spectral imaging and non-imaging sensors are powerful bio- and geo-chemical data acquisition tools that can play a crucial

*Corresponding author: jung@inf.elte.hu

role in the early detection of crop management risk factors, such as soil nutrition supply, pests and diseases, or in the prevention and minimization of field-scale chemical treatments.

The real benefit of proximal- and remote-sensing is the capability to characterize spatial or field variability that cannot be parameterized more effectively any other way. High-resolution spectral sensing provides the opportunity for both research and industry to develop novel approaches and technologies for putting prevention-oriented and site-specific farming into practice.

The main foci of our chapter are the following: i) hyperspectral remote sensing and field spectroscopy technologies also need high spatial and temporal resolution in order to answer the needs of agricultural applications. When looking at the four resolution principles of remote sensing (spectral, temporal, spatial, and radiometric), the temporal is the most under-sampled one in most situations while high demand exists; ii) non-scanning snapshot hyperspectral imaging technology will enable researchers to overcome the scanning limitations and provide flexible sensing solutions in time and space for regular field applications (soil sampling, fertilizing, phytopathology, etc.); iii) vegetation narrow-band indices in the range of 400-1100 nm are anticipated to become the basis of the next generation of agricultural sensors due to their cost-efficiency, non-saturating behavior, and high sensitivity. In particular, the so-called red-edge region has potential for future high-resolution vegetation indices as well; iv) soil spectroscopy over 1100 nm up to 2500 nm has great sensing potential but is very cost-intensive and showing limitations in flexibility and mobility. Here we basically deal with soil spectroscopy under 1100 nm because most hyperspectral drone cameras use this spectral region for detecting and mapping.

Our chapter helps to better understand the matches between field spectroscopy, optical sensors, spectral cameras, hyperspectral sensors, and the biophysical, biochemical properties, and reactions of cultivated plants in organic farming and agroecology.

5.2. Remote Sensing and Spatial Variability

The spectral resolution describes the electromagnetic spectrum to sense material properties and characterizes the number and width of the spectral channels available for spectroscopic sampling. The spectral resolution could also be interpreted as the 'chemical resolution' since the spectral resolution resolves the apparent spectral material properties and links chemistry to spectroscopy. Accordingly, higher spectral resolution provides more detailed chemical insights (Goetz *et al*., 1985).

Conventional spatial resolution in remote sensing provides two-dimensional space information. In many present and future remote- and proximal sensing measurements, three-dimensional spatial measurements will be carried out coupled with the spectral domain. The traditional spatial resolution definition will be extended and supplemented. The fineness of the spatially-distributed data depends on the sensor and platform. There is a technical limitation for the spectral and spatial resolutions of the satellite platforms which show that high spectral resolution and high spatial resolution cannot be achieved at the same time from the same altitude. It has complex technical aspects – one of them is a justifiable signal-to-noise ratio. The signal-to-noise ratio (SNR) compares preferred signal levels to unpreferred ones. It is complex to give an average SNR for a sensor or multispectral data because it depends on wavelengths, radiance levels, and other technical issues. Generally, it is expected that non-imaging systems provide higher SNR values compared to imaging ones. Satellites with less than 1-meter pixel size have typically less than ten broad spectral bands while satellites with more than ten spectral bands have typically larger pixel sizes than 10m on the ground. One way to increase the spatial and spectral resolution is to change the sensor and reduce the altitude of the data capturing. This demand challenges the remote sensing platforms and initiated many different forms of terrestrial and near-ground imaging and non-imaging spectroscopy.

Digital imaging is the capture, storage and display of object information in electronic forms. In color imaging, three broad bands (blue, green, and red) are used to best reproduce real object properties in a virtual form. The RGB (red, green, and blue) bands are spectral channels as well. When the number of spectral channels is increased (over 100) and the spectral range is extended (400-1000 nm, 400-1000 nm or more), imaging spectroscopy or hyperspectral imaging is applied.

Spatial scales of field phenomena are not absolute and are customized to specific needs and applications. From a global (that is, Earth-observing) point of view, scales smaller than 10^4 km^2 are referred to as local scale (IPCC, 2014), which are higher by several magnitudes than the common agricultural management scales in Europe. For site-specific observations, further downscaling is needed. For crop management, the variability on the field and sub-field scale are of interest and the variability at distances of 50 m or less are mainly related to management practices (Adamchuk *et al.*, 2010). For organic agriculture, the average size of organic holdings in the EU-28 amounted to 47 ha in 2013. The largest organic holdings were located in Slovakia (average total holding area of 474 ha/holding), and the smallest organic farms in Malta (less than 1 ha/holding). Accordingly, future remote sensing applications must cope with even less than 1-5 m^2 ground resolution (EC, 2016).

Temporal resolution is a driver in agricultural remote sensing that controls flexibility and data availability. The periodical returns of satellites are typically not customized and the air-borne campaigns with high-temporal resolution are very cost-intensive and complex. Considering the application areas of remote sensing, agriculture is one of the most time-critical. The entire agricultural sector and production are based on time-critical processes, including sowing, plant protection, fertilizing, irrigating, and all management decisions. There is thus a need to sense at higher temporal rates to overcome present limitations and to allow targeted technological interactions.

In spatial down-scaling when the measurement height drops down to 100, 10, and 1 m, the temporal, spatial and spectral resolution can be significantly increased and new demands or application needs, such as mobility (on the fly) and flexibility (vehicle-based), can be considered.

The temporal resolution affects not only the process accuracy but also the imaging process. Recent developments show that a novel kind of imaging technique (the snap-shot spectroscopy, *see below*) enables high-rate spectral images to generate spectral video sequences that are an obvious advantage in on-line process monitoring and controlling of agroecological conditions, both in field and indoor.

The fourth resolution, the radiometric resolution is a technical term that characterizes the sensitivity of the detector or the wavelength-dependent energy resolving power of a sensor. It is typically quantified by bit values. Accuracy and stability are essential in radiometric calibrations in order to calculate radiance and/or reflectance that are the derivatives and representative outputs (information carriers) of the remotely sensed data and the primary inputs for further statistical analyses.

In the present work, we address the spectral sensing needs of sustainable agricultural practices and small-scale holdings, which are generally less involved in the high-tech developments of precision farming. From a practical point of view, high importance was afforded to those research studies that used field spectrometers and/or spectral cameras and attempted to understand the agricultural values, biophysical, and biochemical properties or reactions of cultivated plants. Only outdoor or field-related applications were considered, processed agricultural products, food or other related indoor products were not studied. It is behind the scope of this chapter to comprehensively analyze the situation in animal husbandry, aquaculture, fungi- or viticulture, which are of interest but might be the subject of other or further works. In the context studied here, the main benefit of hyperspectral remote sensing and/or (imaging and non-imaging) spectroscopy is the capability to characterize spatial or field variability that cannot be parameterized more effectively any other way.

5.3. Hyperspectral Remote Sensing and Field Spectroscopy

The use of field-, air- and space-borne hyperspectral data is of increasing importance for innovative agricultural applications nowadays. In contrast to multispectral remote sensing, 'hyperspectral' means that there are very narrow spectral bands throughout the electromagnetic spectrum (Lauden and Bareth, 2006). In spectral imaging the number of available spectral bands is crucial. A normal digital camera has three color channels (RGB). A multispectral system has a two-digit number of spectral bands, and a hyperspectral camera has typically a three-digit number of channels.

The term 'hyperspectral' is not limited to specific spectral ranges. Hyperspectral data sets are generally composed of about 100 to 200 or more spectral channels with relatively narrow bandwidth (1-10 nm). This technology combines imaging with spectroscopy in a single optical system that often produces large data sets and requires high performing processing methods. Thirty years ago, Elvidge (1990) pointed out that the use of spectrometers to characterize vegetation, soil, or environmental parameters would offer new opportunities. In the meantime, many application areas were identified, mostly in science and research. Proven agri-industrial solutions in hyperspectral remote sensing are still under intensive development. In agroecology inter alia, remote sensing topics often focused on different conditions of stress caused by pest or disease incidences, nutrition deficiencies, drought, and frost, etc. Vegetation stress may cause biochemical anomalies in the cellular or leaf structure, affecting the pigment system or the canopy moisture content, which could be detected and mapped by optical sensors.

The demand for out-of-the-lab devices initiated the early field spectroscopy experiments with non-imaging measurements, which originated from laboratory spectroscopy and required respective developments in optics and portable platform techniques. From the beginning, portable or hand-held field spectroradiometers were very popular in geology, soil and vegetation spectroscopy as they assured flexible and rapid field data acquisition (Milton *et al.*, 2009). Thus, spectroscopy in the visible (VIS) and near-infrared (NIR) has been widely used either in the laboratory (Ben-Dor and Banin, 1995) or for in-situ monitoring (Stevens *et al.*, 2008). Non-imaging field spectroradiometers provide the highest spectral resolution and high-quality information content for estimating agroecology traits with multivariate methods. However, using a point spectrometer only integrative measurements can be performed, which hamper the analysis of spatial variability. Field campaigns with portable field spectroscopy are often complemented with data of air- or space-borne imaging spectrometers to cover larger areas; large-area coverage in-flight campaigns often leads to decreased accuracies of estimated soil properties compared to point measurements (due to a lower signal-to-noise ratio and disturbing atmospheric influences, for example). Variable soil and surface properties (as moisture content, roughness, crusting, or texture) induce spectral variability that is critical for large-scale calibration approaches (Wight *et al.*, 2016). There is an obvious gap between integrative point measurements and airborne or even space-borne image data, which may be filled by hyperspectral image data proximally sensed at the field scale. Field imaging line-scanners are less widespread in ground truthing than portable point spectroradiometers, as operating a field line scanner on a tripod with a rotation stage is very time-consuming as compared to the use of a point spectroradiometer. Non-scanning or snapshot hyperspectral imaging is one possible solution to overcome this limitation in-field usability and to bridge the gap in the data chain (Hagen *et al.*, 2012). Snapshot hyperspectral imaging enables rapid data acquisition as the entire image with all spectra is captured at once within a few milliseconds in a hand-held or portable mode (Jung *et al.*, 2015).

Field spectroscopy has been reshaped and extended by new platforms in the last few years. This kind of platform liberalization changes our ground-truthing attitudes, toolboxes, and methods in the fieldwork. Traditionally field spectroscopy was used to support airborne and space-borne campaigns. Devices used here were typically non-imaging spectrometers (*see* Fig. 1)

(A) (B)

Fig. 1: A non-imaging (A) and an imaging spectromter (B) in field use (*Source*: A. Jung)

However, technical working principles among the instruments are very similar, hence some conditions should be considered. Reflectance quantities measured in the field and lab have special and different geometrical situations. Schaepman-Strub *et al.* (2008) give a comprehensive overview on spectroscopic measurement scenarios. Under field conditions, the hemispherical – conical reflectance factor is measured, because optimal solar irradiance is diffuse (solar radiation comes from all directions of the hemisphere) and the reflected light is captured through the field of view (in a conical and directional geometry) of the spectrometer. Laboratory reference measurements are often needed as well to calibrate or verify field spectra. This is another lighting situation that is described by biconical reflectance factors (Schaepman-Strub *et al.*, 2008). Both the illumination and the spectrometer have a given direction represented by two conical geometries. These two cases are the most typical lighting geometries in ground-truthing spectroscopy. In both recent cases, dry-chemistry is conducted, which is often complemented with wet-chemistry measurements to develop quantitative statistical models. For measuring reflectance quantities in the field, two types of techniques are common – one is called the single beam, the other the dual beam (*see* Fig. 2).

Single beam means that prior to the object reflectance measurement, an indirect irradiance measurement is made, using a special white panel. This is the white referencing process. In this case the object reflectance and the white reference measurements are not at the same time to calculate the hemispherical–conical reflectance factor in the field. In such situations, white reference

(A) (B)

Fig. 2: A single-beam (A) and a dual-beam (B) non-imaging spectroscopic measurement (*Source*: A. Jung)

measurements are made in regular time shifts (15-30 minutes) because the sun position is changing and new irradiance conditions are arising. This is also the main disadvantage to be mentioned for single beam spectrometers. The advantage is the better signal-to-noise ratio because all light transmitting fiber optic cables end on the same sensor and are not deviated into two bundles (see dual beam). Dual-beam spectrometers measure the solar irradiance and the object reflectance at the same time to calculate the reflectance factor in real-time. This advantage can be utilized under changing irradiance conditions and if time-saving is needed. Here spectral sensing capability is reduced by the divided transmission bundles – this is more critical in weak signal regions over 1000 and 1700 nm, which are relevant spectral ranges in soil and geological spectroscopy. For this reason, dual-beam spectrometers are more often used in vegetation and water spectroscopy since the spectral power under 1000 nm is sufficient.

It is worth mentioning that in imaging and field spectroscopy, spectral ranges and parts have some conventional labeling. It is informative for beginners or users, who are intending to purchase instruments. For most remote sensing projects in soil, water, and vegetation spectroscopy, the measurable solar spectrum is between 400 and 2500 nm. This is called full range. Spectrometers (imaging or non-imaging) working in this spectral region are full-range devices. If the spectral interval is located between 400 and 1000 nm, it is called half range. There is another nomenclature which splits the spectrum for technical reasons. In this case VNIR stands for visible and near infra-red light (400-1000 nm) and SWIR for short wave infra-red (1000-2500 nm). SWIR is often divided into SWIR I (1000-1700 nm) and SWIR II (1700-2500 nm). It typically accrues when a spectrometer has unique sensors for VNIR, SWIR I, and SWIR II to maximize signal-to-noise ratios.

When using a white reference panel, special care should be taken since they are sensitive optical materials. Two types of white panels are widespread. If there is no moisture risk (unlikely situation in the field) and the budget is low, $BaSO_4$ (barium sulphate) is an option, but unfortunately, it is very hydrophile. The other choice is an optical PTFE (polytetrafluoroethylene), which is very stable, hydrophobic, heat resistant, and non-reactive, which are optimal properties for spectral measurements. Both white panels are easy to get scratched, dirty, and dusty in the field. They can be purified, depending on damaging grades. These panels are used in different sizes, the most common are 20 × 20 cm in field spectroscopy, while 1 × 1 m panels are popular in hyperspectral airborne campaigns. White panels have a typical average reflectance of 90-99 per cent, depending on material thickness and purity. Material reflectance is wavelength dependent and so each panel gets a unique reflectance curve, which can be involved in object reflectance measurements to correct for white reference uncertainties. It can be responsible for 5-10 per cent changes in spectral signals. If the entire reflectance curve of the white panel is involved in the reflectance factor calculation, it is called the 'absolute reflectance'.

5.4. Snapshot Imaging Spectroscopy and Ground-truthing

The proximal and remote sensing spectral detectors are either imaging or non-imaging sensors. Hyperspectral imaging traditionally utilizes whisk- or push-broom scanners mounted on satellites or airplanes, or on the ground. Until recently, light-weighted spectral scanners were not widely used because of technical limitations. One of the first successful fix-wing miniature spectral scanning measurements was achieved by Zarco-Tejada *et al.* (2013). The light-weighed scanners (<1-2 kg) are mainly working in the spectral range between 400-1100 nm. These typically utilize the push-broom spectral imaging. Hyperspectral cameras with scanning principle cannot control random movements and cannot be used as hand-held imagers or on vehicle-based platforms (UAV, multicopter, tractors, etc.). Mobile imaging field spectroscopy requires sensors that are flexible and easy to operate. Non-scanning hyperspectral imaging has been recently introduced for outdoor applications. Non-scanning spectral imaging is called 'snapshot imaging spectroscopy' (Hagen *et al.*, 2012) and it has a different principle from the push- and whiskbroom sensors (*see* Fig. 3).

A hyperspectral non-scanning camera is generally designed to utilize the instant spatio-spectral surplus of real-time data acquisition. It means that all spectra and image pixels are taken at the same time. A snapshot light-splitting architecture integrated on a sensing sensor chip with appropriate spatial resolution captures the full-frame image with a high spectral (> 100 bands) and radiometric resolution (> 14 bit). The image capturing process benefits from a powerful light collection capacity (a simple geometric factor) that exceeds all scanning and all throughput-

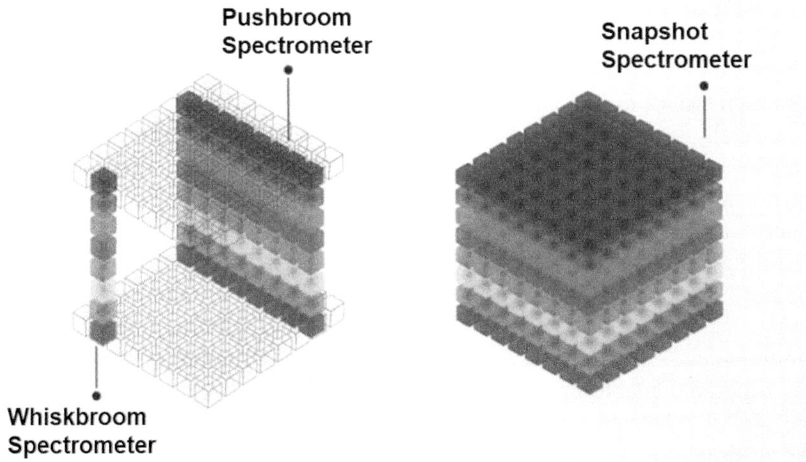

Fig. 3: Working principles of hyperspectral imaging, colors represent different wavelengths (*Source*: Courtesy of Cubert GmbH, Germany)

division snapshot instruments. This is called the 'snapshot advantage' (Hagen *et al.*, 2012) and the primary limitation to constructing high spatial resolution snapshot cameras is the limited number of pixels, the development of much larger detector arrays is needed. For hyperspectral snapshot camera, in a normal sunlight situation, the integration time of taking one hyperspectral data cube is about 1ms. Such a camera is able to capture more than 10 spectral image data cubes per second, which facilitates hyperspectral video recording. The commercially-available snapshot imaging spectrometers record hyperspectral full-frame images with more than 30-100 bands in a spectral range of 400-1000 nm. At the moment the spectral range is limited to VNIR, because SWIR sensors (InGaAs, PbS) are set to detect much lower energy wavelengths (1000-2500 nm), which means a lower number of pixels and a very low spatial and spectral resolution at the end.

The snapshot advantage prefers the time-critical applications, both in the laboratory and in the field. When looking at the four resolutions of remote sensing, the temporal resolution is generally the most under-sampled. This is significant especially for vegetation studies and crop management because of physiological and phenotypical changes (Zarco-Tejada *et al.*, 2011). Knowing more about temporally-resolved spectral crop information is of high importance in agriculture amongst others because of timely and targeted nutrition supply, preventive and precision pest control, and generally for closing the productivity gap of sustainable agriculture systems. Beyond the temporal aspect there is a general and global demand on high-resolution data in agricultural process controlling. The technical paradigm change in imaging field spectroscopy will enhance the effectiveness and availability of the commercial sensors, and will foster applications in agroecology and organic farming.

The real-time image-capturing capability of the proximal snapshot imaging spectroscopy is essential for capturing moving objects (i.e. leaves and canopy) or being on a moving platform (i.e. vehicle, UAV, robots, or human being) at high-resolution scales; maybe you can add a reference? Organic farming applications are typically small-scaled and individual detection and treatments of species or canopies may be of economic, environmental and professional interest; maybe you can add a reference? In the next part, two case studies will be shortly presented which used snapshot hyperspectral imaging technologies – one was implemented in soil spectroscopy, the other in UAV-based vegetation monitoring. The first one used a spectral snapshot camera for soil mapping. For the data collection the hyperspectral imaging sensor was mounted on a single tripod. An ASD Pro-Lamp model (14.5 volt, 50 watt) was used as an illumination source, which is also tripod-mountable for indoor laboratory, diffuse reflectance measurements. The size of the white reference panel (Zenith Polymer®) was 30 cm × 30 cm. The air-dried raw soil samples were prepared for illuminated diffuse reflectance measurements. The distance between sensor and soil sample was set to 35 cm in the nadir position, the illumination zenith angle was 45°. All samples were prepared on a reflection neutral plate (spectrally tested) and covered, prior to the spectroscopic measurement, by a black passepartout (reflectance under 5 per cent over the entire spectral range from 350 to 1100 nm) with a window of 20 cm × 20 cm. This study focused more on soil constituents available to detect in VNIR (350-1100 nm). The authors used partial least squares regression (PLS) as a statistical calibration method to estimate soil organic carbon (OC), hot-water extractable carbon (HWE-C), and nitrogen (N). The obtained results from the camera data were satisfactory with coefficients of determination (R^2) between 0.62 and 0.84 in the cross-validation, but only with crushed samples and when combing PLS with an effective spectral variable selection technique. For in-field studies without any sample preparation, further studies are needed. Approaches considering soil surface roughness and/or the elimination of shadow pixels from the acquired images might both be promising to improve the accuracy of estimates.

Bareth *et al.* (2015) used a snapshot hyperspectral imaging camera in a farming experiment to study its usability on UAV platform to monitor crops. The silicon CCD chips of the camera captured a 1 M pixel grayscale image as well as a 50 by 50 hyperspectral image with over 100 spectral channels. At a flying altitude of 30 m, the grayscale image had a ground resolution of about 1 cm and a pure hyperspectral ground resolution of about 20 cm. However, the latter may be pan-sharpened to the resolution of the grayscale image. This study concluded that the combination of 3D imaging techniques and snapshot hyperspectral imaging enables the precise and accurate monitoring of dynamic crop growth through phenological changes. A multi-temporal crop surface analysis enables the precise tracking of plant height and plant growth while hyperspectral analysis derives physiological vegetation parameters, like chlorophyll or nitrogen content and others. To monitor crop growth behavior, crop vitality, and crop stress snapshot hyperspectral imaging may be an ideal tool (Bareth *et al.*, 2015).

5.5. Vegetation Spectroscopy and Narrow Bands

Remote sensing of biophysical parameters, such as phytomass, leaf area index, and canopy structure have intensively been analyzed (Schellberg *et al.*, 2008; Asner and Heidebrecht, 2002; Numata *et al.*, 2007; Mutanga *et al.*, 2003; Clevers *et al.*, 2007; Beeri *et al.*, 2007). Behind the biophysical parameters, numerous papers have been devoted to biochemical components, such as foliar constituents, chlorophyll a and b, carotenoids, lignin, cellulose, protein, water, and further components (Elvidge, 1990; Van Der Meer, 2004; Almeida and Filho, 2004; Somers *et al.*, 2010). Many of the studies used high-resolution full-range spectra (400-2500 nm) because some foliar chemistry components show indications only over 2000 nm, such as lignin and cellulose (Bannari *et al.*, 2006; Kokaly *et al.*, 2007). Our study focuses on narrow-band indications in the range of 400-1100 nm.

For historical and technical reasons, the multispectral satellite remote sensing initiated many agricultural applications. The spectrum-based methods initially used broad (50-100 nm) spectral bands, which have been narrowed by scientific high-resolution sensors over the last decades (Zhao *et al.*, 2007; Adam *et al.*, 2010). The VNIR spectral range will remain relevant in the next generation of crop sensor developments as well but it will likely be spectrally-enhanced to produce high-resolution crop or soil sensors. The narrow- and broad-band comparisons (Thenkabail *et al.*, 2012; Zhao *et al.*, 2007) highlight best the benefits of the narrow-band indices, such as non-saturating behavior or high sensitivity in vegetation dynamics (phenology). Thenkabail *et al.* (2000, 2004b) gave an excellent overview of using hyperspectral narrow bands for vegetation analysis and agricultural applications. After Thenkabail (2002) the narrow bands can be classified as very narrow bands (1-15 nm), narrow bands (16-30 nm), intermediate bands (31-45 nm) and broad bands (greater than 45 nm). Based on this classification, the first-generation crop sensors (nitrogen-sensors, for example) belong basically to the broadband detectors. For future VNIR crop sensor developments, the following spectral narrow-bands could be of interest (*see* Table 1).

Narrow-band studies (Zarco-Tejada *et al.*, 2012) showed that classification accuracies have been increased. Generally, the hyperspectral narrow bands explain about 10-30 per cent greater variability in quantitative biophysical models as compared to broadband bands and are not sensible to saturation problems in biophysical estimations. These two benefits are to be considered in the design of future high-resolution imaging or non-imaging crop sensors. There is another important part in VNIR spectrum: the so-called red-edge region is probably becoming increasingly important for novel optical field sensors.

The responses of crops to ecological factors and vegetation conditions vary over time. Stress-induced biochemical and biophysical changes in the cellular or leaf structure can affect the pigment system or the canopy moisture content and distribution. A very promising tool to detect vegetation condition is to study the

Table 1: Narrow Band Indices for Biochemical and Biophysical Plant Parameters

Wavelength	Parameter	Indications	References
375 nm	Biochemical	Leaf water content	Thenkabail *et al.*, 2000, 2002, 2004a,b
466 nm	Biochemical	Leaf chlorophyll	
515 nm	Biochemical	Leaf nitrogen	Thenkabail *et al.*, 2004a,b
520 nm	Biochemical	Pigment content	Wrolstad *et al.*, 2005; Gitelson and Merzlyak ,2003; Thenkabail *et al.*, 2004a
525 nm	Biochemical	Leaf nitrogen	Gitelson *et al.*, 2001; Lee *et al.*, 2003; Wessman, 1990
575 nm	Biochemical	Leaf nitrogen	Gunasekaran *et al.*, 1985; Zhao *et al.*, 2003
675 nm	Biochemical	Leaf chlorophyll	Gitelson and Merzlyak, 1994; Thenkabail *et al.*, 2004a,b; Chan and Paelinckx, 2008
700 nm	Biochemical	Nitrogen stress	Thenkabail *et al.*, 2004a,b; Chan and Paelinckx, 2008; Lichtenthaler *et al.*, 1996
720 nm	Biochemical	Nitrogen stress	Sims and Gamon, 2002; Thenkabail *et al.*, 2002; le Maire *et al.*, 2008; Peñuelas, 1995
740 nm	Biochemical	Leaf nitrogen	Merzlyak *et al.*, 1999; Thenkabail *et al.*, 2004a,b; Chan and Paelinckx, 2008
490 nm	Biophysical	Crop yield	Thenkabail *et al.*, 2004a,b
550 nm	Biophysical	Biomass	Buschmann and Nagel, 1993; Sims and Gamon, 2002; Yang *et al.*, 2010; Thenkabail *et al.*, 2002; Chan and Paelinckx, 2008
682 nm	Biophysical	Crop yield	Thenkabail *et al.*, 2004a,b;
845 nm	Biophysical	Biomass	Peñuelas *et al.*, 1994; le Maire *et al.*, 2008
915 nm	Biophysical	Crop yield	Thenkabail *et al.*, 2002; Peñuelas, 1995
975 nm	Biophysical	Leaf moisture	Danson, *et al.*, 1992; Yao *et al.*, 2010
1100 nm	Biophysical	Biomass	Ustin *et al.*, 2004; Abdel-Rahman *et al.*, 2010

sharp rise of the reflectance curve between 670-780 nm. This segment is called the red-edge region. Both the position and the slope of the red-edge region change due to physiological conditions and can result in a blue- or red shift of the red-edge position. The red-edge index is defined as the position of the inflexion point of the red-NIR slope of a vegetation reflectance curve. A reliable detection of this index requires spectral sampling at about 10 nm intervals or higher, which requires high-resolution spectral measurements (Filella and Peñuelas, 1994). There are well-known methods to define red-edge position (REP) (Smith *et al.*, 2004). The REP is strongly correlated with foliar chlorophyll content, and hence it provides a very sensitive indicator for a variety of environmental factors affecting leaves, such as nutrition deficiency, drought, senescence, etc. The REP is also present in spectra for vegetation recorded by remote or proximal sensing sensors. Due to its importance in vegetation mapping, a number of techniques have been developed to best determine REP for foliar spectral reflectance. The numeric derivation and interpolation techniques of the reflectance curve are also widely used. Comprehensive spectral analysis has been conducted on fruits and other agricultural products (Zude *et al.*, 2006) in scientific studies. Recent developments in REP oriented hyperspectral imaging and non-imaging spectroscopy offer new perspectives and approaches for field heterogeneity mapping and spectral mobile services.

5.6. VNIR Field Spectroscopy and Soil Constituents

Soil heterogeneity is a crucial factor in agriculture as variable site characteristics strongly influence the growth and yield of crops and may also affect the incidence of pests (Patzold *et al.*, 2008). Its development is particularly influenced by the soil-forming factors, which contribute to the natural and management-induced soil heterogeneity that can be interpreted at different spatial scales – from global and continental scales down to farm and field scales (Jenny, 1941). Nevertheless, soil properties vary specifically in space and time, to which soil monitoring strategies should be adapted. For instance, nitrate in soils shows spatial dynamics being more pronounced than that of soil carbon content. Soil pH or soil colors have low temporal dynamics and do not change significantly within weeks or months, while soil temperature or soil moisture have very high temporal dynamics within short periods, such as hours.

Beyond this, soil management differs between organic and conventional farming systems. This is reflected by greater biological activity of organically managed soils, while soil chemical and physical parameters show less pronounced but still detectable differences. Soil microbial biomass, enzyme activities, and the metabolic quotient are thus parameters of certain interest for organically farmed soils; the balance of N_2 emissions and nitrate losses is shifted in comparison with conventionally systems and soluble fractions of some other nutrients were also found to be different (Mäder *et al.*, 2002). Despite these differences, there is a suite of main parameters that get monitored for soils of all farming systems (soil organic carbon SOC, inorganic C, dry bulk density, pH, total nitrogen N, soil texture).

For monitoring issues, soil spectroscopy in the VNIR-SWIR wavelength domain (visible to near- and short-wave infrared, defined as the 400-2500 nm region) has been established as an efficient method to quantify various soil properties (*see* e.g. Stenberg *et al.*, 2010; Soriano-Disla *et al.*, 2014), primarily applied in the laboratory, but also on-site with portable instruments (Stevens *et al.*, 2008; Mouazen *et al.*, 2010; Kuang and Mouazen, 2011). Typically, field reflectance spectra are collected by portable field spectrometers (1D high-resolution spectra), larger areas may be sensed by air- or space-borne imaging spectrometers with a more limited spatial resolution. The latter is far less common, as data availability is still a bottleneck. Lausch *et al.* (2019) give a comprehensive overview of the state-of-the-art approaches for monitoring soil characteristics with air- and space-borne remote sensing techniques.

Proximally sensed hyperspectral image data seem to offer the opportunity to close the gap between in-situ field spectrometer and airborne image data, but field studies with this kind of data and with the purpose of regular soil monitoring are missing. Nevertheless, there is a comprehensive list of studies and works conducted with line-scanners in the laboratory. Different from point measurements, imaging techniques also allow for a complete scanning of soil profiles to characterize vertical heterogeneities, to identify diagnostic horizons, and to quantify soil properties (e.g. concentrations of carbon, nitrogen, aluminium, iron, and manganese, *see* Steffens and Buddenbaum (2013), who utilized a hyperspectral scanner between 400-1000 nm to characterize a stagnic Luvisol profile. Profile data can then be used to model soil functions and to improve the process-related understanding of chemical and biological transformations (Hengl and MacMillan, 2019; Ogen *et al.*, 2018).

Key chemical, biological, and physical soil properties that can be estimated with spectroscopy in the 400-2500 nm domain are listed by a couple of overview articles (e.g. Cécillon *et al.*, 2009; Stenberg *et al.*, 2010; Soriano-Disla *et al.*, 2014). With a view to biological soil properties (that may be of interest, as mentioned, especially for organically farmed soils), Cécillon *et al.* (2009) state a good spectral predictivity for microbial biomass, soil respiration, and the ratio of microbial to organic C. Successful application of spectroscopy for the estimation of these properties is, however, not solely based on their optical activity (if any), but rather due to correlations with other and spectrally active constituents, above all quantity and quality of soil organic matter (Soriano-Disla *et al.*, 2014). Plant-available nutrients and pH, for example, are also not considered spectrally active. Correlations to other soil properties that may be used for their estimation are usually of local nature, so that calibrations are per se not transferable in space and time but have to be defined or at least to be re-calibrated locally (Hill *et al.*, 2010; Stenberg *et al.*, 2010). In contrast, SOC, clay, N, carbonates, water, iron (Fe) oxides, and some C and N fractions are considered optically active, i.e. spectral principles usable for their estimation are (largely) known and diagnostic wavelengths have a direct physical meaning. Among these properties, studies modeling carbon (total and organic) and clay content are likely to be most frequent. Whereas SOC has no well-pronounced features in the 400-2500 nm region, we find typical absorption features for clay minerals in the 1900-2400 nm range (Chabrillat *et al.*, 2002). The SWIR domain has also been identified as very relevant for the spectral assessment of OC and N, a series of key wavelengths for both are located in the region beyond 1700 nm (Cécillon *et al.*, 2009; Bellon-Maurel and McBratney, 2011). Nevertheless, in the following, we focus on some exemplary studies using the 400-1100 nm (VNIR) range only, as the development of instruments using this domain is less cost-intensive compared to 'full range' (400-2500 nm)-instruments and thus it could be easily utilized in future agriculture applications for assessing both vegetation and soil properties.

Daniel *et al.* (2003) used in-situ measurements in the 400-1050 nm range and tested different bandwidths to estimate soil organic matter, phosphorus, and potassium in a neural network approach. In the best case, R^2 values greater than 0.8 were obtained for all constituents. Viscarra-Rossel *et al.* (2006) compared systematically how far a range of soil properties could be quantified when using visible (400-795 nm) or NIR-SWIR (810-2400 nm) data. With the latter, an R^2 of at least 0.50 was achieved for OC ($R^2 = 0.60$), aluminium (Al), clay, sand contents, pH, and lime requirement. With data from the visible, this was only achieved for OC (R^2 equalled again 0.60), which can be traced back to an often strong impact of organic matter on soil color. This gives also the ability for relatively simple spectral indices using bands from the visible domain, as proposed for example by Thaler *et al.* (2019). They defined a new SOC index (SOCI) based on a simple ratio of reflectances in the visible blue and the product of reflectances in the visible green and red. SOC predictions with this index were similarly accurate as a prediction from the SWIR/NIR ratio. Similarly, Aitkenhead *et al.* (2018) used RGB photography and retrieved good predictions for organic matter. Furthermore, they also found Al, Fe and magnesium (Mg) to be well predictable when present in oxidized form and in a sufficient concentration to impact soil color. An additional integration of site characteristics (as topography, climate, soil type, geology) was helpful to improve prediction accuracies. Gholizadeh *et al.* (2020) compared various soil color spaces and color indices derived from reflectance spectroscopy (400-700 nm) and from data of an RGB digital camera; SOC results for both datasets were similar.

Compared to plant parameters (*see* Table 1), narrow-band indices in the 400-1100 nm range relying on reflectances measured at one single sensitive wavelength or on simple band combinations are less common in spectroscopy or remote sensing of soils (*see* overview in Table 2), as well-defined absorption features are rare. The SOCI of Thaler *et al.* (2019), for example, indeed realizes a simple combination of discrete wavelengths (478, 546, and 659 nm), but actually refers to the continuous behavior of soil reflectance curves in the visible domain, which is a decreasing slope and a typical change of the shape from concave to convex with increasing SOC contents (Bartholomeus *et al.*, 2008). Bartholomeus *et al.* (2008) thus identified '1/Slope 400-600 (nm)' to be an appropriate spectral index, linearly related to $SOC^{1/4}$. Nevertheless, Stenberg *et al.* (2010) document that spectral effects in the visible may also be induced by many other soil properties, such as texture, structure, moisture, and mineralogy, so that soil darkness would only be a useful discriminator for different SOC contents within a limited geological variation.

Table 2: Examples of Spectral Band Indices and Sensitive Wavelengths in the 400-1100 nm Range for the Estimation of Soil Variables

Soil Parameter	Used Spectral Variable/Index	References
Organic carbon	SOCI (= B/R/G) with 478, 546, and 659 nm for blue (B), green (G), and red (R) spectral bands	Thaler *et al.*, 2019
Organic carbon	1/Slope 400-600 (nm)	Bartholomeus *et al.*, 2008
Organic carbon, total nitrogen, oxalate-extractable iron	CIE color system with L* and b* as predictors (L* = lightness to darkness; b* = blue- to yellowness)	Moritsuka *et al.*, 2014; Stiglitz *et al.*, 2018
Hematite, goethite, Hematite : goethite ratio	Absorption features at 0.48-0.55 µm, 0.64-0.73 µm, 0.85-1.0 µm	Morris *et al.*, 1985; Cudahy and Ramanaidou, 1997
Hematite	Redness indices, based on R, G, and partly B spectral bands, e.g. $R^2/(B \times G^3)$	Madeira *et al.*, 1997; Mathieu *et al.*, 1998
Electrical conductivity	• Salinity indices (including R, G or B, partly near-infrared NIR spectral bands) • Vegetation Soil Salinity Index: $VSSI = 2 \times G - 5 \times (R + NIR)$	Dehni and Lounis, 2012; Nguyen *et al.*, 2020

5.7. Conclusion

Hyperspectral remote sensing and field spectroscopy in the visible and near-infrared (VNIR) has been widely used in vegetation and soil sensing, both in the laboratory and in-situ soil monitoring. Following the general goal of this book, we discussed spectral imaging systems that are relevant for UAV-s and field spectroscopy. A selected list of variables will be provided in Table 3, which focuses on indicators in agroecology and organic farming.

Typically, field reflectance spectra are collected by portable field spectrometers, which are often complemented by UAV-, air- or space-borne imaging spectrometers with a more limited spectral resolution. Compared to portable field spectroscopy, drone-based imaging spectroscopy has a great potential to cover larger areas during a flight campaign, but spectral accuracies of estimated vegetation and soil

Table 3: Main Detectable Proxies for Agroecology and Organic Farming Using Field Spectroscopy for Mapping and Quantifying Plant and Soil Parameters

Vegetation measures	**Biochemical indicators:** chlorophyll, xanthophyll, carotene, lignin, cellulose, starch, sugar, carbon, etc.
	Physiological processes: photosynthesis, stoma conductivity, transpiration, nutrient cycles, etc.
	Phenology: leaf , flowering phenology, ripening, senescence, scale dependent phenology, etc.
	Volumetric and geometric parameters: density, size, yield, shape, area, spatial distribution, heterogeneity, relationships, connectivity, dimensionality, etc.
	Plant stress and pathology: biotic, abiotic sources, invasive species, disturbance and hazards, etc.
Soil measures	**Geochemical indicators:** inorganic components, calcium, magnesium, sulfur, micro nutrients, etc.
	Soil functions and dynamics: cation-exchange capacity (CEC), evaporation, moisture regulation, pH, heat flux, emissivity, etc.
	Soil taxonomy: clay minerals, silt and sand content, carbonates, iron, sulphate content, etc.
	Soil stress and soil pathology: erosion, land degradation, desertification, soil acidity, over use, etc.
	Structure and geometric parameters: fragmentation, density, size, shape, area, spatial distribution, heterogeneity, relationships, connectivity, 3D architecture, etc.
	Soil life and organic: soil organic carbon (SOC) and matter (SOM), total nitrogen, microbial biomass and activity, etc.

properties are usually lower (due to a lower signal-to-noise ratio and disturbing atmospheric influences, for example). With traditional instrument concepts (i.e. ground truthing linked to the data of air- and space-borne scanners), there is a gap in the 'point-pixel-image' approach as proximally sensed hyperspectral image data were less available until recently. To overcome this limitation and to bridge the gap in the hyperspectral image data chain, snapshot hyperspectral imaging has been introduced, which enables rapid spectral image data acquisition in the field. The recent technological advantages of field spectroscopy enable researchers and users in agroecology and organic farming to adapt remote sensing know-how and experiences to small-scale or small-sized agricultural situations. For further reading, it is recommended to study comprehensive and extended reviews of remote sensing applications in soil and vegetation sciences (Lausch *et al.*, 2018, 2019; Weiss *et al.*, 2020). These papers try to consider most of the remote sensing technologies (optical, thermal, radar, lidar, etc.) and associated aspects. There are many available commercial sensors and systems available on the market. The spectral ranges will be extended and more and more miniature spectral cameras for drones are constructed to cover the solar spectrum. This spectral extension will open new perspectives, especially in mineralogy and soil mapping.

Bibliography

Abdel-rahman, E.M., F.B. Ahmed and M. van den Berg (2010). Estimation of sugarcane leaf nitrogen concentration using *in-situ* spectroscopy, *Int. J. Appl. Earth Obs. Geoinf.*, 12: 52-57.

Adam, E., O. Mutanga and D. Rugege (2010). Multispectral and hyperspectral remote sensing for identification and mapping of wetland vegetation: A review, *Wetl. Ecol. Manag.*, **18**(3): 281-296.

Adamchuk, V.I., R.B. Ferguson and G.W. Hergert (2010). Soil heterogeneity and crop growth, pp. 3-16. *In*: E.C. Oerke, R. Gerhards, G. Menz and R.A. Sikora (Eds.). *Precision Crop Protection – The Challenge and Use of Heterogeneity*, Springer, New York, USA.

Aitkenhead, M., C. Cameron, G. Gaskin, B. Choisy, M. Coull and H. Black (2018). Digital RGB photography and visible-range spectroscopy for soil composition analysis, *Geoderma*, 313: 265-275.

Almeida, T.I.R. and D.S. Filho (2004). Principal component analysis applied to feature-oriented band ratios of hyperspectral data: A tool for vegetation studies, *Int. J. Remote Sens.*, **25**(22): 5005-5023.

Asner, G.P. and K.B. Heidebrecht (2002). Spectral unmixing of vegetation, soil and dry carbon cover in arid regions: Comparing multispectral and hyperspectral observations, *Int. J. Remote Sens.*, **23**(19): 3939-3958.

Bannari, A., A. Pacheco, K. Staenz, H. McNairn and K. Omari (2006). Estimating and mapping crop residues cover on agricultural lands using hyperspectral and IKONOS data, *Remote Sens. Environ.*, **104**(4): 447-459.

Bareth, G., H. Aasen, J. Bendig, M.L. Gnyp, A. Bolten, A. Jung, R. Michels and J. Soukkamäki (2015). Low-weight and UAV-based hyperspectral full-frame cameras for monitoring crops: Spectral comparison with portable spectroradiometer measurements, PFG-J, *Photogramm. Rem.*, 1: 69-79.

Bartholomeus, H.M., M.E. Schaepman, L. Kooistra, A. Stevens, W.B. Hoogmoed and O.S.P. Spaargaren (2008). Spectral reflectance-based indices for soil organic carbon quantification, *Geoderma*, 145: 28-36.

Beeri, O., R. Phillips, J. Hendrickson, A.B. Frank and S. Kronberg (2007). Estimating forage quantity and quality using aerial hyperspectral imagery for northern mixed-grass prairie, *Remote Sens. Environ.*, **110**(2): 216-225.

Bellon-Maurel, V. and A. McBratney (2011). Near-infrared (NIR) and mid-infrared (MIR) spectroscopic techniques for assessing the amount of carbon stock in soils – Critical review and research perspectives, *Soil Biol. Biochem.*, 43: 1398-1410.

Ben-Dor, E. and A. Banin (1995). Near-infrared analysis as a rapid method to simultaneously evaluate several soil properties, *Soil Sci. Soc. Am. J.*, 59: 364-372.

Buschmann, C. and E. Nagel (1993). In vivo spectroscopy and internal optics of leaves as basis for remote sensing of vegetation. *Int.J. Remote Sens.*, **14**(4): 711-722.

Cécillon, L., B.G. Barthès, C. Gomez, D. Ertlen, V. Genot, M. Hedde, A. Stevens and J.J. Brun (2009). Assessment and monitoring of soil quality using near-infrared reflectance spectroscopy (NIRS), *Eur. J. Soil Sci.*, 60: 770-784.

Chabrillat, S., A.F.H. Goetz, L. Krosley and H.W. Olsen (2002). Use of hyperspectral images in the identification and mapping of expansive clay soils and the role of spatial resolution, *Remote Sens. Environ.*, 82: 431-445.

Chan, J.C.W. and D. Paelinckx (2008). Evaluation of Random Forest and Adaboost tree-based ensemble classification and spectral band selection for ecotope mapping using airborne hyperspectral imagery, *Remote Sens. Environ.*, **112**(6): 2999-3011.

Clevers, J.G.P.W., G.W.A.M Van Der Heijden, S. Verzakov and M.E. Schaepman (2007). Estimating grassland biomass using SVM band shaving of hyperspectral data, *Photogramm. Eng. Remote Sensing*, **73**(10): 1141-1148.

Cudahy, T.J. and E.R. Ramanaidou (1997). Measurement of the hematite: Goethite ratio using field visible and near-infrared reflectance spectrometry in channel iron deposits, Western Australia, *AGSO J. Aust. Geol. Geophys.*, **44**(4): 411–420.

Daniel, K.W., N.K. Tripathi and K. Honda (2003). Artificial neural network analysis of laboratory and *in situ* spectra for the estimation of macronutrients in soils of Lop Buri (Thailand), *Soil. Res.*, **41**(1): 47-59.

Danson, F.M., M.D. Steven, T.J. Malthus and J.A. Clark (1992). High-spectral resolution data for determining leaf water content, *Int. J. Remote Sens.*, **13**(3): 461-470.

Dehni, A. and M. Lounis (2012). Remote sensing techniques for salt affected soil mapping: Application to the Oran region of Algeria, *Procedia Eng.*, 33: 188-198.

EC (European Commission) (2016). Facts and Figures on Organic Agriculture in the European Union, European Commission; Retrieved from: https://ec.europa.eu/info/sites/info/files/food-farming-fisheries/farming/documents/organic-agriculture-2015_en.pdf; accessed on 17 January, 2020.

Elvidge, C.D. (1990). Visible and near infrared reflectance characteristics of dry plant materials, *Int. J. Remote Sens.*, **11**(10): 1775-1795.

Filella, I. and J. Peñuelas (1994). The red edge position and shape as indicators of plant chlorophyll content, biomass and hydric status, *Int. J. Remote Sens.*, **15**(7): 1459-1470.

Gholizadeh, A., M. Saberioon, R.A. Viscarra Rossel, L. Boruvka and A. Klement (2020). Spectroscopic measurements and imaging of soil color for field scale estimation of soil organic carbon, *Geoderma*, 357: 113972.

Gitelson, A. and M.N. Merzlyak (1994). Spectral reflectance changes associated with autumn senescence of *Aesculus hippocastanum* L. and *Acer platanoides* L. leaves: Spectral features and relation to chlorophyll estimation, *J. Plant Physiol.*, **143**(3): 286-292.

Gitelson, A.A and M.N. Merzlyak (2003). Relationships between leaf chlorophyll content and spectral reflectance and algorithms for non-destructive chlorophyll assessment in higher plant leaves, *J. Plant Physiol.*, **160**(3): 271-282.

Gitelson, A.A., M.N. Merzlyak and O.B. Chivkunova (2001). Optical properties and non-destructive estimation of anthocyanin content in plant leaves, *Photochem. Photobiol.*, **74**(1): 38-45.

Goetz, A.F.H., G. Vane, J.E. Soloman and B.N. Rocks (1985). Imaging spectrometry for Earth remote sensing, *Science*, 228: 1147-1153.

Gunasekaran, S., M.R. Paulsen and G.C. Shove (1985). Optical methods for non-destructive quality evaluation of agricultural and biological materials, *J. Agric. Eng.*, **32**(3): 209-241.

Hagen, N., R.T. Kester, L. Gao and T.S. Tkaczyk (2012). Snapshot advantage: A review of the light collection improvement for parallel high-dimensional measurement systems, *Opt. Eng.*, 51: 111702-1.

Hengl, T. and R.A. MacMillan (2019). Predictive Soil Mapping with R; Retrieved from https://soilmapper.org/; Accessed on 15 March, 2021.

Hill, M.J., T. Udelhoven, M. Vohland and A. Stevens (2010). The use of laboratory spectroscopy and optical remote sensing for estimating soil properties, pp. 67-86. *In*: E.C. Oerke, R. Gerhards, G. Menz and R.A. Sikora (Eds.). *Precision Crop Protection – The Challenge and Use of Heterogeneity*, Springer Science, Dordrecht.

IPCC (Intergovernmental Panel on Climate Change). (2014). Working Group I: The Scientific Basis, 10.1.2 The Regional Climate Problem; retrieved from https://www.ipcc.ch/; accessed on 15 March, 2021.

Jenny, H. (1941). *Factors of Soil Formation*, McGraw-Hill, New York, USA.

Jung, A., M. Vohland and S. Thiele-Bruhn (2015). Use of a portable camera for proximal soil sensing with hyperspectral image data, *Remote Sens.*, **7**(9): 11434-11448.

Kokaly, R.F., B.W. Rockwell, S.L. Haire and T.V. King (2007). Characterization of post-fire surface cover, soils, and burn severity at the Cerro Grande Fire, New Mexico, using hyperspectral and multispectral remote sensing, *Remote Sens. Environ.*, **106**(3): 305-325.

Kuang, B. and A.M. Mouazen (2011). Calibration of visible and near infrared spectroscopy for soil analysis at the field-scale on three European farms, *Eur. J. Soil Sci.*, 62: 629-636.

Lauden, R. and G. Bareth (2006). Multitemporal hyperspectral data analysis for regional detection of plant diseases by using a tractor- and an airborne-based spectrometer, *PFG-J. Photogramm. Rem.*, 3: 217-227.

Lausch, A., O. Bastian, S. Klotz, P.J. Leitão, A. Jung, D. Rocchini, M.E. Schaepman, A.K. Skidmore, L. Tischendorf and S. Knapp (2018). Understanding and assessing vegetation health by *in situ* species and remote-sensing approaches, *Methods Ecol. Evol.*, 9: 1799-1809.

Lausch, A., J. Baade, L. Bannehr, E. Borg, J. Bumberger, S. Chabrilliat, P. Dietrich, H. Gerighausen, C. Glässer, J.M. Hacker, D. Haase, T. Jagdhuber, S. Jany, A. Jung, A. Karnieli, R. Kraemer, M. Makki, C. Mielke, M. Möller, H. Mollenhauer, C. Montzka, M. Pause, C. Rogass, O. Rozenstein, C. Schmullius, F. Schrodt, M. Schrön, K. Schulz, C. Schütze, C. Schweitzer, P. Selsam, A.K. Skidmore, D. Spengler, C. Thiel, S.C. Truckenbrodt, M. Vohland, R. Wagner, U. Weber, U. Werban, U. Wollschläger, S. Zacharias and M.E. Schaepman (2019). Linking remote sensing and geodiversity and their traits relevant to biodiversity – Part I: Soil characteristics, *Remote Sens.*, **11**(20): 2356.

le Maire, G., C. François, K. Soudani, D. Berveiller, J.Y. Pontailler, N. Bréda, H. Genet, H. Davi and E. Dufrêne (2008). Calibration and validation of hyperspectral indices for the estimation of broadleaved forest leaf chlorophyll content, leaf mass per area, leaf area index and leaf canopy biomass, *Remote Sens. Environ.*, **112**(10): 3846-3864.

Lee, D.W., J. O'Keefe, N.M. Holbrook and T.S. Field (2003). Pigment dynamics and autumn leaf senescence in a New England deciduous forest, eastern USA, *Ecol. Res.*, **18**(6): 677-694.

Lichtenthaler, H.K., A. Gitelson and M. Lang (1996). Non-destructive determination of chlorophyll content of leaves of a green and an aurea mutant of tobacco by reflectance measurements, *J. Plant Physiol.*, **148**(3): 483-493.

Madeira, J., A. Bedidi, J. Pouget, B. Cervelle and N. Flay (1997). Spectrometric indices (visible) of hematite and goethite contents in lateritic soils. Application to a TM image for soil mapping of Brasilia area, *Int. J. Remote Sens.*, **18**(13): 2835-2852.

Mäder, P., A. Fließbach, D. Dubois, L. Gunst, P. Fried and U. Niggli (2002). Soil Fertility and Biodiversity in Organic Farming, *Science*, 296: 1694.

Mathieu, R., M. Pouget, B. Cervelle and R. Escadafal (1998). Relationships between satellite-based radiometric indices simulated using laboratory reflectance data and typic soil color of an arid environment, *Remote Sens. Environ.*, 66: 17-28.

Merzlyak, M.N., A.A. Gitelson, O.B. Chivkunova and V.Y. Rakitin (1999). Non-destructive optical detection of pigment changes during leaf senescence and fruit ripening, *Physiol. Plant*, **106**(1): 135-141.

Milton, E.J., M.E. Schaepman, K. Anderson, M. Kneubühler and N. Fox (2009). Progress in field spectroscopy, *Remote Sens. Environ.*, 113: 92-109.

Moritsuka, N., K. Matsuoka, K. Katsura, S. Sano and J. Yanai (2014). Soil color analysis for statistically estimating total carbon, total nitrogen and active iron contents in Japanese agricultural soils, *Soil Sci. Plant Nutr.*, 60: 475-485.

Morris, R.V., H.V. Lauer, Jr., C.A., Lawson, E.K. Gibson, Jr., G.A. Nace and C. Stewart (1985). Spectral and other physicochemical properties of submicron powders of hematite (alpha-Fe_2O_3), maghemite (gamma-Fe_2O_3), magnetite (Fe_3O_4), goethite (alpha-FeOOH), and lepidocrocite (gamma-FeOOH), *J. Geophys. Res.*, **90**(B4): 3126-3144.

Mouazen, A.M., B. Kuang, J. De Baerdemaeker and H. Ramon (2010). Comparison among principal component, partial least squares and back propagation neural network analyses for accuracy of measurement of selected soil properties with visible and near infrared spectroscopy, *Geoderma*, 158: 23-31.

Mutanga, O., A.K. Skidmore and S. van Wieren (2003). Discriminating tropical grass (*Cenchrus ciliaris*) canopies grown under different nitrogen treatments using spectroradiometry, *ISPRS J. Photogramm. Remote Sens.*, **57**(4): 263-272.

Nguyen, K.A., Y.A. Liou, H.P. Tran, P.P. Hoang and T.H. Nguyen (2020). Soil salinity assessment by using near-infrared channel and vegetation soil salinity Index derived from Landsat 8 OLI data: A case study in the Tra Vinh Province, Mekong Delta, Vietnam, *Prog. Earth Planet. Sci.*, **7**(1): 1-16.

Numata, I., D.A. Roberts, O.A. Chadwick, J. Schimel, F.R. Sampaio, F.C. Leonidas and J.V. Soares (2007). Characterization of pasture biophysical properties and the impact of grazing intensity using remotely sensed data, *Remote Sens. Environ.*, **109**(3): 314-327.

Ogen, Y., C. Neumann, S. Chabrillat, N. Goldshleger and E. Ben-Dor (2018). Evaluating the detection limit of organic matter using point and imaging spectroscopy, *Geoderma*, 321: 100-109.

Patzold, S., F.M. Mertens, L. Bornemann, B. Koleczek, J. Franke, H. Feilhauer and G. Welp (2008). Soil heterogeneity at the field scale: A challenge for precision crop protection, *Precis. Agric.*, 9: 367-390.

Peñuelas, J., I. Filella, P. Lloret, F. Munoz and M.Vilajeliu (1995). Reflectance assessment of mite effects on apple trees, *Int. J. Remote Sens.*, **16**(14): 2727-2733.

Peñuelas, J., J.A. Gamon, Al. Fredeen, J. Merino and C.B. Field (1994). Reflectance indices associated with physiological changes in nitrogen- and water-limited sunflower leaves, *Remote Sens. Environ.*, **48**(2): 135-146.

Schaepman-Strub, G., M.E. Schaepman, T.H. Painter, S. Dangel and J.V. Martonchik (2006). Reflectance quantities in optical remote sensing – Definitions and case studies, *Remote Sens. Environ.*, **103**(1): 27-42.

Schellberg, J., M.J. Hill, R. Gerhards, M. Rothmund and M. Braun (2008). Precision agriculture on grassland: Applications, perspectives and constraints, *Eur. J. Agron.*, **29**(2-3): 59-71.

Sims, D.A. and J.A. Gamon (2002). Relationships between leaf pigment content and spectral reflectance across a wide range of species, leaf structures and developmental stages, *Remote Sens. Environ.*, **81**(2): 337-354.

Smith, K.L., M.D. Steven and J.J. Colls (2004). Use of hyperspectral derivative ratios in the red-edge region to identify plant stress responses to gas leaks, *Remote Sens. Environ.*, **92**(2): 207-217.

Somers, B., S. Delalieux, W.W. Verstraeten, A.V. Eynde, G.H. Barry and P. Coppin (2010). The contribution of the fruit component to the hyperspectral citrus canopy signal, *Photogramm. Eng. Rem. S.*, **76**(1): 37-47.

Soriano-Disla, J.M., L.J. Janik, R.A. Viscarra Rossel, L.M. Macdonald and M.J. McLaughlin (2014). The performance of visible near- and mid-infrared reflectance spectroscopy for prediction of soil physical, chemical and biological properties, *Appl. Spectrosc. Rev.*, 49: 139-186.

Steffens, M. and H. Buddenbaum (2013). Laboratory imaging spectroscopy of a stagnic Luvisol profile. High resolution soil characterisation, classification and mapping of elemental concentrations, *Geoderma*, 195: 122-132.

Stenberg, B., V. Rossel, A.M. Mouazen and J. Wetterlind (2010). Visible and Near-infrared Spectroscopy in Soil Science, pp. 163-215. *In*: D.L. Sparks (Ed.). *Advances in Agronomy*, Academic Press, Burlington, USA. Stevens, A., B. van Wesemael, H. Bartholomeus, D. Rosillon, B.Tychon and E. Ben-Dor (2008). Laboratory, field and airborne spectroscopy for monitoring organic carbon content in agricultural soils, *Geoderma*, 144: 395-404.

Stiglitz, R.Y., E. A. Mikhailova, J.L., Sharp, C.J. Post, M.A. Schlautman, P.D. Gerard and M.P. Cope (2018). Predicting Soil Organic Carbon and Total Nitrogen at the Farm Scale Using Quantitative Color Sensor Measurements, *Agronomy*, 8: 212.

Thaler, E.A., I.J. Larsen and Q. Yu (2019). A New Index for Remote Sensing of Soil Organic Carbon Based Solely on Visible Wavelengths, *Soil Sci. Soc. Am. J.*, 83: 1443-1450.

Thenkabail, P.S., R.B. Smith and E. De Pauw (2000). Hyperspectral vegetation indices and their relationships with agricultural crop characteristics, *Remote Sens. Environ.*, **71**(2): 158-182.

Thenkabail, P.S., R.B. Smith and E. De Pauw (2002). Evaluation of narrowband and broadband vegetation indices for determining optimal hyperspectral wavebands for agricultural crop characterization, *Photogramm. Eng. Rem. S.*, **68**(6): 607-622.

Thenkabail, P.S., E.A. Enclona, M.S. Ashton, C. Legg and M.J. De Dieu (2004a). Hyperion, IKONOS, ALI, and ETM+ sensors in the study of African rainforests, *Remote Sens. Environ.*, **90**(1): 23-43.

Thenkabail, P.S., E.A. Enclona, M.S. Ashton and B. Van Der Meer (2004b). Accuracy assessments of hyperspectral waveband performance for vegetation analysis applications, *Remote Sens. Environ.*, **91**(3): 354-376.

Thenkabail, P.S., J.J.G. Lyon and A. Huete (2012). *Hyperspectral Remote Sensing of Vegetation*, CRC Press, Florida, Boca Raton, USA.

Ustin, S.L., D.A. Roberts, J.A. Gamon, G.P. Asner and R.O. Green (2004). Using imaging spectroscopy to study ecosystem processes and properties, *Bioscience*, **54**(6): 523-534.

Van Der Meer, F. (2004). Analysis of spectral absorption features in hyperspectral imagery, *Int. J. Appl. Earth Obs. Geoinf.*, **5**(1): 55-68.

Viscarra Rossel, R.A., D.J.I. Walvoort, A.B. McBratney, L.J. Janik and J.O. Skjemstad (2006). Visible, near infrared, mid infrared or combined diffuse reflectance spectroscopy for simultaneous assessment of various soil properties, *Geoderma*, 131: 59-75.

Weiss, M., F. Jacob and G. Duveiller (2020). Remote sensing for agricultural applications: A meta-review, *Remote Sens. Environ.*, 236: 111402.

Wessman, C.A. (1990). Evaluation of canopy biochemistry, pp. 135-156. *In*: R.J. Hobbs and H.A. Mooney (Eds.). *Remote Sensing of Biosphere Functioning*, Springer, New York, USA.

Wight, J.P., A.J. Ashworth and F.L. Allen (2016). Organic substrate, clay type, texture, and water influence on NIR carbon measurements, *Geoderma*, 261: 36-43.

Wrolstad, R.E., R.W. Durst and J. Lee (2005). Tracking color and pigment changes in anthocyanin products, *Trends in Food. Sci. Technol.*, **16**(9): 423-428.

Yang, F., J. Li, X. Gan, Y. Qian, X. Wu and Q. Yang (2010). Assessing nutritional status of *Festuca arundinacea* by monitoring photosynthetic pigments from hyperspectral data, *Comput. Electron. Agric.*, **70**(1): 52-59.

Yao, X., Y. Zhu, Y. Tian, W. Feng and W. Cao (2010). Exploring hyperspectral bands and estimation indices for leaf nitrogen accumulation in wheat, *Int. J. Appl. Earth Obs. Geoinf.*, **12**(2): 89-100.

Zarco-Tejada, P.J., V. González-Dugo and J.A.J. Berni (2011). Fluorescence, temperature and narrow-band indices acquired from a UAV platform for water stress detection using a micro-hyperspectral imager and a thermal camera. *Remote Sens. Environ.*, 117: 322-337.

Zarco-Tejada, P.J., V. González-Dugo and J.A. Berni (2012). Fluorescence, temperature and narrow-band indices acquired from a UAV platform for water stress detection using a micro-hyperspectral imager and a thermal camera, *Remote Sens. Environ.*, 117: 322-337.

Zarco-Tejada, P.J., M.L. Guillén-Climent, R. Hernández-Clemente, A. Catalina, M.R. González and P. Martín (2013). Estimating leaf carotenoid content in vineyards using high resolution hyperspectral imagery acquired from an unmanned aerial vehicle (UAV), *Agric. For. Meteorol.*, 171: 281-294.

Zhao, D., K.R. Reddy, V.G. Kakani, J.J. Read and G.A. Carter (2003). Corn (*Zea mays* L.) growth, leaf pigment concentration, photosynthesis and leaf hyperspectral reflectance properties as affected by nitrogen supply, *Plant Soil*, **257**(1): 205-218.

Zhao, D., L. Huang, J. Li and J. Qi (2007). A comparative analysis of broadband and narrowband derived vegetation indices in predicting LAI and CCD of a cotton canopy, *ISPRS J. Photogramm. Remote Sens.*, **62**(1): 25-33.

Zude, M., B. Herold, J.M. Roger, V. Bellon-Maurel and S. Landahl (2006). Non-destructive tests on the prediction of apple fruit flesh firmness and soluble solids content on tree and in shelf-life, *J. Food Eng.*, **77**(2): 254-260.

CHAPTER

6

Drones for Good: UAS Applications in Agroecology and Organic Farming

Salvatore Eugenio Pappalardo[1]* and Diego Andrade[2]

[1] Department of Civil, Environmental and Architectural Engineering, University of Padova, Italy

[2] Drone & GIS, Quito, Ecuador

6.1. Introduction: Flying Robots on Agroecosystems

More than "flying robots". Drones. They are more properly defined as 'unmanned aerial systems' (UAS) and today, they embody different data acquisition tools and approaches together: geo-information and communication technologies (GeoICT), MEMs and sensors, robots, people, artificial intelligence, social intelligence, Internet of Things (IoT), Big Data. Today, small, low-cost quadcopters with 'special eyes' or mimetic bionic-birds fly almost everywhere: on river corridors, on forests, on the city, on farmlands (Pajares, 2015; Tang and Guofan, 2015; White *et al.*, 2016; Baena *et al.*, 2018; Merkert and James Bushell, 2020). Drones for civil and environmental applications – or Drones for Good – are becoming even more diffused, assuming a key role especially within the domain of agriculture by supporting actual challenges of increasing sustainability in cropping and agro-food production systems (Sylvester, 2018). In fact, UAS recently seduced and entered many fields of cropping systems, particularly through the framework of Agriculture 4.0, within the different declinations of precision agriculture, smart farming, and climate-resilient faming systems (Radoglou-Grammatikis, 2020; Tsouros *et al.*, 2019). They are mainly deployed for monitoring crop yields, assessing nutrie nt and water stress, mapping weed distribution, and for pest management (Radoglou-Grammatikis, 2020).

The epoch of the 'flying robots' for agriculture and agro-environmental monitoring started a decade ago when drones 'slipped away' from the military

*Corresponding author: salvatore.pappalardo@unipd.it

aviation technologies fences, by entering into the domain of civil applications (Kim *et al.*, 2019). Through a huge emphasis from the worlds of academy, national and international institutions, and agro-industries, drones rapidly broke into the 'collective imagery' as the flying robots which will make the difference in pursuing sustainability in agriculture. This emphasis is well synthetized by mainstream articles from *MIT Technology Review*, which enormously sponsored the forthcoming entrance of 'agricultural drones' (2014) and, later, highlighted the ways they are revolutionizing agriculture (2016). In fact, as reported by Goldman Sachs research (2021), the expansion of drones in agriculture seems to be confirmed also in terms of growth of drone industry and services; agriculture is the second one after construction sector, with a total addressable market worth USD 6,000 million. Globally, the drone market size was USD 4,400 million in 2018 and it is expected to grow to USD 63,600 million in 2025, with a compound annual growth rate of 55.9 per cent during such temporal range (*Market Insider*, 2021). It is estimated that agricultural drones will grow to about USD 32,400 million by 2050 which will represent almost 25 per cent of UAS global market (Kim and Kim, 2019). At present a wide range of UAS are available on the global market. If on one hand, different 'ready to fly' UAS are produced by big manufacturers (DJI, AGEagle, Parrot, Trimble Navigation, Precisonhawk); on the other, once UAS open hardware and open software notably increased, giving the opportunity to assemble and to build an operational drone for aerial surveys (Gayathri Devi *et al.*, 2020).

This chapter will explore the world of UAS and their applications in different domains of agriculture; it is structured in the following sections:

- From the space to the near surface: UAS in agriculture
- Agricultural UAS: platforms, sensors, components
- UAS applications in sustainable agriculture and agroecology
- UAS for preserving spider monkey and agroecosystem management: experiences from tropical forests of Chocò (Ecuador)
- Opportunities and perspectives for the agroecology transition

6.2. From the Space to the Near Surface: UAS in Agriculture

In the past, GIScience was widely characterized by an increasing massive use of remote sensing technologies and platforms, mainly equipped on aircraft and satellites, to acquire spatial information about Earth surface processes through specific sensors (Goodchild, 2007). At present, a wide range of satellite-derived images are available: public aerospace programs, such as from USGS/NASA Landsat (US) and ESA Copernicus (European Union) or commercial satellites (WorldView, Planet among others). For a deeper understanding, see detailed explanations in Chapter 11.

However, due to their spatial resolutions – which usually range from 30 to 10m per pixel for public aerospace programs, or up to 0.2 m per pixel for commercial platforms – remotely-sensed data from satellite is generally scarcely suitable for application at agroecosystem or at a detailed scale (Tsouros *et al.*, 2019). Moreover, satellite temporal resolution – or namely frequency of revisiting time over the same area of interest – may represent a critical constraint in terms of image acquisition. In fact, satellite platforms are not generally suitable to capture images in a required time-frame, as often needed for acquiring remotely sensed information from agroecosystem dynamics and cropping cycle (Kim *et al.*, 2019; Zhang *et al.*, 2021). Moreover, some environmental conditions, such as cloud cover and atmospheric factors, may drastically affect the quality of imagery, making difficult or, in some cases, impossible, to extract data and useful information (Kim *et al.*, 2019).

Airborne cameras have a long track in the history of remote sensing; however, the use of aircraft for aerial surveys is at present economically onerous as it would require a strong coordination between farmers to acquire large portions of agricultural territory to make it cost-effective.

On the other hand, the lately rapid and extensive spreading of UAS is currently offering new opportunities for a deeper understanding of agroecosystem complexity and for supporting a paradigm shift in agriculture. In fact, according to specific national regulations, UAS can fly at much lower altitudes compared to aircraft/satellite, usually from few meters up to 120-600 meters above the ground (Zhang *et al.*, 2021). Such flight altitude combined with the actual available technology of sensors considerably increases spatial resolution up to 0.01 m per pixel, or even higher. Some authors refer to this characteristic as the 'ultra-high' spatial resolution of UAS-derived images (Tsouros *et al.*, 2019). Moreover, different UAS can be equipped with a wide range of image-acquisition devices, from optical to multi and hyper-spectral sensors (Kim and Kim, 2019).

One of the advantages of integrating UAS for spatial analysis in agriculture is related to the low latency represented by on-demand repeatability of acquisition flight, which makes ultra-high resolution aerial surveys more suitable for agroecosystem monitoring and management. In fact, drones may survey farmland every week, every day or even every hour, given the chance to perform on-demand multi-temporal time-series, able to detect changes, and to unveil new opportunities in agrosystem management (Radoglou-Grammatikis, 2020; Marino and Alvino, 2018). Therefore, direct control of temporal resolution of aerial surveys may give both to researchers and to farmers an integrated technical and operative support for studying ecosystem dynamics and for rapid interventions on the field.

6.3. Agricultural UAS: Platforms, Sensors, Components

This section describes aerial platforms and the main components of an UAS. UAS are structured in different components and elements interacting with each other. Key elements and components are five: i) one (or more) aerial platform (commonly named Unmanned Aerial Vehicles, UAV); ii) a payload constituted by one (or more) sensor for spatial data acquisition or mechanical devices; iii) an UAV remote controller combined with a ground control station; iv) a human operator; and v) a GIS-based software for image processing and output maps.

6.3.1. Aerial Platforms: Multi-rotors, Fixed-wing and Hybrids

Firstly, we refer to the term platform in relation to the underlying aerial-vehicle structure which is the physical support for mounting extra tools and peripherals, such as MEM, GPS, and sensors. At present, different typologies of aerial platforms are available and can be adopted for agricultural purposes, according to the specific aims, the operational conditions, and the context.

They include rotorcraft and fixed-wing aircraft on one side; aerostatic balloons, blimps, and kites on the other (Fig. 1). Even if, at present, the most diffused platforms for agrosystems monitoring and management are rotorcraft and fixed-wing aircraft, the adoption of the long-stand aerial photography represented by balloons or kites should not be excluded *a priori* for photogrammetry surveys, as they still represent important alternatives for particular contexts and needs (Bryson *et al.*, 2013; Lorenz and Scheidt, 2014).

In general, the main elements which characterize an aerial platform and, therefore, its operational functions and range, are the aerodynamic features represented by the wings. Indeed, there are two types of primary aerial platforms: rotary- and fixed-wing (Radoglou-Grammatikis, 2020).

Rotary-wing platforms are usually multi-rotor models which are classified according to the number of propellers. With the exception of the traditional unmanned helicopters (one propeller), multi-rotors platforms are divided in the following categories: tricopters (three propellers); quadcopters (four propellers); exacopters (six propellers); octocopters (eight propellers) (Kim and Kim, 2019; Radoglou-Grammatikis, 2020). Generally, increase in the number of propellers corresponds to largest payload capacity (up to 9.5 kg for octocopters) and size of UAS. Quadcopters and hexacopters are usually smaller and are adapted to carry a payload ranging above 1.25-2.6 kg (Hayat *et al.*, 2016; Vergouw *et al.*, 2016).

Major advantages of employing multi-rotor platforms in agriculture are the following: i) ease of use compared to fixed-wing platforms (no runaway is needed), ii) the capability of taking-off and landing vertically, and iii) the ability of hovering on a given area for detailed inspection (Chapman *et al.*, 2014; Hassler and Baysal-Gurel, 2019).

Fixed-wing platforms are similar, both in shape and in aerodynamics, to an airplane. They require a reserved space as runway or a catapult (i.e. Trimble UX5), according to their size (from 90 to 300 cm wingspan). The main advantages are related to their longer flight autonomy and faster velocity compared to multi-rotors platforms. In fact, they are capable of covering vast areas of land rapidly, and to support high temporal and spatial resolution data acquisition; in addition, some fixed-wing platforms can carry heavier payloads for extended routes (Hogan *et al.*, 2017). However, they are not adapted for aerial survey in narrow spaces or for tasks which require operation of hovering or manoeuvring. They are generally preferred for application in wide field-mapping tasks for large portions of areas. With the exception of some assembled UAV (Moudrý *et al.*, 2018), fixed-wing UAV are generally more expensive and in some countries they are limited due to internal regulation of keeping the aircraft in visual line of sight (VLOS) with the pilot (Torresan *et al.*, 2017).

Finally, an interesting technological solution among modern platforms is represented by the hybrid-wing which integrates propellers for taking-off and landing, but also fixed-wing for large field-mapping tasks (Kim and Kim, 2019).

Aerial platforms vary in weight, size, flight autonomy, payload, and power. Aerial platforms are generally classified according to their weight, specifically named maximum take off mass (known as MTOM); hence, they are commonly divided in 'small' (≤15 kg), 'light' (≤7kg) and 'ultra-light' (≤0.250 kg) (Zhang *et al.*, 2021).

A less explored opportunity for low-cost aerial surveys is today represented by aerostatic balloons, blimps, and kites (Lorenz and Scheidt, 2014). Different platforms, which do not integrate any propellers or electric engines, are at present available. Generally, they are more suitable for semi-static or punctual aerial surveys or data acquisition for small areas. They are adapted for different geographical contexts, especially for non-invasive aerial surveys in sensitive ecosystems (Bryson *et al.*, 2013). Main characteristics and categories of aerial platforms are summarized in Table 1. It is worth noting that each typology of the above-mentioned aerial platforms presents the corresponding pros and cons.

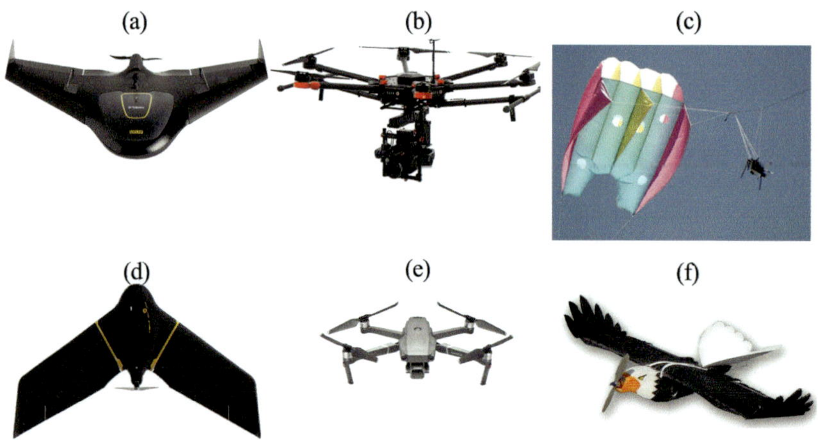

Fig. 1: Main typologies of unmanned aerial vehicles: (a) Fixed-wing UAV Trimble® UX5 (100 cm wingspan); (b) Multi-rotor hexacopter DJI® Matrice600; (c) Kite platform and camera; (d) Fixed-wing UAV EbeeSensfly® (115 cm wingspan); (e) Multi-rotor quadcopter DJI® Mavic Pro 2; (f) Bionic bird, Drone Bird®
(*Source*: Author's elaboration)

Table 1: Main Characteristics of Aerial Platforms for Agricultural Monitoring and Management (*Source*: Author's elaboration)

Aerial Platform Category	Advantages	Disadvantages
Rotary-wings (quadcopter, hexacopter, octocopter)	Ease of use Take-off/landing vertically Hovering on a specific spot Capture detailed images Suitable for narrow spaces	Low flight autonomy (15-25′) Limited payload Not suitable for extreme environments (tropical context, high temperatures)
Fixed-wings	High flight autonomy (20-40′) Data acquisition on vast areas Large payload	Runway or catapult for take-off Requirement of flight ability and control No hovering
Kite & balloon	Extremely low-cost Handmade assembly Limited legal regulations	Not suitable for large mobile mapping Limitations in stability Requirement of technical skills

6.3.2. Payload: Sensors and Peripherals

The component that gives 'special eyes' or other specific functions to UAVs is represented by the payload. Generally, it is constituted by different types of sensors for spatial data acquisition, but it could be implemented by other mechanical or electronic peripherals (grippers, discharger devices, biological and chemical sensors, weather sensors). By mounting these equipments, UAS are turning into powerful observation-and-sensing systems which may speed up a more comprehensive understanding of agroecosystem processes and functions, by interlinking ground sensors and stations based on IoT technologies (Gupta *et al.*, 2015; Hayat and Yanmaz, 2016).

Kind and number of elements of payload that can be installed on a UAV depend on their size and weight; the main aspect to be considered is the UAV's payload lift capability. Therefore, every aerial platform will have a maximum payload which limits size and weight of equipment that can be adopted. Similarly, general performances of UAV, such as flight time, stability, and velocity are strongly affected by the payload. It is noteworthy that many UAV manufacturers, such as DJI or Parrots provide on-board sensors which comply with the mentioned characteristics (Kim and Kim, 2019; Easterday *et al.*, 2019). UAS applications in agriculture usually require adoption of small and lightweight payload to ensure performance, both in data acquisition and flight range (Zhang *et al.*, 2021).

Typically, UAV sensors can be classified in the following types:

- Visible light sensors (RGB)
- Multispectral sensors
- Hyperspectral
- Thermal sensors
- Light detection and ranging sensors (LiDAR)

6.3.2.1. Visible Light Sensors (RGB)

Undoubtedly, visible light sensors – or commonly named RGB cameras – are the most used optical devices integrated into UAVs. These cameras produce the image most typically recognized in photography, by using red, green, and blue bands (or channels) within the range of visible light for image composition. Different typology of RGB cameras are at present available for aerial surveys: from reflex to mirror-less, from bridge to compact cameras (Yonah *et al.*, 2018). They are generally capable of acquiring images from high to ultra-high spatial resolution, according to pixel count and sensor size. The main advantage of RGB cameras is the relative ease of use, both in terms of image acquisition and data processing, by using common photogrammetry software (Zheng *et al.*, 2018; Tewes and Schellberg, 2018). Moreover, aerial surveys can be performed in different skylight conditions, both with cloud cover or cloudiness; however, weather changes during the UAV survey time-frame may extremely affect the quality of mosaic composition, due to changes in light conditions and, therefore, the different image exposures (Roth and Streit, 2017).

Downsides of using only RGB cameras are mainly due to their incapability of detecting different parameters which are not included the visible range. Consequently, RGB cameras are often coupled with multispectral sensors (Gruner *et al.*, 2019; Hassler and Baysal-Gurel, 2019).

6.3.2.2. Multispectral Sensors

Multispectral sensors expand the capability to obtain information beyond the visible spectrum of human eyes. As vegetation absorbs and reflects light in a wider range of spectrum, a larger amount of information can be, therefore, derived from multispectral images. Particularly, this spectral information is essential to assess, to monitor, and to manage different components and dynamics of agroecosystems: from physiological, biological, and physical characteristics of vegetation, to biodiversity and water management (Patrick *et al.*, 2017; Iqbal *et al.*, 2018).

The most diffused use of multispectral sensors in agriculture is related to the generation of several vegetation indices by the use of combinations of specific bands, commonly located in the near infrared (NIR) region of spectrum, within 750 nm and 2,500 nm wavelength. Therefore, multispectral sensors are designed to acquire information in multiple channels of light spectrum (typically, from 4 to 12 bands) and they cover large wavelength ranges (from 50 to 100 nm wide). Undoubtedly, the most important and adopted vegetation index for analyses on vegetation is the normalized difference vegetation index (NDVI) (Zaman-Allah *et al.*, 2015; Zhang *et al.*, 2018; Hassler and Baysal-Gurel, 2019); however, many variants based on bands in the NIR region were developed to increase performances of multispectral analyses. It is worth noting that, as most of the vegetation has higher spectral response within a slight portion between Red and NIR, different sensors are implemented with a dedicated channel around 717 nm wavelength, called Red-Edge (Hassler and Baysal-Gurel, 2019).

Disadvantages of multispectral sensors are mainly linked to the complexity of data to be acquired and processed for deriving useful information. In fact, use of multispectral sensors requires corrections in different phases of the processing workflow: i) on site before the aerial survey for image acquisition (radiometric calibration); ii) during pre-processing (image enhancement and mosaicking); iii) during the calculation of vegetation indices (Zhang *et al.*, 2021).

In terms of accessibility, multispectral sensors for UAS are usually much more expensive as compared to RGB cameras. It is not rare that RGB cameras are hacked and modified by stakeholders to extend the capability to acquire information in NIR and Red-Edge as well. This improvement is technically possible by complete substitution of the original RGB optical filter with another one, turning the original camera into a multispectral sensor in NIR region. Commonly, the result from hacking the camera is a hybrid sensor which acquires invisible RGB and NIR together. Clearly, hacked sensor will no longer work in visible light acquisition mode; hence, the use of original RGB camera together with the modified NIR camera is documented in many cases (Zhang *et al.*, 2021).

6.3.2.3. Hyperspectral Sensors

Likewise multispectral cameras are capable of detecting information beyond the visible light spectrum. The main significant differences are related to the number of available bands and the bandwidths. In general, hyperspectral cameras can capture specific and independent spectral information by hundreds, or even thousands, of bands which cover narrow wavelength windows, ranging from 10 to 20 nm (Hunt and Daughtry, 2018). Detailed explanations of hyperspectral sensors and image-processing techniques are described in Chapter 4 of the present book.

The adoption of such cameras on UAS seems to be very promising in agriculture as they can be adopted for different purposes: mapping plant species and phytocenosys dynamics by detecting specific spectral signatures, measuring physiological processes of vegetation, plant phenotyping and modeling (Hunt and Daughtry, 2018; Tsouros *et al.*, 2019). Unfortunately, lightweight hyperspectral sensors which are suitable for UAS platforms are currently in full technological development and, therefore, they are still very expensive, both for public institutions and farmers; hence, they are not commonly adopted in agricultural applications.

In addition, these sensors require a huge amount of computational resources as hyperspectral imaging typically generates an enormous volume of data to be processed and managed.

6.3.2.4. Thermal Sensors

Thermal sensors are specific cameras which are able to detect the temperature of surfaces and objects. As all bodies with temperature > 0 K (–273°C - –459°F) have the physical property of emitting energy in the infrared spectrum, these sensors are capable of capturing and – after calibration processes – return an output in terms of thermal imaging (Hassel and Baisal-Gurel, 2019). They usually detect infrared energy within a wavelength range from 750 to 10^6 nm (REF). In general, thermal sensors are adopted for mapping and assessing spatial variability of evapo-transpiration rate of vegetation and water stress associated with other physical factors, such as morphology, pedology, and micrometeorology (Granum *et al.*, 2015; Ribeiro-Gomes and Hernández-López, 2017).

The main constraint of thermal cameras is related to the low spatial resolution as compared to the other mentioned sensors (Ribeiro-Gomes and Hernández-López, 2017). This typology of sensors is not commonly adopted in agriculture as it is particularly expensive on one hand, and requires advanced skills and competences in data pre- and post-processing, on the other. Thermal sensors are often combined with RGB and multispectral sensors for UAS survey (Lioy *et al.*, 2021).

6.3.2.5. Light Detection and Ranging Sensors (LiDAR)

Light detection and ranging (LiDAR) devices are active sensors which are able to acquire 3D information (x,y,z) by emitting a beam of light pulses which hit surfaces and objects; light is reflected back and recorded by the sensor as spatial information (Maltamo *et al.*, 2014).

In general, LiDAR sensors are consolidated technologies commonly adopted as laser scanners for on-ground surveys. Since more than twenty years, airborne LiDAR is widely used for different environmental applications, such as geomorphological and topographic applications. High-resolution digital surface models (DSM) and digital terrain models (DTM) are the first-level output of using LiDAR data. By analyzing and integrating DSM and/or DTM with other information, it is possible to exploit LiDAR data in various applications (Vepakomma *et al.*, 2004; Lombard *et al.*, 2019).

Only recently, by the rapid advances in technology development, LiDAR sensors are integrated into UAS platforms, gaining even more attention in a wide range of applications. Due to their effective capability to accurately measure 3D structures, LiDAR technology provides different opportunities, especially in forest ecology, agriculture, soil and water management (Bagaram *et al.*, 2018). Common applications in agroforestry refer to canopy height and density measurements, fractional vegetation coverage, above-ground mass estimations, and land mapping (Zhang *et al.*, 2021).

The main constraint of deploying LiDAR survey is today represented by the extremely high costs of sensors which also may require an adequate UAV in terms of payload and safe aerial operations.

6.3.3. Ground Control Station and UAV Controller

To deploy an effective aerial survey, dedicated flight planning, a real-time flight control, and drone monitoring are required. The ground control station – commonly named GCS – is a computer (tablet, smartphone or laptop) by which the human operator is able to monitor, in real-time, UAV data acquisition during the flight (Kim and Kim, 2019). In addition, GCS continuously communicates to UAV controller, which is commonly the remote control device working in two-way data link for managing both flight operations and the autopilot system. With the UAV control system, different information acquired by the set of sensors integrated on to the drone allows control over important parameters, such as the flight altitude, the planimetric distance from the take-off/landing base (home point), the inside and outside temperature, the presence of obstacles, and air force (Kim and Kim, 2019). All the acquired information from UAV sensors is therefore displayed on the GCS which allows direct monitoring of the flight, both for real-time assessment of the aerial survey-data acquisition and for possibly performing recovery or safety operations.

Usually, GCS is based on dedicated proprietary software or applications provided by the UAV manufacturers or by other software houses, such as UgCS (universal ground control station), DroneDeploy; on the other hand, according to UAV hardware compatibility, different open-source software is available and is currently under development, such as mission planner ground station, MAV Pilot, APM Planner 2.0, MAVProxy, QGroundControl.

6.3.4. Human UAS Operator

The human control in UAS is crucial in all phases: from flight planning to the aerial survey. Firstly, it is necessary, and in most of countries mandatory by law, to pilot the UAV during the flight. Even if most of aerial surveys are performed automatically by the GCS by accomplishing the pre-planned flight for spatial data acquisition, a pilot is always required to assist all the aerial operations. Normally, a second operator is often required to support the pilot in all flight operations, in order to assist possible recovery manoeuvrings.

Beyond the UAV pilot, the human component is essential also in upstream and downstream phases of the aerial survey. In the preliminary phase, a geographical analysis of the area of interest by using GIS-based software is strongly recommended, in order to: (i) set up an optimized flight scenario which is able to maximize capability of data acquisition; (ii) identify possible physical limitations to flight (obstacles, accessibility, topography, infrastructures, sensitive places); (iii) examine critical factors that may affect data acquisition (water bodies, weather conditions, vegetation). In the post-flight phase, all data acquired by aerial survey must be processed, visualized, and analyzed.

6.4. UAS Applications in Sustainable Agriculture and Agroecology

Thanks to the wide range of UAS platform typologies, sensors, and possible interlinks with agro-environmental ground-based sensor networks, a broad set of applications in the domain of agriculture are at present experienced. Moreover, by considering the current speed of UAS technology development, areas of application may be further consolidated as well as other potential uses in the future will be tested and implemented (Hunt and Daughtry, 2017). However, UAS applications are mainly developed in different domains and sub-domains of farming, with particular emphasis within the Agriculture 4.0 framework: precision farming, smart farming, and sustainable agriculture (Hunter *et al.*, 2017). Unfortunately, scientific literature does not report UAS applications in the field of agroecology as such.

As an intrinsic function of most remote sensing technologies, land-cover mapping and classification are the main achievements of using UAS in agriculture. By the multi-scalar geometric resolution provided by UAS (from sub-meter to sub-centimeter resolution) which may fly at different altitudes, such information becomes crucial to understand spatial distribution, variability, and dynamic changes of land-cover features. Therefore, classification can be performed by discriminating, within large portions of surface, different land cover macro classes, i.e. forests, agricultural patches, grazing lands, bare soil and build-up areas; on the other hand, UAS ultra-resolution acquisition capability gives the opportunity to perform extremely detailed land cover/land use classification, enabling recognition of specific habitat types, phytochenosys, and individual plants (Ahmed *et al.*, 2017; Strong *et al.*, 2017; Librán-Embid *et al.*, 2020).

In addition, they might be exploited to produce high-resolution three-dimensional maps of forests or individual tree. This is made possible by photogrammetric elaborations, such as structure from motion techniques (known as SfM), by using stereoscopic images acquired by RGB cameras or LiDAR data.

In general, UAS applications help to obtain useful diagnostic information of different agroecosystem components and dynamics, derived from image acquisition and processing. It includes, among others: vegetation growth and yield, above-ground biomass, nutrients and chlorophyll contents, water stress, plant and animal diversity, plant species density, presence of pollinators, soil characteristics, soil water, and terrain morphology (Jay *et al.*, 2019; Cruzan *et al.*, 2016). Diagnostic information may be acquired in different phases of vegetation growth by different aerial surveys, making UAS a powerful tool for monitoring at multiple temporal and spatial scales. Continuous high-resolution monitoring gives to farmers the possibility to know where and when to deploy action during the growing period of vegetation (Nonni *et al.*, 2018).

One interesting approach to clarify and to summarize UAS applications which are diffused in precision agriculture is presented by Hunt and Daughtry (2017). This work proposes to divide UAS employments in three niches, according to the objectives and costs: 'scouting' for problems, monitoring to prevent yield losses, and planning agricultural management operations. Firstly, UAS can be used for 'praecox diagnosis' to rapidly detect emerging issues by real-time image acquisition and, therefore, to support decision making for interventions. Secondly, it can be employed for monitoring crop changes by advanced adoption of different sensors which require calibration, pre- and post-data processing from GIScientists or professionals. Finally, the third niche is related to the use of UAS for planning and management, which is today mainly oriented only for nutrient applications (2017).

As the most diffused applications are related to mapping, classifying, and monitoring land cover, we present common UAS employments simplified by areas of interest, which may have intersections at each other: vegetation, soil, agrosystems, and biodiversity.

6.4.1. Vegetation Monitoring

This activity represents the most diffused UAS applications to support agricultural practices. It usually combines the use of RGB cameras with multispectral sensors to identify possible critical issues on the land cover (Marcial-Pablo *et al.*, 2019). The main purposes are to detect and to map, at a very detailed scale, the health status of plants by analyzing different vegetation stresses: nutrients deficits, water stress, and plant diseases (Zhang *et al.*, 2021).

To perform these tasks, several vegetation indices based on multispectral bands have been adopted in remote sensing analyses, according to the specific objectives. Most common vegetation indices exploited in agriculture are the following: NDVI, difference vegetation index (DVI), enhanced vegetation index (EVI), ratio vegetation index (RVI), Red-edge vegetation stress index (RVTI), green normalized vegetation index (GNVI), chlorophyll absorption ratio index (CARI), nitrogen nutrition index (NNI), and photochemical reflectance index (PRI) (Liu *et al.*, 2018; Galiano *et al.*, 2012). It is worth noting that the combination of NIR with red bands is often adopted for above-ground biomass estimation, canopy structure, and calculation of the leaf area index (Gruner *et al.*, 2019). A complete overview of vegetation indices, operating bands, and applications in agriculture is summarized by Padua *et al.* (2017) in Table 3.

Another emerging application is represented by exploiting the ultra-resolution of UAS images to identify individual or clustered specific plant species, commonly defined in conventional agriculture as weeds. This application has found notable interest in precision agriculture technology, by the site-specific weed management framework (Peña *et al.*, 2013; Castaldi *et al.*, 2017). This approach aims to control weed and to drastically reduce the use of herbicides within the crop by detecting weed in early stages and by deploying a strict site-specific herbicide distribution. To pursuit this goal, a detailed weed map is required for precise operations and actions. Spatial analysis can be performed, both by image photo-interpretation techniques and by automatic extraction for weed detection. The first choice does not require advanced skills or expertise but, according to the size of the surveyed area, it can be time-consuming; the second one is time-efficient but requires skills and competences in GIS-analyses and modeling. In the latter case, use of machine-learning techniques and computer -ision analyses are required. The most common automatic classification techniques are the following: object-based image analysis (OBIA), artificial neural network (ANN), and maximum likelihood classifier (MLL) (Tamouridou *et al.*, 2017; De Castro *et al.*, 2012; Bechtel *et al.*, 2008). Generally, computer vision techniques are based on the use of both RGB and multispectral bands. However, RGB cameras can be used alone for automatic land-cover classification, simplifying calibration, and data processing (Ayhan and Kwan, 2020).

One promising application of automatic mapping specific plant species in organic farming and in agroecology is the use of low-cost commercial drones, equipped with a standard RGB camera. A representative case study in the framework of organic farming is reported by Mattivi *et al.* (2021). In this experimental research, a Parrot Anafi UAV was adopted to automatically extract presence of *Sorghum halepense, Chenopodium* and *Amaranthus retroflexus* in a maize-crop field. Results showed good performances of detecting weed by testing ANN, OBIA, and MLL (Figure 2). Moreover, this study also showed the feasibility of adopting a completely open-source workflow for RGB image processing (OpenDroneMap software) and automatic weed extraction by using open algorithms and packages available in SAGA and QGIS software (Mattivi *et al.*, 2021).

It is noteworthy that even if weed mapping is mainly developed within precision farming, the use of such information offers to organic farming and agroecology the opportunity to scout farmers for geovisualizing components of biodiversity and for improving agrosystems management.

6.4.2. Soil Monitoring

Assessing general conditions and physico-chemical characteristics of soil system in agroecosystem is paramount. Soil texture, soil moisture contents (SMC), soil organic matter (SOM), soil water, soil temperature, electrical conductivity, and biological activity are the most important aspects that can be assessed by using UAS (Jorge *et al.*, 2019; Sobayo *et al.*, 2018; Krížová *et al.*, 2018). To monitor soil-related characteristics, multi-spectral, hyper-spectral and thermal sensors are generally required, often combined together.

According to experimental studies of Wang (2016) and Guo *et al.* (2020), SOM, which is an important indicator of soil fertility, can be modeled and estimated by combining multi-spectral with hyper-spectral images. UAS equipped with thermal infrared sensor can be exploited to assess the spatial distribution of crop water deficit (Chisholm *et al.*, 2013; Chen *et al.*, 2019). In addition, thermal imaging can be also used for estimating the soil moisture, the water temperature comprehensive index, as well as the SMC, at different soil depths (Zhang *et al.*, 2019;

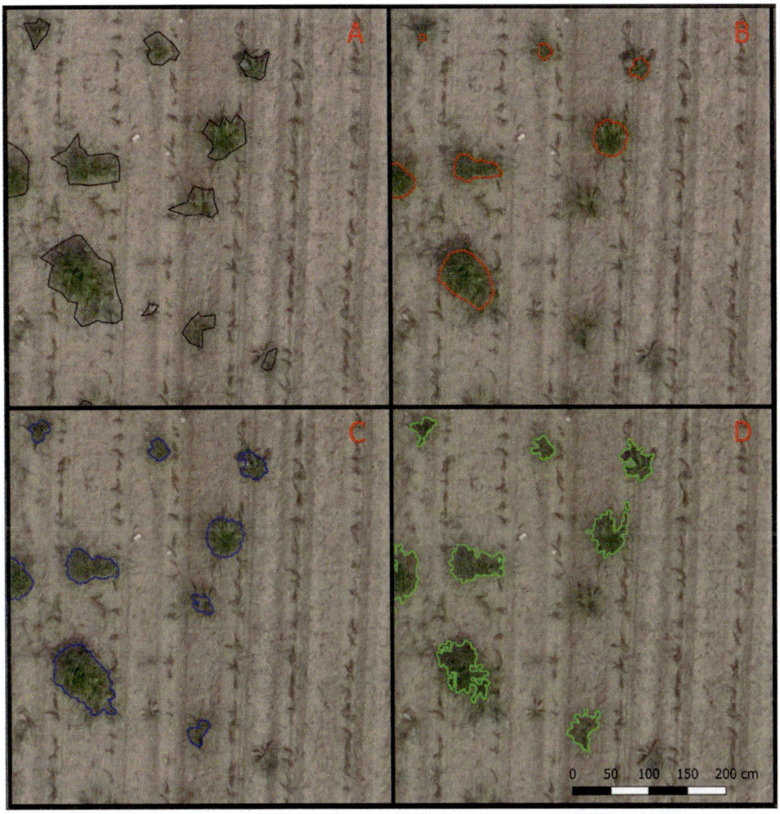

Fig. 2: Details of the weed map obtained with: (A) expert photo interpretation (reference data), (B) MLC method, (C) ANN method, and (D) OBIA method (Mattivi *et al.*, 2021)

Zhang *et al.*, 2021). To deploy such applications, usually adopted in the domain of precision agriculture, it is necessary to manage a set of specific hardware (UAV and sensors), dedicated software, and expertise which might represent critical elements that make scarcely accessible UAS to medium/small farms.

On the other hand, more user-friendly and affordable systems are at present adopted, especially for scouting farmers on a specific site and for supporting decision-making processes. It is the case of water stagnation in low-lying areas from intense precipitation, which is due to the lack of proper drainage or infiltration processes (Hunt *et al.*, 2018). By using a small low-cost UAS equipped with RGB cameras it is possible to map in detail the flooded areas and to deploy rapid interventions.

In general, soil monitoring by the use of UAS and different kinds of sensors is mainly oriented to increase efficiency of water management and irrigation, in the framework of smart farming.

6.4.3. Agroecosystems and Biodiversity Monitoring

Only recently, some efforts and successful attempts to bring UAS technologies within an agroecological framework to manage agricultural lands and agroforest ecosystems were accomplished (Xavier *et al.*, 2018; Padua *et al.*, 2017; Libràn-Embid *et al.*, 2020). The role of integrating biodiversity conservation with habitat management for agricultural-landscape diversification is widely documented. In fact, different strategies to improve and manage ecosystem services through agrobiodiversity, such as pollination and pest control, are at present experimented (Gurr *et al.*, 2017; Landis, 2017). They substantially require a shift in geographic scales – from crop to farm and to landscape – in agroecosystem and natural-resources management. The main effort is oriented to consolidate the relationship between plant and animal diversity and to pursuit in beneficial effects on productivity of agroecosystems (Snyder and Tylianakis, 2012; Gurr *et al.*, 2017; Libràn-Embid *et al.*, 2020). These strategies include the use of UAS for different purposes: mapping plant diversity, detecting floral resources and animals, as well monitoring habitat changes (Padua *et al.*, 2011; Libràn-Embid *et al.*, 2020).

In this framework UAS is representing a promising technology to support agroecosystem and biodiversity monitoring and management. For instance, it was adopted to monitor and to assess the implementation of vegetative buffer strips, such as wildflowers, hedgerows or shrubs at (or within) the field margins, in order to increase useful biodiversity, such as beneficial organisms (Tschumi *et al.*, 2017; Balzan *et al.*, 2016).

One among the most common UAS applications is related to manual or automatic discrimination of flowers within agricultural landscapes in order to identify, to assess plant diversity, and to enhance biocontrol processes (Mullerova *et al.*, 2017). For instance, some studies reported good accuracy in mapping and classifying *Heracleum mantegazzianum* (giant hogweed) (Michez *et al.*, 2016), *Robiniapseudo acacia* (black locust) (Mullerova *et al.*, 2017) and *Iris pseudacorus* (yellow flag iris) (Hill *et al.*, 2017). In addition, by combining remote-sensing imaging techniques with ground agro-environmental data, emerging experimentations are showing the capability of using UAS for estimating arthropod populations and understanding agroecosystems dynamics (Carl *et al.*, 2017; Xavier *et al.*, 2018). Related to this, an interesting study, which adopted UAS for agrobiodiversity monitoring, was developed by Xavier *et al.* (2018) in South Georgia (USA). They used an DJI® M100 hexacopter equipped with an RGB ZenmuseX3 camera combined with ground data to monitor and predict the population-beneficial arthropod as pollinators, by mapping flower areas from high-resolution UAV imagery. Their results highlight concrete possible UAS applications for agroecosystem management by showing a positive correlation between greater areas of blooming flowers and higher numbers of pollinators (Xavier *et al.*, 2018).

UAS technologies were also tested for different scopes within integrated agroforestry management (Padua *et al.*, 2017). They were adopted to measure both ecological and structural properties, such as canopy gaps, floristic biodiversity, phytochemical features, and height metrics in forests, shrub, and grass ecosystem (Anderson and Gaston, 2013). Fixed-wing UAS is used to assess canopy gaps and floristic biodiversity in the forest under-storey, indicating that very-high spatial resolution is sufficient to reveal strong dependency between disturbance patterns and plant diversity (Getzin *et al.*, 2012). In addition, by using SfM photogrammetry technique, UAS can be employed to assess growth, both on individual tree or groups of trees (Gatziolis *et al.*, 2015). An important application is also related to forest-fires detection and monitoring by using multiple UAS equipped with infrared and RGB cameras and a central station (Merino *et al.*, 2011).

As concealing food production with biodiversity conservation is one of the key elements of agroecology, some efforts at using UAS to monitor fauna in agricultural landscapes were deployed. By combining the use of RGB with thermal cameras, UAS provides a useful tool to detect and to track movement of many endothermic animals and environmental anomalies in temperatures as well (Costa *et al.*, 2013). These tasks may be useful to quantify and to localize presence of animals in agricultural landscapes, reducing the unintentional kills, and increasing harvest efficiency (Libràn-Embid *et al.*, 2020). Several studies reported important results in optimizing relationships between farming management and the presence of different species of fauna, such as *Circus pygargus* (Mulero-Pázmány and Negro, 2011), *Capreolus capreolus* (Cukor *et al.*, 2019), *Vanellus vanellus* (Israel and Reinhard, 2017).

Only recently, other UAS applications to monitor and to assess animal biodiversity in agroecosystems are offering new opportunities for both optimizing harvests and valorising human-environment relationships. One ongoing experimental research is about detecting and assessing wasps' nests through the use of UAS thermal sensors (Lioy *et al.*, 2021). As wasps' nests might play an important role as they are pest predators in many crops (Prezoto *et al.*, 2019), their precise localization and assessment is essential. Other promising UAS applications are related to the localization and quantification of important vertebrate pollinators and seed dispersers, such as bats and hummingbirds. In fact, it has been demonstrated that their absence can drastically reduce fruit or seed production up to 60 per cent on an average (Ratto *et al.*, 2018). Spatial distribution and behavior about vertebrate pollinators and seed dispersers may represent an important task for improving agroecosystem management and wildlife biodiversity conservation. In addition, the combined use of a multispectral sensor with thermal camera showed interesting performances in detecting birds and mammals, allowing UAS-derived counts and age of colony-nesting (Chretien *et al.*, 2016; Weissensteiner *et al.*, 2015). Hence, detection and tracking of certain species which have mobility in and around farmlands might make an important contribution to agroecosystem planning and biodiversity conservation (Libràn-Embid *et al.*, 2020).

6.5. UAS for Preserving Spider Monkey and for Agroecosystem Management: Experiences from Tropical Forests of Chocò (Ecuador)

The present study is developed in the tropical forest ecosystems of Ecuador, under the Washu Project. The general framework of the project is to develop an integrated model by combining scientific investigation, environmental education, and community education to create empowered, strong, and independent communities for conservation practitioners and for their own forests.

One of the main tasks was to support management and rehabilitation of spider monkey (*Atelesfuscipes fuscipes*), which is one of 25 most threatened primates in the world, listed within the category Critically Endangered (CR) and included in Appendix II of CITES. Moreover, spider monkey is currently the most threatened primate in Ecuador, especially through illegal trafficking and habitat loss. They inhabit the northern and central region of the Ecuadorian coast, and the western foothills. They live in tropics and humid subtropics between 100 and 1700m a.s.l., both in continuous forest and forest patches – principally in primary and older secondary forests. Spider monkeys are vulnerable to ecosystem degradation as their diet is based on mature fruits; therefore, larger areas of healthy forest are required to acquire food. A group formed of 30 individuals occupies approximately 90 to 250 hectares. Their ecological role is crucial as they are, among neotropical primates, the best disperser species due to their digestive system and a mobility range of about 6 kms per day. Moreover, as umbrella species, conservation of spider monkey results in a wider protection of habitat also for other endangered animals, such as jaguars or the green macaws.

Main threats for spider monkeys are deforestation, unsustainable agricultural practices, cattle, and mining. A combination of such factors has led to the loss and fragmentation of spider monkey habitat and a severe reduction in the population size of this primate.

To support conservation programs for spider monkey and its ecosystems, a UAS-based monitoring plan was developed in 2014 by Drone & GIS enterprise (Quito). By considering context and resources, particular attention was dedicated to the hardware and software setup: a low-cost fixed-wing UAV was identified and adopted for aerial surveys (E384 by Event38); it was equipped with a low-cost RGB camera (Samsung NX1000, 16 mm lens). To perform aerial surveys as well spatial analyses, GCS Mission Planner and QGIS open-source software were used; Pix4Dmapper® was selected to perform SfM elaborations (Fig. 3).

In addition, to perform aerial surveys in a morphologically complex area, a DTM (30 m resolution) from the Shuttle Radar Topography Mission was integrated in the flight plans. By using QGIS, different areas of interests of about 500 ha each were analyzed and selected for UAV aerial surveys. Each area of 500 ha is completely covered by three UAV mission plans. For the flight plan, an altitude of 250 m a.s.l. and a speed of 15 m/s were set; to obtain reliable orthophotos

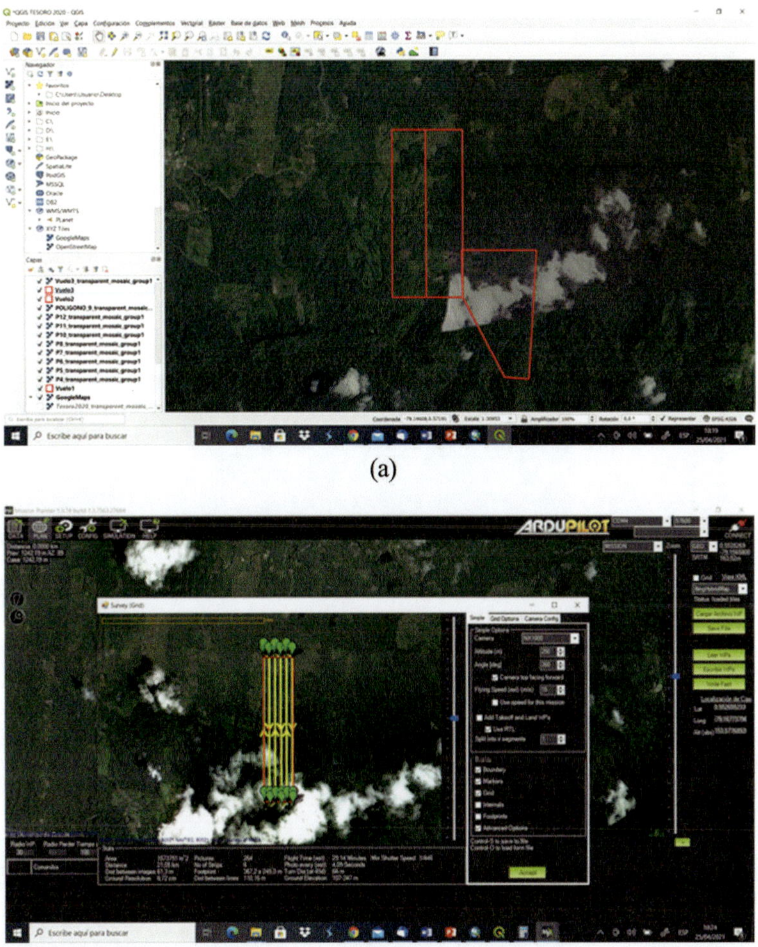

(a)

(b)

Fig. 3: Open-source software showing: (a) geographic analysis and definition of areas of interest in QGIS environment, and (b) specific parameters for UAV survey with Mission Planner

and DSM output by SfM, standard frame overlaps were configured for image acquisition during the flight (sidelap 70 per cent; overlap 75 per cent). By setting these parameters, three UAV surveys were performed obtaining about 6.7 cm of ground sampling resolution, during 30 minutes of flight. The main characteristics are summarized in Table 2.

Results from processing and analyzing UAV dataset allowed to clearly identify and to map important deforestation hotspots and important processes of ecosystem degradation within the study area (Fig. 4).

Table 2: Main Settings and Parameters for UAV Aerial Survey

	Flight 1	Flight 2	Flight 3
Ground resolution:	6.72 cm	6.72 cm	6.72 cm
Distance between images:	61.3 m	61.3 m	61.3 m
Pictures:	264	264	302
Flight time:	29:14 minutes	30:04 minutes	38:06 minutes
Photo interval (est):	4.09 sec	4.09 sec	4.09 sec

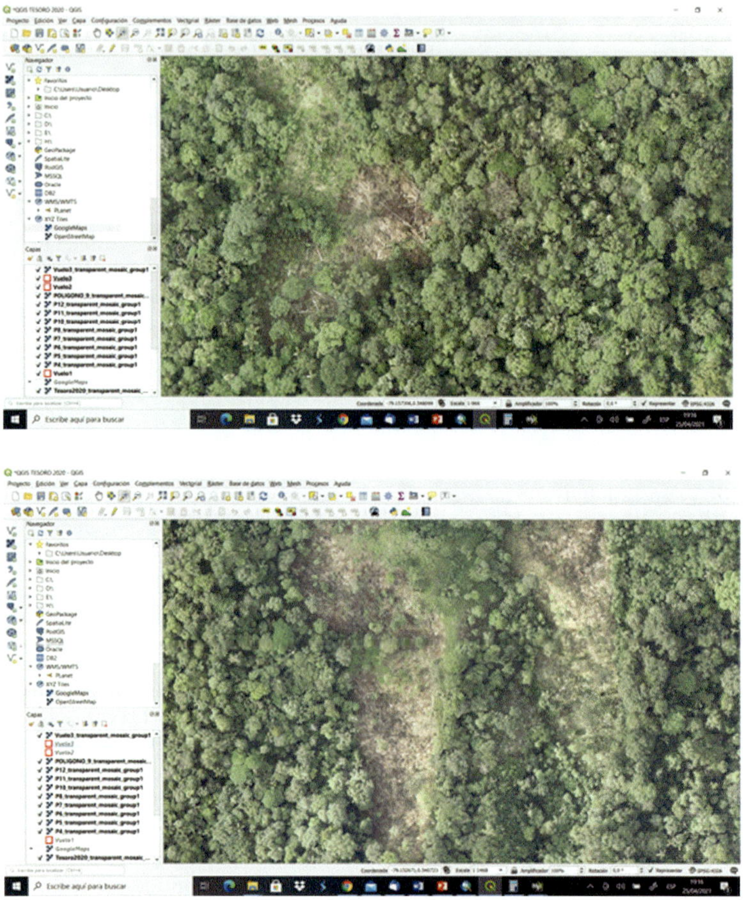

Fig. 4: High-resolution ortophoto obtained after photrammetric analysis, showing deforestation hotspots

Such results are paramount for spider monkey habitat conservation as well for agroecosystem management to be shared with local indigenous farmers. It is noteworthy that by using a fixed-wing UAV in favorable weather conditions, it was possible to perform 1,200 ha of data acquisition in one single day, at ultra-high spatial resolution imaging. On the other hand, by considering the weather conditions, such as cloud cover over tropical forests of Ecuador, usable high-resolution images (0.3 m) from commercial satellites are rare. Therefore, the use of a fixed-wing UAV capable of acquiring spatial data of a large portion of surface represents an opportunity for biodiversity and ecosystem monitoring.

6.6. Opportunities and Perspectives for the Agroecology Transition

Despite the recent and the actual proliferation of UAS for different applications in farming systems, it seems there are important further steps to globally fulfill, or to make substantial advances, in new pathways towards agroecological transition. At present, agricultural activities are drastically shaping about 37.4 per cent (56.1 M km^2) of all land surfaces on Earth (150 M km^2), making farmlands the widest human-modified ecosystem (FAO, 2016; 2017). Magnitude and extension of multi-scalar impacts of agriculture are widely documented in scientific literature: land use and land-cover changes, contamination and degradation of soil and freshwater systems, loss of genetic and functional diversity (biosphere integrity), alteration of global biogeochemical flows, and increase in anthropogenic greenhouse gases (Campbell *et al.*, 2017; Kissinger *et al.*, 2012; Shindell, 2016; Steffen *et al.*, 2016). To face the global challenges and to significantly increase sustainability of agriculture at different geographic scales – from ecosystem to landscape as far as the biosphere scale – dramatic changes to approach and to manage agrosystems are required. At present, a unique opportunity window for driving agriculture toward a sustainable model of farming and natural resources management is embodied by the agroecological approach (Altieri *et al.*, 2017). It represents a paradigm shift of conceiving agriculture by adopting a holistic approach for food production, supporting and valorising ecological functions and processes, and bio-cultural diversity and socio-economic values of agroecosystems (Wezel, 2009; Altieri, 1989). By such a conceptual and applicative framework, agroecology is ever more marking new pathways for investigating complexity of agroecosystems worldwide, in order to increase functional diversity, to control biogeochemical fluxes into a close-loop system, and to pursue socio-economic sustainability of agricultural production as well (Altieri, 1989; Wezel *et al.*, 2009).

In this framework, the systemic approach of geographical information science (GIScience) combined with the use of GeoICT and UAS offers a twofold opportunity for understanding ecological complexity and, therefore, to design and manage agroecosystems: firstly, it is able to integrate different biophysical, ecological, hydrological, anthropic, and socio-economic dynamics into spatially explicit analyses and modeling about the complex interactions of socio-environmental systems; secondly, it includes participatory methodologies which may represent powerful tools in supporting local community empowerment, public decision making processes, policy support, and planning in agroecosystem design and management (Walsh, S.J., Crews-Meyer, 2002; Goodchild, 2007; Goodchild *et al*., 2007).

Bibliography

Ahmed, O.S., A. Shemrock, D. Chabot, C. Dillon, G. Williams, R. Wasson and S.E. Franklin (2017). Hierarchical land cover and vegetation classification using multispectral data acquired from an unmanned aerial vehicle, *Int. J. Remote Sens*., **38**(8-10): 2037-2052.

Altieri, M.A. (1989). Agroecology: A new research and development paradigm for world agriculture, *Agriculture, Ecosystems and Environment*, 27: 37-46.

Altieri, M.A,, C.I. Nicholls and R. Montalba (2017). Technological approaches to sustainable agriculture at a crossroads: An agroecological perspective, *Sustainability*, 9: 349.

Anderson, K. and K.J. Gaston (2013). Lightweight unmanned aerial vehicles will revolutionize spatial ecology, *Frontiers in Ecology and the Environment*, **11**(3): 138-146.

Ayhan, B. and C. Kwan (2020). Tree, shrub, and grass classification using only RGB images, *Remote Sens*., **12**(8): 1333.

Baena , S., D.S. Boyd and J. Moat (2017). UAVs in pursuit of plant conservation – Real world experiences, *Ecological Informatics*, 47: 2-9.

Bagaram, M.B., D. Giuliarelli, G. Chirici, F. Giannetti and A. Barbati (2018). UAV remote sensing for biodiversity monitoring: Are forest canopy gaps good covariates? *Remote Sens*., 10: 1397.

Balzan, M.V., G. Bocci and A.C. Moonen (2016). Utilization of plant functional diversity in wildflower strips for the delivery of multiple agroecosystem services, *Entomol. Exp. Appl*., 158: 304-319.

Bechtel, B., A. Ringeler and J. Böhner (2008). Segmentation for object extraction of trees using MATLAB and SAGA, Hamburg, *Beiträgezur Phys. Geogr. und Landschaftsökologie*, 19: 1-12.

Bryson, M., M. Johnson-Roberson, R.J. Murphy and D. Bongiorno (2013). Kite aerial photography for low-cost, ultra-high spatial resolution multi-spectral mapping of intertidal landscapes, *PLoS One*, **8**(9): e73550.

Drone market outlook in 2021: Industry growth trends, market stats and forecast. https://www.businessinsider.com/drone-industry-analysis-market-trends-growth-forecasts?r=US&IR=T Visited on April 23rd 2021

Campbell, B.M., D.J. Beare, E.M. Bennett, J.M. Hall-Spencer, J.S.I. Ingram, F. Jaramillo, R. Ortiz, N. Ramankutty, J.A. Sayer and D. Shindell (2017). Agriculture production as a major driver of the Earth system exceeding planetary boundaries, *Ecology and Society*, **22**(4).

Carl, C., D. Landgraf, M. van der Maaten-Theunissen, P. Biber and H. Pretzsch (2017). *Robiniapseudo acacia* L. flowers analyzed by using unmanned aerial vehicle (UAV), *Remote Sens*., 9: 1091.

Castaldi, F., F. Pelosi, S. Pascucci and R. Casa (2017). Assessing the potential of images from unmanned aerial vehicles (UAV) to support herbicide patch spraying in maize, *Precis. Agric*., 18: 76-94.

Chapman, S., T. Merz, A. Chan, P. Jackway, S. Hrabar, M. Dreccer, E. Holland, B. Zheng, T. Ling and J. Jimenez-Berni (2014). Pheno-copter: A low-altitude, autonomous remote-sensing robotic helicopter for high-throughput field-based phenotyping, *Agronomy*, 4: 279-301.

Chen, Z., C.Y. Ma, H. Sun, Q. Cheng and F.Y. Duan (2019). The inversion methods for water stress of irrigation crop based on unmanned aerial vehicle remote sensing, *China Agric. Inf.*, 31: 23-35.

Chisholm, R.A., J. Cui, S.K.Y. Lum and B.M. Chen (2013). UAV LiDAR for below-canopy forest surveys, *Journal of Unmanned Vehicle Systems*, **1**(1): 61-68.

Chrétien, L.-P., J. Théau and P. Ménard (2016). Visible and thermal infrared remote sensing for the detection of white-tailed deer using an unmanned aerial system, *Wildl. Soc. Bull.*, **40**(1): 181-191.

Costa, J.M., O.M. Grant and M.M. Chaves (2013). Thermography to explore plant-environment interactions, *J. Exp. Bot.*, 64: 3937-3949.

Cruzan, M.B., B.G. Weinstein, M.R. Grasty, B.F. Kohrn, E.C. Hendrickson, T.M. Arredondo and P.G. Thompson (2016). Small unmanned aerial vehicles (micro-UAVs, drones) in plant ecology, *Appl. Plant Sci.*, **4**(9): apps.1600041.

Cukor, J., J. Bartoška, J. Rohla, J. Sova and A. Machálek (2019). Use of aerial thermography to reduce mortality of roe deer fawns before harvest, *Peer J.*, 7: e6923.

De Castro, A.I., M. Jurado-Expósito, J.M. Peña-Barragán and F. López-Granados (2012). Airborne multi-spectral imagery for mapping cruciferous weeds in cereal and legume crops, *Precis. Agric.*, 13: 302-321.

Easterday, K., C. Kislik, T.E. Dawson, S. Hogan and M. Kelly (2019). Remotely sensed water limitation in vegetation: Insights from an experiment with unmanned aerial vehicles (UAVs), *Remote Sens.*, **11**(16): 1853.

FAO (Food and Agriculture Organization). (2013). The State of Food and Agriculture: Food System for Better Nutrition, Rome, Italy.

FAO (Food and Agriculture Organization). (2017). FAOSTAT: Land Use Indicators; Retrieved from http://www.fao.org/faostat/en/#data/EL; accessed on 21 April, 2021.

García Galiano, S.G. (2012). Assessment of vegetation indexes from remote sensing: Theoretical basis, pp. 65-75. *In:* Erena, M., López-Francos, A., Montesinos, S. and J.-P. Berthoumieu (Eds.). *The Use of Remote Sensing and Geographic Information Systems for Irrigation Management in Southwest Europe*, CIHEAM / IMIDA / SUDOE Interreg IVB (EU-ERDF), Zaragoza, Spain.

Gatziolis, D., J.F. Lienard, A. Vogs and N.S. Strigul (2015). 3D tree dimensionality assessment

using photogrammetry and small unmanned aerial vehicles, *PLoS One*, **10**(9): e0137765.

Gayathri Devi, K., N. Sowmiya, K. Yasoda, K. Muthulakshmi and B. Kishore (2020). Review on application of drones for crop health monitoring and spraying pesticides and fertilizer, *JCR*, 7(6): 667-672.

Getzin, S., K.Wiegand and I. Schöning (2012). Assessing biodiversity in forests using very high-resolution images and unmanned aerial vehicles, *Methods in Ecology and Evolution*, **3**(2): 397-404.

Goldman Sachs (2021). Drones: Reporting for Work; Retrieved from https://www.goldmansachs.com/insights/technology-driving-innovation/drones/; accessed on 21 April, 2021.

Goodchild, M.F., M. Yuan and T.J. Cova (2007). Towards a general theory of geographic representation in GIS, *International Journal of Geographical Information Science*, **21**(3): 239-260.

Goodchild, M.F. (2017). Citizens as sensors: The world of volunteered geography, *GeoJournal*, 69: 211-221.

Granum, E., M.L. Pérez-Bueno, C.E. Calderón, C. Ramos, A. de Vicente, F.M. Cazorla and M. Barón (2015). Metabolic responses of avocado plants to stress induced by *Rosellinia necatrix* analyzed by fluorescence and thermal imaging, *Eur. J. Plant Pathol.*, 142: 625-632.

Grüner, E., T. Astor and M. Wachendorf (2019). Biomass prediction of heterogeneous temperate grasslands using an SfM approach based on UAV imaging, *Agronomy*, 9: 54.

Guo, H., X. Zhang, Z. Lu, T. Tian, F.F. Xu, M. Luo, Z.G. Wu and Z.J. Sun (2020). Estimation of organic matter content in southern paddy soil based on airborne hyperspectral images, *J. Agric. Sci. Tech.*, 22: 60-71.

Gupta, L., R. Jain and G. Vaszkun (2015). Survey of important issues in UAV communication networks, *IEEE Commun. Surv. Tut.*, **18**(2): 1123-1152.

Gurr, G.M., S.D. Wratten, D.A. Landis and M. You (2017). Habitat management to suppress pest populations: Progress and prospects, *Annu. Rev. Entomol.*, 62: 91-109.

Hassler, S.C. and F. Baysal-Gurel (2019). Unmanned aircraft system (UAS) technology and applications in agriculture, *Agronomy*, 9: 618. https://doi.org/10.3390/agronomy9100618

Hayat, S., E. Yanmaz and R. Muzaffar (2016). Survey on unmanned aerial vehicle networks for civil applications: A communications viewpoint, *IEEE Commun. Surv. Tut.*, **18**(4): 2624-2661.

Hill, D.J., C. Tarasoff, G.E. Whitworth, J. Baron, J.L. Bradshaw and J.S. Church (2017). Utility of unmanned aerial vehicles for mapping invasive plant species: A case study on yellow flag Iris (*Iris pseudacorus* L.), *Int. J. Remote Sens.*, 38: 2083-2105.

Hodgson, J.C. and L.P. Koh (2016). Best practice for minimising unmanned aerial vehicle disturbance to wildlife in biological field research, *Curr Biol.*, **26**(10): R404-5. doi: https://doi.org/10.1016/j.cub.2016.04.001

Hogan, S.D., M. Kelly, B. Stark and Y.Q. Chen (2017). Unmanned aerial systems for agriculture and natural resources, *Calif. Agric.*, 71: 5-14.

Hunt, E.R. and C.S.T. Daughtry (2018). What good are unmanned aircraft systems for agricultural remote sensing and precision agriculture? *International Journal of Remote Sensing*, **39**(15-16): 5345-5376.

Hunter, M.C., R.G. Smith, M.E. Schipanski, L.W. Atwood and D.A. Mortensen (2017). Agriculture in 2050: Recalibrating targets for sustainable intensification, *Bioscience*, **67**(4): 386-391.

Iqbal, F., A. Lucieer and K. Barry (2018). Simplified radiometric calibration for UAS-mounted multispectral sensor, *Eur. J. Remote Sens.*, 51: 301-313.

Israel, M. and A. Reinhard (2017). Detecting nests of lapwing birds with the aid of a small unmanned aerial vehicle with thermal camera, pp. 1199-1207. *In:* International Conference on Unmanned Aircraft Systems (ICUAS), Miami, USA.

Jay, S., F. Baret, D. Dutartre, G. Malatesta, S. Héno, A. Comar, M. Weiss and F. Maupass (2019). Exploiting the centimeter resolution of UAV multispectral imagery to improve remote-sensing estimates of canopy structure and biochemistry in sugar beet crops, *Remote Sens. Environ.*, 231: 110898.

Jorge, J., M. Vallbé and J.A. Soler (2019). Detection of irrigation in homogeneities in an olive grove using the NDRE vegetation index obtained from UAV images, *Eur. J. Remote Sens.*, 52: 169-177.

Kim, J., S. Kim, C. Ju and H.I. Son (2019). Unmanned aerial vehicles in agriculture: A review of perspective of platform, control, and applications, *IEEE*, 7: 105100-105115.

Kissinger, G., M. Herold and V. De Sy (2012). *Drivers of Deforestation and Forest Degradation: A Synthesis Report for REDD+ Policymakers*, Lexeme Consulting, Vancouver, Canada.

Krížová, K., M. Kroulík, J. Haberle, J. Lukáš and J. Kumhálová (2018). Assessment of soil electrical conductivity using remotely sensed thermal data, *Agron. Res.*, 16: 784-793.

Landis, D.A. (2017). Designing agricultural landscapes for biodiversity-based ecosystem services, *Basic Appl. Ecol.*, 18: 1-12.

Librán-Embid, F., F. Klaus, T. Tscharntke and I. Grass (2020). Unmanned aerial vehicles for biodiversity-friendly agricultural landscapes – A systematic review, *Sci. Total Environ.*, 732: 139204.

Lioy, S., E. Bianchi, A. Biglia, M. Bessone, D. Laurino and M. Porporato (2021). Viability of thermal imaging in detecting nests of the invasive hornet *Vespa velutina*, *Insect Sci.*, 28: 271-277.

Liu, Z., W. Wan, J.Y. Huang, Y.W. Han and J.Y. Wang (2018). Progress on key parameters inversion of crop growth based on unmanned aerial vehicle remote sensing, *Trans. Chin. Soc. Agric. Eng.*, **34**(24): 60-71.

Lombard, L., R. Ismail and N. Poona (2019). Modeling forest canopy gaps using LiDAR-derived variables, *Geocarto Int.*, 6049: 1-15.

Lorenz, R.D. and S.P. Scheidt (2014). Compact and inexpensive kite apparatus for geomorphological field aerial photography, with some remarks on operations, *Geo. Res. J.*, 3-4: 1-8. https://doi.org/10.1016/j.grj.2014.06.001

Maltamo, M., E. Naesset and J. Vauhkonen (2014). Forestry applications of airborne laser scanning, *Concepts Case Studies*, Springer, Dordrecht, The Netherlands.

Marcial-Pablo, M.D.J., A. Gonzalez-Sanchez, S.I. Jimenez-Jimenez, R.E. Ontiveros-Capurata and W. Ojeda-Bustamante (2019). Estimation of vegetation fraction using RGB and multispectral images from UAV, *Int. J. Remote Sens.*, 40: 420-438.

Marino, S. and A. Alvino (2018). Detection of homogeneous wheat areas using multi-temporal UAS images and ground truth data analyzed by cluster analysis, *Eur. J. Remote Sens.*, 51: 266-275.

Markets Insider (2021). Global Drone Service Market Report 2019: Market is Expected to Grow from USD 4.4 Billion in 2018 to USD 63.6 Billion by 2025, at a CAGR of 55.9 per cent; Retrieved from https://markets.businessinsider.com/news/stocks/global-drone-service-market-report-2019-market-is-expected-to-grow-from-usd-4-4-billion-in-2018-to-usd-63-6-billion-by-2025-at-a-cagr-of-55-9-1028147695; accessed on 21 April, 2021.

Mattivi, P., S.E. Pappalardo, N. Nikolić, L. Mandolesi, A. Persichetti, M. De Marchi and R. Masin (2021). Can commercial low-cost drones and open-source GIS technologies be suitable for semi-automatic weed mapping for smart farming: A case study in NE Italy, *Remote Sensing* (in press).

Merkert, R. and J. Bushell (2020). Managing the drone revolution: A systematic literature review into the current use of airborne drones and future strategic directions for their effective control, *Journal of Air Transport Management*, 89: 101929.

Merino, L., F. Caballero, J.R. Martínez-De-Dios, I. Maza and A. Ollero (2011). An unmanned aircraft system for automatic forest fire monitoring and measurement, *Journal of Intelligent & Robotic Systems*, **65**(1-4): 533-548.

Michez, A., H. Piégay, L. Jonathan, H. Claessens and P. Lejeune (2016). Mapping of riparian invasive species with supervised classification of unmanned aerial system (UAS) imagery, *Int. J. Appl. Earth Obs. Geoinf.*, 44: 88-94.

MIT Technology Review (2014). Agricultural Drones; Retrieved from https://www.technologyreview.com/technology/agricultural-drones/; accessed on 21 April, 2021.

MIT Technology Review (2016). Six Ways Drones Are Revolutionizing Agriculture; Retrieved from https://www.technologyreview.com/2016/07/20/158748/six-ways-drones-are-revolutionizing-agriculture/; accessed on 21 April, 2021.

Moudrý, V., R. Urban, M. Štroner, J. Komárek, J. Brouček and J. Prošek (2018). Comparison of a commercial and home-assembled fixed-wing UAV for terrain mapping of a post-mining site under leaf-off conditions, *International Journal of Remote Sensing*, **13**(12): 1672-1694.

Mulero-Pázmány, M. and J.J. Negro (2011). *Small UAS for Montagus Harriers (Circus pygargus) Nests Monitoring*, RED UAS International Congress University of Engineering, Seville, Spain.

Müllerová, J., T. Bartaloš, J. Bruna, P. Dvořák and M. Vítková (2017). Unmanned aircraft in nature conservation: An example from plant invasions, *Int. J. Remote Sens.*, 38: 2177-2198.

Nonni, F., D. Malacarne, S.E. Pappalardo, D. Codato, F. Meggio and M. De Marchi (2018). Sentinel-2 data analysis and comparison with UAV multispectral images for precision viticulture, *GI_Forum*, 1: 105-116.

Pádua, L., J. Vanko, J. Hruška, T. Adão, J.J. Sousa, E. Peres and R. Morais (2017). UAS, sensors, and data processing in agroforestry: A review towards practical applications, *International Journal of Remote Sensing*, **38**(8-10): 2349-2391.

Pajares, G. (2015). Overview and current status of remote sensing applications based on unmanned aerial vehicles (UAVs), *Photogrammetric Engineering & Remote Sensing*, **81**(4): 281-330.

Patrick, A., S. Pelham, A. Culbreath, C. Corely Holbrook, I.J. De Godoy and C. Li (2017). High throughput phenotyping of tomato spot wilt disease in peanuts using unmanned aerial systems and multispectral imaging, *IEEE Instrum. Meas. Mag.*, 20: 4-12.

Peña, J.M., J. Torres-Sánchez, A.I. de Castro, M. Kelly and F. López-Granados (2013). Weed mapping in early-season maize fields using object-based analysis of unmanned aerial vehicle (UAV) images, *PLoS One*, 8: 1-11.

Prezoto, F., T.T. Maciel, M. Detoni, A.Z. Mayorquin and B.C. Barbosa (2019). Pest control potential of social wasps in small farms and urban gardens, *Insects*, **10**(7): 192.

Radoglou-Grammatikis, P., P. Sarigiannidis, T. Lagkas and I. Moscholios (2020). A compilation of UAV applications for precision agriculture, *Computer Networks*, 172: 107148.

Ratto, F., B.I. Simmons, R. Spake, V. Zamora-Gutierrez, M.A. MacDonald, J.C. Merriman, C.J. Tremlett, G.M. Poppy, K.S.-H. Peh and L.V. Dicks (2018). Global importance of vertebrate pollinators for plant reproductive success: A meta-analysis, *Front. Ecol. Environ.*, **16**(2): 82-90.

Ribeiro-Gomes, K., D. Hernández-López, J.F. Ortega, R. Ballesteros, T. Poblete and M.A. Moreno (2017). Uncooled thermal camera calibration and optimization of the photogrammetry process for UAV applications in agriculture, *Sensors*, 17: 2173.

Roth, L. and B. Streit (2017). Predicting cover crop biomass by lightweight UAS-based RGB and NIR photography: An applied photogrammetric approach. *Precis. Agric.* 19: 93-114. doi: 10.1007/s11119-017-9501-1

Shindell, D.T. (2016). Crop yield changes induced by emissions of individual climate-altering pollutants, *Earth's Future*, 4: 373-380.

Snyder, W.E. and J.M. Tylianakis (2012). The ecology of biodiversity-biocontrol relationships, pp. 21-40. *In:* Gurr, G.M., Wratten, S.D. and W.E. Snyder (Eds.). *Biodiversity and Insect Pests: Key Issues for Sustainable Management*, Wiley Blackwell, Chichester, UK.

Sobayo, R., H.H. Wu, R. Ray and L. Qian (2018). Integration of convolutional neural network and thermal images into soil moisture estimation, pp. 207-210. *In:* Proceedings of the 2018 1st International Conference on Data Intelligence and Security (ICDIS), South Padre Island, USA.

Steffen, W., K. Richardson, J. Rockström, S.E. Cornell, I. Fetzer, E.M. Bennett, R. Biggs, S.R. Carpenter, W. de Vries, C.A. de Wit, C. Folke, D. Gerten, J. Heinke, G.M. Mace, L.M. Persson, V. Ramantan, B. Reyers and S. Sörlin (2015). Planetary boundaries: Guiding human development on a changing planet, *Science*, **347**(6223): 1259855.

Strong, C.J., N.G. Burnside and D. Llewellyn (2017). The potential of small-unmanned aircraft systems for the rapid detection of threatened unimproved grassland communities using an enhanced normalized difference vegetation index, *PLoS One*, **12**(10): e0186193.

Sylvester, G. (2018). *E-agriculture in Action: Drones for Agriculture*, FAO. Bangkok, Thailand.

Tamouridou, A.A., T.K. Alexandridis, X.E. Pantazi, A.L. Lagopodi, J. Kashefi and D. Moshou (2017). Evaluation of UAV imagery for mapping *Silybummarianum* weed patches, *Int. J. Remote Sens.*, 38: 2246-2259.

Tang, L. and G. Shao (2015). Drone remote sensing for forestry research and practices, *J. For. Res.*, 26: 791-797.

Tewes, A. and J. Schellberg (2018). Towards remote estimation of radiation use efficiency in maize using UAV-based low-cost camera imagery, *Agronomy*, **8**(2): 16.

Torresan, C., A. Berton, F. Carotenuto, S.F. Di Gennaro, B. Gioli, A. Matese, F. Miglietta, C. Vagnoli, A. Zaldei and L. Wallace (2017). Forestry applications of UAVs in Europe: A review, *Int. J. Remote Sens.*, **38**(8-10): 2427-2447.

Tschumi, M., M. Albrecht, C. Bärtschi, J. Collatz, M.H. Entling and K. Jacot (2016). Perennial, species-rich wildflower strips enhance pest control and crop yield, *Agric. Ecosyst. Environ.*, 220: 97-103.

Tsouros, D.C., S. Bibi and P.G. Sarigiannidis (2019). A review on UAV-based applications for precision agriculture, *Information*, **10**(11): 349.

Vergouw, B., H. Nagel, G. Bondt and B. Custers (2016). Drone technology: Types, payloads, applications, frequency spectrum issues and future developments, pp. 21-45. *In:* Custers, B. (Ed.). *The Future of Drone Use*, T.M.C. Asser Press, The Hague, The Netherlands.

Vepakomma, U., B. St-Onge and D. Kneeshaw (2008). Spatially explicit characterization of boreal forest gap dynamics using multi-temporal lidar data, *Remote Sens. Environ.*, 112: 2326-2340.

Walsh, S.J. and K.A. Crews-Meyer (2002). *Linking People, Place and Policy: A GIScience Approach,* Klumer Academic Publisher, Norwell, USA.

Wang, L. (2016). A Research About Remote Sensing Monitoring Method of Soil Organic Matter Based on Imaging Spectrum Technology, M.S. thesis, Henan Polytechnic University, Zhengzhou, China.

Weissensteiner, M.H., J.W. Poelstra and J.B.W. Wolf (2015). Low-budget ready-to-fly unmanned aerial vehicles: An effective tool for evaluating the nesting status of canopy-breeding bird species, *J. Avian Biol.*, **46**(4): 425-430.

Wezel, A. (2009). A quantitative and qualitative historical analysis of the scientific discipline of agroecology, *International Journal of Agricultural Sustainability*, **7**(1): 3-18.

Wezel, A., S. Bellon, T. Dore, C. Francis, D. Vallod and C. David (2009). Agroecology as a science, a movement and a practice. A review, *Agron. Sustain. Dev.*, **29**(4): 503-515.

White, J.C., N.C. Coops, M.A. Wulder, M. Vastaranta, T. Hilker and P. Tompalski (2016). Remote Sensing Technologies for Enhancing Forest Inventories: A Review, *Canadian Journal of Remote Sensing*, **42**(5): 619-641.

Xavier, S.S., A.W. Coffin, D.M. Olson and J.M. Schmidt (2018). Remotely estimating beneficial arthropod populations: Implications of a low-cost small unmanned aerial system, *Remote Sens.*, 10: 1485.

Yonah, I.B., S.K. Mourice, S.D. Tumbo, B.P. Mbilinyi and J. Dempewolf (2018). Unmanned aerial vehicle-based remote sensing in monitoring smallholder, heterogeneous crop fields in Tanzania, *Int. J. Remote Sens.*, 39: 5453-5471.

Zaman-Allah, M., O. Vergara, J.L. Araus, A. Tarekegne, C. Magorokosho, P.J. Zarco-Tejada, A. Hornero, A.H. Albà, B. Das, P. Craufurd, M. Olsen, B.M. Prasanna and J. Cairns (2015). Unmanned aerial platform-based multi-spectral imaging for field phenotyping of maize, *Plant Methods*, 11: 35.

Zhang, D., X. Zhou, J. Zhang, Y. Lan, C. Xu and D. Liang (2018). Detection of rice sheath blight using an unmanned aerial system with high-resolution color and multispectral imaging, *PLoS One*, **13**(5): e0187470.

Zhang, H., L. Wang, T. Tian and J. Yin (2021). A review of unmanned aerial vehicle low-altitude remote sensing (UAV-LARS) use in agricultural monitoring in China, *Remote Sens.*, 13: 1221.

Zhang, X., Y. Bao, D. Wang, X. Xin, L. Ding, D. Xu, L. Hou and J. Shen (2021). Using UAV LiDAR to extract vegetation parameters of inner Mongolian grassland, *Remote Sens.*, 13: 656.

Zheng, H., T. Cheng, D. Li, X. Zhou, X. Yao, Y. Tian, W. Cao and Y. Zhu (2018). Evaluation of RGB, color-infrared and multispectral images acquired from unmanned aerial systems for the estimation of nitrogen accumulation in rice, *Remote Sens.*, 10: 824.

Part III

Landscapes and Ecosystem Services, Technologies for Agroecological Transitions

CHAPTER

7

WebGIS: Status, Trends and Potential Uptake in Agroecology

Luca Battistella[1]*, Federico Gianoli[2], Marco Minghini[3] and Gregory Duveiller[3]

[1] Interfaculty Department of Geoinformatics, University of Salzburg, Hellbrunnerstrasse 34 5020 Salzburg, Austria

[2] Department of Civil, Environmental and Architectural Engineering, University of Padova

[3] European Commission[a], Joint Research Centre (JRC), 21027 Ispra, Italy

7.1. Introduction

As shown in previous chapters, geographic information systems (GIS) are tools which allow collection, visualization, organization, analyses, and process geospatial data as a collection of thematic layers related to each other through their spatial dimension. This simple yet extremely powerful and versatile concept is currently used to address several real-world problems in a variety of disciplines and domains, including agroecology. Whilst desktop GIS tools were developed from the late 1960s (Haklay and Skarlatidou, 2010), web-based GIS applications only appeared in the 1990s and have since established a shift in the way geospatial information is accessed and analyzed, allowing users to interact with geospatial content without having to install any software on their own computers. This shift also allowed the enrichment of maps with the interactive and dynamic dimension that is peculiar to the web, which holds the potential to express many cartographic design principles, and represent a modern way of displaying and analyzing information.

The web dimension of GIS, also referred to as the geospatial web, has made it possible to reduce the traditionally steep learning curve of GIS (Abdalla and Esmail, 2019) and increase its accessibility to the general public. Web-based GIS

[a] The views expressed are purely those of the authors and may not in any circumstances be regarded as stating an official position of the European Commission.

*Corresponding author: luca.battistella.geo@gmail.com

systems can be used by everyone (including non-experienced users), at any time (even simultaneously) and without the need to use specialized software or to possess specific skills. Searching, accessing, visualizing, and processing geospatial resources on the Internet happens through web communication protocols, such as HTTP (hypertext transfer protocol) (Dragićević, 2004) but also through a local intranet if the tools are meant to be used within a single organization. GIS tools running on the web vary, based on the purpose of the application and on the level of interactivity. For the purpose of this chapter, we can group them into two main categories: Web mapping tools or webGIS and spatial data infrastructures (SDIs).

The term webGIS was first introduced to refer to those applications developed around the beginning of 2000, which progressively transferred GIS functionality to the web, especially for gathering, storing, retrieving, analyzing, and visualizing geospatial data (Peng and Tsou, 2003). Hence, the term webGIS refers to web-based systems somehow resembling desktop GIS in terms of functionalities, including data manipulation and processing. In this regard, most of today's GIS applications available on the web – which only display simple maps (composed of basemaps, layers and legends) and offer a limited level of interaction – should be more appropriately called web maps. However, in the context of this chapter webGIS and web maps will be used synonymously, and the term web mapping will be used to indicate the generic process of developing GIS solutions that run on the web. In such a context, geospatial information is mostly delivered through web services and formats compliant with the web standards for geospatial interoperability issued by the open geospatial consortium (OGC). In the broad context of web mapping, a special role is played by spatial data infrastructures (SDIs) which are more complex frameworks, usually government-driven and bounded to specific administrative areas and legal arrangements, aiming to make geospatial data available in a consistent and interoperable way.

In this chapter we present an overview of the main trends, concepts and types of web mapping and we offer an insight into the available technologies for sharing, analyzing and managing geospatial data in web mapping applications and SDIs. This forms the basis for a discussion on the current status and the potential uptake of web mapping in the field of agroecology, which is increasingly employing digital and data-driven business models (Ajena, 2018) and can highly benefit from the opportunities arising from publishing and sharing geospatial resources on the web. Such resources, particularly those collected at the local level, represent key information to expand and upscale agroecological principles and practices, as well as bridging the gap between local communities and policymakers.

The remainder of the chapter is structured as follows. The next section provides some background notions on web mapping, tracing its history and trends, outlining the increasing dimension of citizen involvement and comparing the proprietary and open-source business models. From the more technological perspective, a distinction between standard web mapping applications and SDIs is introduced. This is further explored in the section 'Spatial Data Infrastructures', which details the traditional software components of, respectively, web mapping and SDI architectures and provides real-world examples of such systems. 'WebGIS and Agroecology' discusses the potential role of web mapping in the field of agroecology, in particular the benefits associated with data dissemination and a potential experience exchange network.

7.2. Background and Concepts

Humans have a natural affinity to think spatially, perceive the world in a geographical-oriented way, and use maps to convey spatial ideas about places and phenomena. Interactive web maps and augmented reality applications based on geospatial data are nothing but the natural evolution of classical printed maps and atlases in the modern multimedia world. In this section we briefly illustrate the history of webGIS and the directions in which it has progressively evolved, focusing also on the main differences between proprietary and open-source business models, and introducing the two main categories of web mapping applications and SDIs.

7.2.1. Web Mapping Trends and Types

Since the creation of the World Wide Web in 1989, serving geospatial data has been a challenge. The first web viewer, called Xerox PARC map viewer, which consisted of an HTML page embedding a static image of the map, was developed in 1993 (Putz, 1994), while the first geospatial server, called MapServer, was started a few months later in 1994 at the University of Minnesota with the support of NASA. Today, after a quarter of a century, advancements of web technologies (along with the web itself) have grown exponentially, leading to features and functionalities that were initially unimaginable. A synthetic timeline of webGIS developments up to the present days, showing some of the most significant events, is provided in Fig. 1. The reader will notice that many software tools mentioned in the chapter are included in this timeline.

Fig. 1: Timeline of the most significant mapping events since the creation of the World Wide Web (*Source*: Authors' elaboration from Veenendaal *et al.*, 2017)

Veenendaal *et al.* (2017) identified nine web mapping eras. These are not bound to specific and distinct time frames; on the contrary, they feature some temporal overlapping and should be seen as subsequent evolutions which can very well summarize the geospatial web development and its main trends:

- *Static*: The initial era inaugurated by the Xerox PARC map viewer, grounded on basic HTTP and HTML technologies and characterized by static maps and poor user interaction;
- *Dynamic*: The era where the term webGIS was introduced to refer to map-based applications resembling the desktop ones, where users started to have some degree of control on map layers but the overall interaction remained limited;
- *Services*: Enabled by the rise of OGC standards, the era characterized by the emergence of service-oriented architectures (SOA), used for example in SDIs, and mapping application programming interfaces (APIs), e.g. the Google Maps API launched in 2005;
- *Interactive*: The era based on the AJAX (asynchronous JavaScript and XML) technology, which offered an improved user experience and allowed the creation of map mashups, i.e. combinations of multi-source data into the same web map;
- *Collaborative*: Pulled by Web 2.0 and the emergence of user-generated content over the web, the era characterized by the birth of OpenStreetMap in 2004 (*see* the section 'Web Mapping 'Without' Coding') and the appearance of the words crowdsourcing and volunteered geographic information (VGI);
- *Digital globe*: The era marked by the massive use of virtual globes that expanded the web mapping user base from the specialist to the global community; among the first virtual globes include Google Earth, Microsoft Virtual Earth (currently Bing) and NASA World Wind;
- *Mobile*: The era characterized by the diffusion of hand-held devices, enabled with multiple sensors (including the GPS), and the development of location-based services (LBS), and augmented reality applications; Pokémon Go, released in 2016, is one of the best examples;
- *Cloud*: The era focused on the cloud to provide storage, software, services, and infrastructure to scale traditional applications and manage big geospatial data, including satellite imagery and data from the Internet of Things (IoT); examples of software as a service (SaaS) platforms in the web mapping domain are Google Earth Engine, ArcGIS Online, CARTO and Mapbox;
- *Intelligent*: Driven by Web 3.0 or semantic web, the era characterized by the intelligent discovery, extraction, and contextualization of multi-source geospatial information and transformation into knowledge; the term location intelligence was coined exactly to describe this.

Given its potential relevance for the agroecology domain, the next subsection focuses in more detail on the collaborative dimension of web mapping, which – despite originating from the collaborative era mentioned above – embraces the developments and technologies of many of the other eras.

7.2.1.1. Collaborative Web Mapping

As mentioned above, Web 2.0 and its geospatial extension GeoWeb 2.0 (Maguire, 2007) have paved the way for web mapping applications characterized not only by increased performance and user interaction (*see* the section 'Geoportals') but also the inclusion of user-generated geospatial content. It was exactly in this context that many new terms were coined, such as neogeography, geotag, and the most well-known VGI (Goodchild, 2007). The latter refers to geospatial information contributed by volunteers with a variety of possible motivations (Coleman *et al.*, 2009). This phenomenon has become commonplace in the last ten years thanks to a number of technological drivers, such as the diffusion of LBS and the open access availability of high-resolution satellite imagery. The most popular VGI project to date is OpenStreetMap, which aims to create a free and open map of the world and has so far attracted some millions of users (Mooney and Minghini, 2017). Integrating citizen-generated content into web mapping applications has become a common practice which may be adopted for very different purposes, ranging from the simple web visualization and analysis of such user-collected data, to the production of new geospatial datasets enabled within the map viewer (e.g. by allowing users to digitize objects on top of satellite imagery). In some cases, web mapping applications are developed to raise community participation in reporting local environmental or social issues (e.g. pollution, crime or mugging) to stimulate political discussion and trigger decision making (Brovelli *et al.*, 2015). Many terms have been coined in the literature to indicate both the user-generated geospatial data and the processes leading to their creation, the latter including crowdsourcing[1], participatory sensing[2], geocollaboration and citizen science[3]. See *et al.* (2016) provide a comprehensive classification of all such terms and introduce the overarching expression crowdsourced geographic information. Nevertheless, the functionality of these collaborative web mapping applications and the software tools exploited to achieve them are those described in 'Web Mapping Applications' section. There is no doubt that crowdsourced geographic information projects will continue to gain momentum in the future and modern web and webGIS tools will not only provide a technological enabler, but also increase the role and contributions of citizens in a wide variety of both scientific and societal domains.

7.2.2. Business Models: Proprietary vs Open Source

In terms of business models, webGIS software is split into two domains: open source and proprietary. It is noteworthy that a high cost does not necessarily translate into high value. The main pros of proprietary solutions are related to the high number of ready-to-use tools and the fact that they are extremely user-friendly, allowing users to create web mapping applications with no programming capabilities. This comes at the cost of the licence for the deployment and maintenance of these systems, which might represent an obstacle for both single users and small businesses. Open-source approaches, on the other hand, boast

[1] The act of outsourcing a task to an undefined network of people.

[2] Approach focused on data collected through the device sensors.

[3] Activities in which citizens are involved in scientific projects.

no costs for licensing and grant the possibility to develop highly customizable applications. However, such systems also come with some cons. Programming skills are typically essential for the development and maintenance of open-source web mapping applications, which often require a steep learning curve that may dissuade users from creating them. However, open source web mapping has recently gained an increased popularity thanks to a mature and proactive community that makes it easier to build professional solutions in a knowledge-sharing context. This philosophy has created a virtuous circle, which allows the channeling of investments on both the individual professional growth and, as a consequence, the advancement of the related open-source project development. In this way, over the years, the collaborative nature of open-source communities (formed not only by developers but also users, project managers, researchers, educators, etc.) has driven such a high degree of innovation that today open-source geospatial technology is able to fully compete with the proprietary counterpart (Coetzee *et al.*, 2020). Indeed, today's open-source software products cover all geo-technology areas and geospatial applications, and as a whole form a rich, mature, and modular ecosystem of tools addressing any user need (Brovelli *et al.*, 2017). As highlighted by Minghini *et al.* (2020), open-source geospatial software is just one component within the wider movement of geospatial openness, which is also formed by open geospatial data, open geospatial standards, and their interconnected communities.

Since the early 2000s, open-source web mapping technology has been drawing the attention of businesses and governments on a global scale. In contrast to open-source desktop tools which have become commonplace, we believe that the main barrier still preventing the massive adoption of open-source web mapping solutions is the small number of ready-to-use tools for implementing functional webGIS applications without programming skills. Some examples of such tools are described in web mapping 'without' coding, although all of them require at least a provider to host the maps and/or some technical skills to configure and manage a server. Additionally, from the pure management perspective, organizations and users usually assume that something they have paid for works better than something free of cost. These two factors, in many cases, are those influencing the decision to opt for proprietary rather than open-source software. However, over the last decades, open-source webGIS solutions have developed and matured so rapidly that they have become a powerful and cost-effective solution to provide adequate dissemination, sharing and management of geospatial information over the web.

At the moment, the state-of-the-art of proprietary GIS software is mainly represented by the ecosystem of tools by the Environmental Systems Research Institute (ESRI), which was founded in 1969 and has been developing and selling popular GIS solutions since then. On the other hand, open-source GIS software witnessed a different evolution. While the first open-source web mapping solution was only started in 1994, the origin of open-source desktop GIS software dates back to the 80s with the first release of GRASS GIS (Neteler and Mitasova, 2013).

7.2.3. Geospatial Web Components

To create a webGIS, it is first necessary to define its main purpose. Should the webGIS provide or collect data? What is the main target audience? Is the information going to be public or restricted to specific users? How will users interact with the webGIS? Should it be primarily accessed from desktop or mobile devices? All such questions are crucial to frame the development workflow and to lay down objectives, development efforts and costs of the system.

Once all these questions are answered, it is of primary importance to focus on the data. The real success of a webGIS relies on the quality of the data that it delivers to the public: a good-looking web application in terms of design and usability would be a failure if it does not also provide good quality data. In the following, we classify geospatial web applications into two main types: standard web mapping applications and SDIs (*see* Fig. 2). These are then described in detail in the following sections.

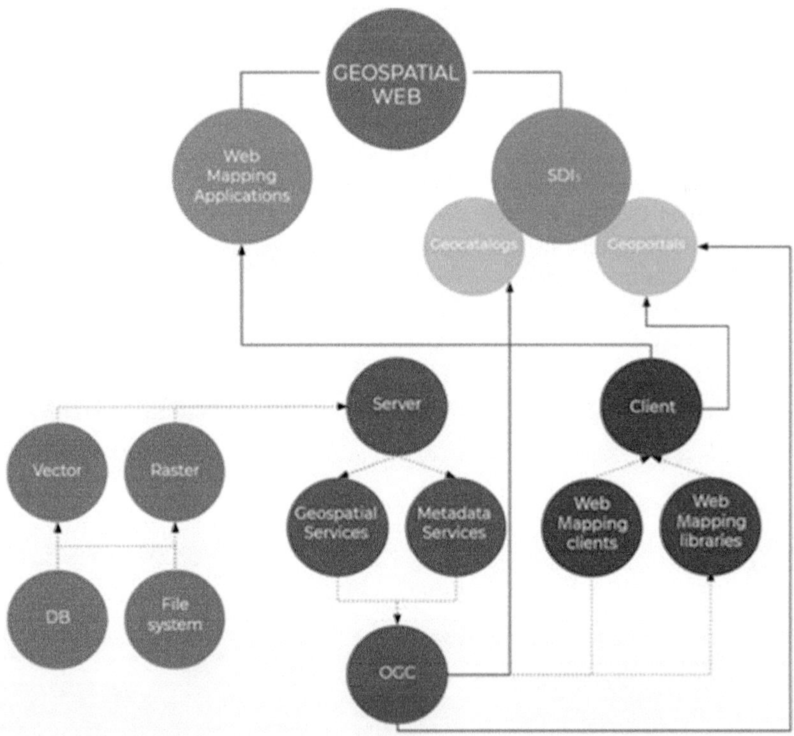

Fig. 2: The geospatial web components, organized into web mapping applications and SDIs (*Source*: Authors' elaboration)

7.3. Web Mapping Applications

Web mapping applications nowadays span from simple visualizations to sophisticated interactive tools. They can integrate and manage data from a wide spectrum of disciplines, using geospatial web services as a common way of sharing and integrating geospatial information. These services are defined by the OGC standards. A key function of those standards is the integration of different and readily-available systems to serve and consume geospatial data, which ultimately allow to geo-enable the web (Skoulikaris *et al.*, 2014). Interoperable OGC web services providing different functionalities can be used simultaneously to combine data from different sources (Dunfey *et al.*, 2006) and this practice of 'service-chaining' allows addition of value to existing services. Exploiting web services based on OGC standards is a fundamental condition to combine distributed data from multiple organizations (Skoulikaris *et al.*, 2014).

A web map could be defined as an interactive display of geospatial information, which is accessible and queryable from the users through a web browser. A web map could be static (i.e. it simply displays data), or dynamic. In addition to displaying data, the latter also allows the user to interact with them through the client, e.g. by retrieving information, calculating data-driven statistics, etc. An example of a dynamic web map is the one from the digital observatory for protected areas (DOPA) (Dubois *et al.*, 2016), that allows the exploration of the world database of protected areas (WDPA) and dynamically retrieves metrics and indicators over the selected protected area (PA), country or ecoregion (*see* Fig. 3). In this specific case, the use of a web map allows interaction with the complexity of the dynamics that underlie a given PA.

A web map is made of specific elements. The first one is a basemap, i.e. a background layer covering the extension of the whole world and providing basic

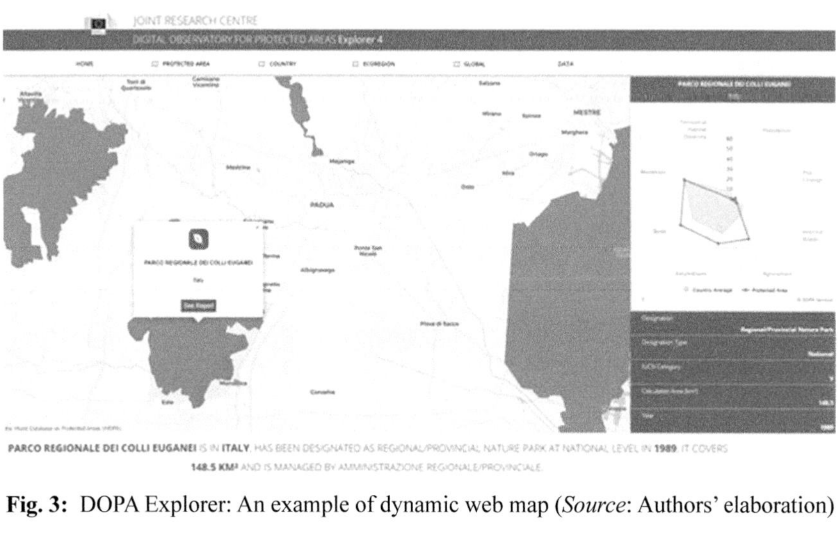

Fig. 3: DOPA Explorer: An example of dynamic web map (*Source*: Authors' elaboration)

information, such as boundaries, rivers, and cities. On top of the basemap, there are layers that represent the actual geospatial information presented to the end-users. Additionally, a web map can offer some functionalities typical of desktop GIS, such as zoom, measure tools, and legend. The layers are displayed in the web map through standard protocols defined by the OGC. The main protocols used in standard web maps are web map service (WMS), web feature service (WFS), and web coverage service (WCS). The WMS, which allows serving geospatial data as georeferenced map images, is the most widely used and has very simple interfaces: *GetCapabilities*, *GetMap*, and *GetFeatureInfo*. When a WMS is accessed, the *GetCapabilities* is the first invoked to obtain the service metadata. The client can get detailed capability information on this service (the list of available layers, the reference systems allowed, the conditions for access and use, etc.) by parsing this 2capability document. This information is used to invoke the *GetMap* operation to request the map with a given resolution and extent, as well as to query specific map features through the *GetFeatureInfo* operation (Wu *et al.*, 2011). Aside from the OGC protocols, it is worth mentioning the Vector Tiles service, which is an emerging method for transferring geospatial data over the web. Although Vector Tiles are pre-rendered as map images using caching systems, the server returns the vector attributes clipped to the boundaries of each tile. Hence, the use of Vector Tiles allows to style layers on-the-fly, perform high-performance spatial queries, and rapidly compute client-side statistics.

The development of a web map passes through several steps and associated software technologies (*see* Fig. 4). The first step is the web server set-up. The most common solutions are based on a Linux distribution, such as Ubuntu or CentOS, with additional software and libraries for web development (Apache2, Tomcat, PHP, etc.). Second, it is necessary to deploy the software to manage geospatial services: typically a map server (e.g. GeoServer, QGIS Server, Deegree or MapServer) and a database, for instance, PostgreSQL with the PostGIS extension. The map server is the software that exposes the geospatial data, both rasters and vectors, to the web through the OGC protocols (mainly WMS, WFS, and WCS) or cached services, such as WMTS or Vector Tiles through the use of caching technologies like GeoWebCache, MapCache and MapProxy. It is possible to describe the function of a map server using the testaurant service analogy. The customer *(map client)* makes a series of requests for a specific set of services (beverage, main course, coffee, etc.). These requests are all made to one person, the waiter *(map server)*, who is responsible for performing the service and returning the information to the client. The map server is like the waiter in a restaurant responsible for handling the client's requests and delivering the end product to the client, in this case serving the geospatial data through OGC protocols. The described workflow on the server set-up and deployment corresponds to the so-called back-end architecture of a web mapping application (*see* Fig. 4).

In addition to the back-end, web mapping applications consist of a front-end architecture. This includes all the software solutions to present geospatial data to end users on a web page. A web page is written in HTML, a markup

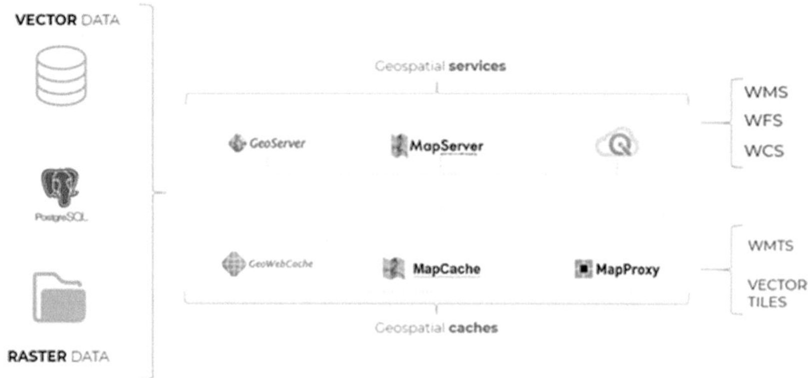

Fig. 4: Examples of software products to set-up the back-end architecture of a web mapping application (*Source*: Authors' elaboration)

language designed to be displayed in a browser, and CSS, a language describing how elements should be displayed (colors, position, etc.). To add elements to a web page and allow users to interact with them, it is necessary to use JavaScript (JS). Many JS libraries to create dynamic maps are freely available; examples are Leaflet JS, OpenLayers and MapBoxGL JS. Below is an example of the HTML and JS code to create a simple web map (shown in Fig. 5) consisting of an OpenStreetMap basemap and a layer of Italian provinces (served by GeoServer via WMS) on top of it.

The web map could be further enriched by adding elements and functionalities through additional JS libraries in order to, e.g. display charts, popups, and other dynamic elements. For example, Fig. 3 shows an example of a dynamic web map with a complex front-end architecture developed with such elements. Figure 7 shows the technologies and the most popular software products used to develop the front-end architecture of a web map, including the mapping libraries mentioned before as well as specific libraries to achieve additional functionalities: Highcharts, Google Charts, and D3 for interactive charts, and LeafletJS, OpenLayers, and Mapbox GL JS as web mapping libraries. These components can be wrapped using Vanilla JS or through the use of a framework, such as React. All these JS libraries are supported by extremely active development communities that keep them up-to-date and often add additional features.

7.3.1. Web Mapping 'Without' Coding

The solutions for creating customized web maps require a significant effort in terms of configuration and programming. However, there is also an entry level to web mapping that does not require any code (or in some cases only a few lines) to be written. Many software products have developed tools and systems which hide the underlying complexity and allow common users to create web maps in an easy and quick way. For example, maps created in the open-source desktop GIS

```html
<html>
  <head>
    <!--Import leaflet library-->
    <script src="libraries/leaflet-src.js"></script>
    <link rel="stylesheet" href="css/leaflet.css" />
    <link rel="stylesheet" href="css/custom.css" />
    <title>WebGIS</title>
  </head>
  <body>
    <!-- banner -->
      <div id= 'banner'>
        <h3>Main Title</h3>
        <hr>
        <p> Paragraph</p>
      </div>
      <!-- end banner -->
        <!-- div map -->
              <div id="map"></div>
        <!-- end of map -->
  </body>
  <script src="main.js"></script>
</html>
```

HTML Example

```javascript
//Initialise map
var max_zoom = 16;
// Width and height of tiles (reduce number of tiles and increase tile size)
var tile_size = 512;
// zoom to Italy (lat,lon, zoom)
var map = L.map('map', {
}).setView([42, 10], 6);
// Define base maps OSM
var light  =  L.tileLayer('https://{s}.basemaps.cartocdn.com/light_all/{z}/{x}/{y}{r}.png', {
        attribution: '&copy; OpenStreetMap &copy; CartoDB',
        subdomains: 'abcd',
        maxZoom: 19
}).addTo(map);
// Define WMS layer and add it to the map
var service_base_url= 'http://localhost:8080/geoserver/workspace_name/wms';
var layer=L.tileLayer.wms(url_region, {
                layers: workspace_name:layer_name,
                transparent: true,
                format: 'image/png',
                opacity:'0.8',
                zIndex: 2
        }).addTo(map);
//Available Layers
var baseMaps = {"White" : light};
var overlayMaps = {'provinces': layer};
//Add Layer Control
layerControl = L.control.layers(baseMaps, overlayMaps, null,  {position: 'topleft'}).addTo(map);
```

Javascript Example (main.js)

Fig. 5: Example of the code to create a web map (*Source*: Authors' elaboration)

software QGIS can be seamlessly transformed into web maps (maintaining the same settings and symbologies) thanks to the server extension QGIS server that is able to automatically expose OGC standards. Similarly, client-side applications for QGIS server have been developed in the form of QGIS plugins (e.g. Lizmap) or external applications (e.g. QGIS Web Client 2 and G3W-SUITE) to allow QGIS users to easily publish web maps without writing a single line of code. This happens through the use of the same web mapping libraries, such as LeafletJS and

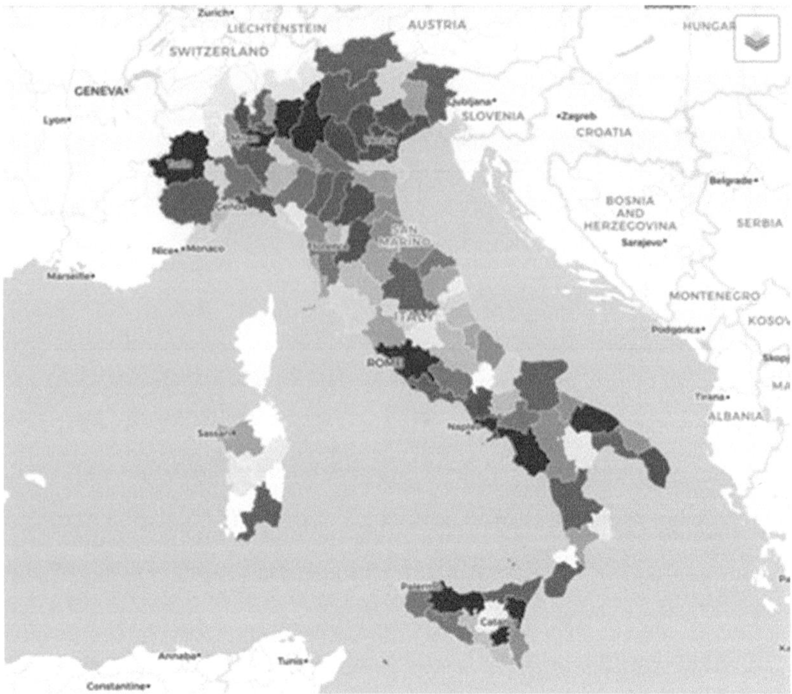

Fig. 6: Web map showing Italian provinces on top of a basemap, created with the code presented above. Basemap: © OpenStreetMap contributors (*Source*: Authors' elaboration)

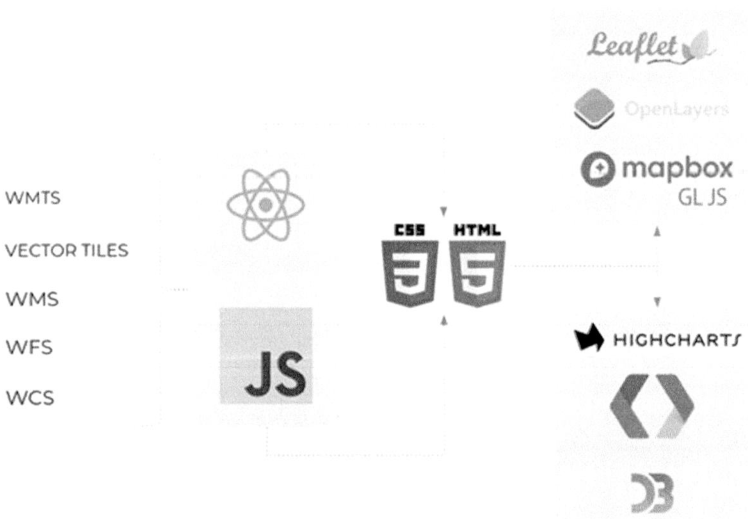

Fig. 7: Examples of software products and languages to develop the front-end architecture of a web mapping application (*Source*: Authors' elaboration)

OpenLayers. Similarly, the ESRI ecosystem offers its own tools to support users in deploying a desktop map to the web through the ArcGIS online platform. The connection of such tools with the ArcGIS desktop application allows the creation of web maps without added costs and complex IT (information technology) architectures (Chmielewski *et al.*, 2018). GeoNode is another open-source technology to publish and share geospatial data on the web and create web maps as well as SDIs with no programming skills required. GeoNode offers a simple, intuitive, and user-friendly web interface to upload data and create web maps and is widely used by NGOs and international organizations worldwide.

7.3.2. Story Maps

In recent years, story maps have become popular ways to make use of web mapping technologies. A story map is nothing more than a web slideshow that combines multimedia with web mapping content in a dynamic and interactive way to present any kind of map-based narrative, e.g. journalistic inquiries, documentaries, project results, etc. ESRI was the first company to provide its customers with a very usable online tool (ArcGIS StoryMaps) to create geographic storytelling. This pioneering work has then inspired the open-source movement, which picked up the idea and developed a plethora of software solutions to create story maps, such as StoryMapJS, TalkingMaps and MapStory. Despite the obvious differences, all these projects allow users to link the communicative power of maps to stories, images, and multimedia. All the tools described in this section are suitable for novice users to approach the world of web mapping by publishing and managing map-based web projects even without having to know how the web works.

7.4. Spatial Data Infrastructures

There are multiple definitions of the term spatial data infrastructure (SDI). One of the earliest was given by the US National Research Council, according to which an SDI is ‘a framework of technologies, policies, and institutional arrangements that together facilitate the creation, exchange, and use of geospatial data and related information resources across an information-sharing community’ (Jabbour *et al.*, 2019, p. 69). In a few words, an SDI allows the sharing of geospatial data between specific stakeholders through the use of dedicated technologies and within a specific legal and political setting. SDIs are usually implemented by governmental organizations at a specific scale, which can range from municipal to regional, national or continental, and they aim to provide access to reliable datasets on a specific domain or area of interest. In some cases, SDIs are used by these organizations to collect data directly from users. One of the pioneering examples of SDIs is INSPIRE (infrastructure for spatial information in the European community), started in 2007 and is still engaging European Union (EU) Member States in the creation of a pan-European infrastructure to support environmental policies within the EU (Cetl *et al.*, 2019).

Given the interest of this chapter on the aspects related to web mapping, the following discussion will not address the legal and organizational dimensions of SDIs but will only focus on their technical and technological features, which are roughly classified in Fig. 8. SDIs mainly consist of two components, a geoportal and a geocatalog, which are described in the following session.

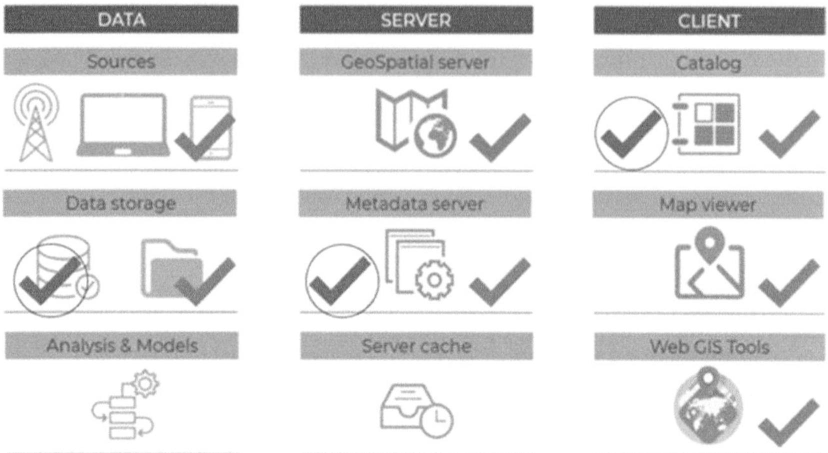

Fig. 8: Main components of SDIs, showing elements included in a geoportal (the check image) and in a geocatalog (components with a circle check).
(*Source*: Authors' elaboration)

7.4.1. Geoportals

Started in 1994, the US National Spatial Data Infrastructure (NSDI) is considered the earliest concept of a geoportal (Gaile and Willmott, 2005). Usually based on a content management system (CMS)[4], a geoportal offers the possibility to browse, access, and visualize geospatial datasets through a web interface. The components of an SDI, which are covered by a geoportal, are highlighted using a check icon in Fig. 7 above. When combined together, and thanks to the interaction with a geocatalog, they form the main gateway for discovering and visualizing the geospatial data of an SDI. Often, a geoportal also allows to filter, query, and download geospatial data through the direct interaction with a map viewer.

The main functionalities provided by a geoportal are the following:

- *User administration*: Different permission levels may be assigned to distinct users to access the geoportal functionalities. For instance, users may or may not be allowed to upload new geospatial data, add/edit metadata, modify the layer layout, access specific web maps, or add new users and assign permissions;

[4] A content management system is a computer software used to manage the creation and modification of digital content. https://en.wikipedia.org/wiki/Content_management_system

- *Data publishing*: Geospatial datasets, both vector (shapefiles, GeoJSON, CSV, KML, KMZ, etc.) and raster (GeoTIFF, GeoPackage, NetCDF, etc.), can be published using OGC protocols (e.g. WMS, WFS and WCS) in order to share them with other people. This feature is managed by the geospatial server in the back-end;
- *Map management*: Combining different layers and base layers in a single web map is one of the key features of a heoportal; users may also be able to change layer styles (e.g. colors, symbols and labels) on the fly. Creating web maps implies that the system is able to reproject all the layers in one single projection, which in many cases is Web Mercator;
- *WebGIS tools*: Geoportals may also include additional, specialized tools for analyzing geospatial data. These can make use of third-party services and in some cases are seamlessly integrated into the geoportal features, for example in the map viewer for functions such as routing and calculation of isochrones.

7.4.2. Geocatalogs

The second key component of an SDI is a geocatalog, which complements the features of a geoportal by offering a metadata repository for geospatial data and the related functions to search and discover them. In a nutshell, a geocatalog does not store the SDI geospatial resources under its data storage but has the task to link them with the metadata (see the elements highlighted with a round check icon in Fig. 7 above). In particular, the geocatalog makes use of the OGC catalogue service for the web (CSW) protocol to query the database, allowing users to store metadata, and retrieve their related geospatial datasets using a single-entry point. The most used software tools to implement geocatalogs are ESRI geoportal server in the proprietary domain and GeoNetwork and pycsw in the open-source realm. In an SDI, the geocatalog is usually integrated with the geoportal using CSW clients.

The main functionalities provided by a geocatalog are the following:

- *Administration*: Enabled functionalities of the geocatalog may be different based on the permission level of each user: for example, users may or may not be allowed to manage (create, edit, and remove) metadata or start a new harvest of the geocatalog. Usually, there are only two types of users of a geocatalog, i.e. administrators and end users;
- *Metadata publishing and management*: A geocatalog allows to handle the metadata of the geospatial data and to store them in a database or in XML files. Metadata structures should be designed to be extensible and modifiable and should allow users to choose between different standards including those from ISO (and profiles thereof, such as the INSPIRE one), Federal Geographic Data Committee (FGDC) or Dublin Core Metadata Initiative (DCMI);
- *Metadata search*: Geocatalogs are equipped with a powerful search engine (e.g. the open-source Apache Solr or Elasticsearch), which allows to search

metadata and related datasets at various levels, e.g. using keywords or other specific metadata fields or drawing a bounding box directly on the map.

7.4.3. An SDI Example: The BIOPAMA Regional Reference Information System

Although it is not strictly related to the agroecological context, an insightful example of an effective adoption of an SDI is the reference information dystem of the biodiversity and rotected areas management (BIOPAMA) project, developed by the International Union for Conservation of Nature (IUCN) and the Joint Research Centre (JRC) of the European Commission. It publishes a wide number of geospatial datasets in African, Caribbean, and Pacific (ACP) countries in the domain of natural resources management (*see* Fig. 8). Through the use of this SDI, BIOPAMA has built a solid information base for decision making on ACP protected areas, which currently count a collection of reliable and up-to-date datasets to improve the long-term conservation and sustainable use of natural resources in ACP countries.

The BIOPAMA SDI is based on GeoNode, which integrates the geoportal and geocatalog components through a combination of open-source technologies:

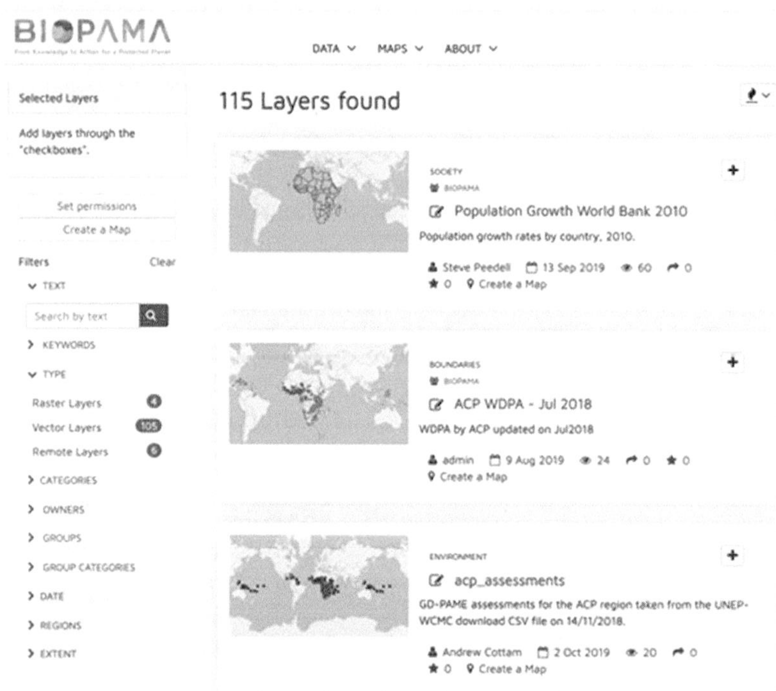

Fig. 9: Geoportal of the BIOPAMA reference information system (*Source*: Authors' elaboration)

the Django web framework, the PostgreSQL/PostGIS database, the GeoServer map server, a map viewer based on MapStore, and the pycswcatalog engine. Combined together, these tools form a powerful stack that provides a gateway for exchanging geospatial data between the administrative bodies and public users collaboratively involved in the project.

7.5. WebGIS and Agroecology

As the concept of agroecology is becoming increasingly popular, along with the recognition that it is a serious way forward to ensure the increase of food production in a sustainable way (Rosset and Altieri, 2017), there is a pressing need to understand how to further expand and upscale agroecological principles from the farm to the regional scale, and perhaps even to the global scale. As put by Rosset and Altieri (2017), there is here a double component of 'scaling out', meaning to spread geographically by involving more people and communities, but also 'scaling up', from local organizations to national or international policy-makers. Two particular aspects of agroecology make this a much more complicated endeavor than expanding conventional agricultural practices. First, in contrast to monoculture systems in conventional agriculture, agroecology thrives on the system's complexity, where the intricate inter-relationships between species from different functional groups ensure a higher resilience of the system. Second, the practical knowledge of how to implement these systems is often highly localized and regionally variable. Together, these constraints call for detailed information to be adequately spatialized, visualized, analyzed, and managed, which carves a clear role for GIS technology in general, and webGIS in particular, in contributing towards scaling agroecology from farm to region.

The uptake of webGIS in this specific sector of agroecology is currently very limited. However, it is very likely to pick up pace as the technology becomes more widely available and the data layers suitable for scaling agroecological principles become more consolidated. Whatever the adopted technology might be, here we elaborate more of the possible functions webGIS can have in the coming years.

The first function of such a web mapping platform is the collection and spatialization of the locations of farms that have adopted certain agroecological practices within a region. Although this function can seem to be simply informative, it should not be neglected as it can serve to further stimulate the adoption of new practices by establishing networks of like-minded people and facilitating the exchange of knowledge, which translates into enhanced connectivity among farmers who are adopting agro-ecological principles and with people keen on purchasing their products. When including open-access and copy-left licensing in the design of such a platform, it is further possible to stimulate the preservation of traditional agroecological knowledge, as is the case of the CONECT-e platform, which collects community contribution of local land races in Spain under a digital commons framework (Calvet-Mir *et al.*, 2018).

Scaling agroecology could also benefit from the function of determining land suitability for expansion. GIS has already been used for this purpose on local scales, e.g. for indigenous Maori medicinal plants by Moore *et al.* (2016), but an application and deployment on larger scales would probably necessitate distributed capacities embedded in a web mapping platform. A precondition is to have consolidated and informative baseline layers from which to determine land suitability. Since agroecosystems rely on the availability of functional biodiversity, species distribution maps and phenology records of both plants and insects would be highly valuable. A concrete example where geospatial web technology is used to generate such information is the iNaturalist platform, a citizen science project that allows mapping and sharing of (geospatial) biodiversity observations across the globe through a dedicated web portal and mobile app. The observations consist of geolocated photographs taken by mobile devices, that enable species identification through a combination of artificial intelligence technology with crowd-sourcing verification from the community. When consensus is reached, the data is embedded in a 'research-grade' dataset of species occurrence (Ueda, 2020). Because of the wide coverage and vast amount of data involved, this offers a highly valuable tool to provide information on both species distribution and phenology that could serve to identify hot-spots where certain agroecological practices could be promoted. Indeed in web mapping platforms for agroecology, the role of Citizen Science, and more in general crowd-sourced geographic information, would be key to promote farmers' engagement and make sure their experience and knowledge of local agroecological practices – whose scale and level of detail have no equivalent in official data – is properly shared. Along the same lines, a webGIS connected to adequate databases of meteorological and environmental variables can further serve to characterize the suitability of crop and insect species based on climatic and edaphic criteria. This can be extended further to seeking the possibilities of adapting to climate change by including data from future climate scenarios. In this sense, a valuable source of both actual and future climate data would be the climate data store (CDS) of the Copernicus Climate Change Service (C3S) of the European Commission. Combining information on successful experiences and suitability in the same platform could allow users to identify practices most adapted to their conditions. A concrete example could be adding spatial layers of environmental conditions to either: (1) the world overview of conservation approaches and technologies (WOCAT) sustainable land management (SLM) database , an effort to compile, document, evaluate, share, disseminate, and apply SLM knowledge; or (2) the agriculture and biodiversity solutions of PANORAMA, a partnership initiative to document and promote examples of inspiring, replicable solutions across a range of conservation and sustainable development topics.

7.5.1. Projects References

In Table 1 are listed all the projects cited in this chapter.

Table 1: The Project Cited in this Chapter (Listed in Order of Appearance)

Organization/Project	WebSite	Last Access
Open Geospatial Consortium (OGC)	https://ogc.org	March 2021
OpenStreetMap (OSM)	https://www.openstreetmap.org	March 2021
Google Earth	https://earth.google.com/web	March 2021
NASA WorldWind	https://worldwind.arc.nasa.gov	March 2021
Pokemon Go	https://pokemongolive.com	March 2021
Google Earth Engine	https://earthengine.google.com	March 2021
ESRI ArcGIS	https://www.arcgis.com	March 2021
CARTO	https://carto.com	March 2021
MapBox	https://www.mapbox.com	March 2021
GRASS GIS	https://grass.osgeo.org	March 2021
Digital Observatory Protected Areas (DOPA)	https://dopa-explorer.jrc.ec.europa.eu	March 2021
GeoServer	http://geoserver.org	March 2021
QGIS	https://qgis.org	March 2021
Deegree	http://www.deegree.org	March 2021
MapServer	https://mapserver.org	March 2021
PostgreSQL	https://www.postgresql.org	March 2021
PostGIS	https://postgis.net	March 2021
GeoWebCache	https://www.geowebcache.org	March 2021
MapCache	https://mapserver.org/mapcache	March 2021
MapProxy	https://mapproxy.org	March 2021
Leaflet JS	https://leafletjs.com	March 2021
OpenLayers	https://openlayers.org	March 2021
MapBoxGL	https://docs.mapbox.com/mapbox-gl-js	March 2021
HighCharts JS	https://www.highcharts.com	March 2021
Google Charts	https://developers.google.com/chart	March 2021
D3	https://d3js.org	March 2021
Vanilla JS	http://vanilla-js.com	March 2021

(Contd.)

Table 1: (*Contd.*)

Organization/Project	WebSite	Last Access
React	https://reactjs.org	March 2021
Lizmap	https://www.3liz.com/en/lizmap.html	March 2021
QGIS Web Client 2	https://github.com/qgis/qwc2-demo-app	March 2021
G3W-SUITE	https://g3wsuite.it/en/g3w-suite-publish-qgis-projects	March 2021
GeoNode	http://geonode.org	March 2021
ArcGIS StoryMaps	https://storymaps.arcgis.com	March 2021
StoryMapJS	https://storymap.knightlab.com	March 2021
TalkingMaps	https://www.talkingmaps.eu	March 2021
MapStory	https://mapstory.org	March 2021
INSPIRE	https://inspire.ec.europa.eu	March 2021
ESRI Geoportal Server	https://www.esri.com/en-us/arcgis/products/geoportal-server/overview	March 2021
GeoNetwork	https://geonetwork-opensource.org	March 2021
Pycsw	https://pycsw.org	March 2021
Apache Solr	https://lucene.apache.org/solr	March 2021
ElasticSearch	https://www.elastic.co/elasticsearch	March 2021
BIOPAMA	https://www.iucn.org/theme/protected-areas/our-work/projects/biopama	March 2021
DJANGO	https://www.djangoproject.com	March 2021
Mapstore	https://mapstore.readthedocs.io/en/latest	March 2021
iNaturalist	https://www.inaturalist.org	March 2021
Climate Data Store (CDS)	https://cds.climate.copernicus.eu/cdsapp	March 2021
World Overview of Conservation Approaches and Technologies (WOCAT)	https://qcat.wocat.net/en/wocat	March 2021
PANORAMA	https://panorama.solutions/en/portal/agriculture-and-biodiversity/map	March 2021

7.6. Conclusion

The interest in agroecology is on the rise. This is partly because it offers plausible contributions towards solving some of humanity's great challenges. It is recognized as a mitigation and adaptation strategy for climate change, and as a catalyst to reduce poverty and inequality by contributing to decent work, and addressing a fundamental human need, such as access to food, in line with the *2030 Agenda for Sustainable Development* (United Nations, 2015). Its demand is rising rapidly and there is currently a remarkable opportunity to strengthen its potential and advance it globally. As outlined in the other chapters of this book, several examples of GIS-based systems in the agroecological domain exist but none of them is structurally designed on the capabilities offered by the web. In this chapter an overview of geospatial web and the main practices, business models, and technologies adopted in web mapping applications was provided to offer baseline knowledge to both experienced and novice users. Although readers should have it clear that the evolution of web technologies is still happening at a fast pace – *see* Kotsev *et al.* (2020) for a recent overview of web-based, data-driven ecosystems and underlying trends – the chapter has hopefully made it clear that the introduction of webGIS principles and technology in the agroecology field might connect farmers as site-specific data generators to regional and global systems. We envision that this process, supported by adequate capacity-building initiatives, will lead to the establishment of fruitful knowledge networks between farmers, researchers, and policymakers, emphasising their mutual dialogue through participatory learning processes. These innovative measures will ultimately need to be placed in a larger policy-driven context to foster a bottom-up approach for collecting and sharing geospatial data, establishing a horizontal integration, and a complete freedom of information to support the agroecological transition from the farm to the regional level.

Bibliography

Abdalla, R. and M. Esmail (2019). *WebGIS for Disaster Management and Emergency Response: Advances in Science, Technology & Innovation*. IEREK Interdisciplinary Series for Sustainable Development.

Ajena, F. (2018). Agriculture 3.0 or (Smart) Agroecology? *Green European Journal*. Retrieved from: https://www.greeneuropeanjournal.eu/agriculture-3-0-or-smart-agroecology. Accessed May 3, 2020.

Brovelli, M.A., M. Minghiniand and G. Zamboni (2015). Public participation GIS: A FOSS architecture enabling field-data collection, *International Journal of Digital Earth*, 8(5): 345-363.

Brovelli, M.A., M. Minghini and R. Moreno-Sanchezand Oliveira (2017). Free and open source software for geospatial applications (FOSS4G) to support Future Earth, *International Journal of Digital Earth*, **10**(4): 386-404.

Calvet-Mir, L., P. Benyei, L. Aceituno-Mata, M. Pardo-de-Santayana, D. López-García, M. Carrascosa-García, A. Perdomo-Molina and V. Reyes-García (2018). The contribution of traditional agroecological knowledge as a digital commons to agroecological transitions: The case of the CONECT-e platform, *Sustainability*, **10**(9): 3214.

Cetl, V., R. Tomas, A. Kotsev, V.N. de Lima, R.S. Smith and M. Jobst (2018). Establishing common ground through INSPIRE: The legally-driven European spatial data infrastructure, pp. 63-84. *In:* J. Döllner, M. Jobst and P. Schmitz. (Eds.). Service-Oriented Mapping: Changing Paradigm in Map Production and Geoinformation Management. Springer, Berlin, De.

Chmielewski, S., M. Samulowska, M. Lupa, D.J. Lee and B. Zagajewski (2018). Citizen science and webGIS for outdoor advertisement visual pollution assessment. Computers, Environment and Urban Systems, 67: 97-109.

Coetzee, S., I. Ivánová, H. Mitasova and M.A. Brovelli (2020). Open geospatial software and data: A review of the current state and a perspective into the future, *ISPRS Int. J. Geo-Inf.*, **9**(2): 90.

Coleman, D., Y. Georgiadou and J. Labonte (2009). Volunteered geographic information: The nature and motivation of producers, *International Journal of Spatial Data Infrastructures Research*, **4**(4): 332-358.

Dragićević, S. (2004). The potential of web-based GIS, *Journal of Geographical Systems*, **6**(2): 79-81.

Dubois, G., L. Bastin, B. Bertzky, A. Mandrici, M. Conti, S. Saura, A. Cottam, L. Battistella, J. Martínez-López, M. Boni and M. Graziano (2016). Integrating multiple spatial datasets to assess protected areas: Lessons learnt from the Digital Observatory for Protected Area (DOPA), *International Journal of Geo-Information*, **5**(12): 242.

Dunfey, R.I., B.M. Gittings and J.K. Batcheller (2006). Towards an open architecture for vector GIS, *Computers & Geosciences*, **32**(10): 1720-1732.

Gaile, G.L. and C.J. Willmott (Eds.) (2005). *Geography in America at the Dawn of the 21st Century*, Oxford University Press, Oxford, UK.

Goodchild, M.F. (2007). Citizens as sensors: The world of volunteered geography, *GeoJournal*, **69**(4): 211–221.

Haklay, M. and A. Skarlatidou (2010). Human-computer interaction and geospatial technologies context, pp. 3-18. *In*: M. Haklay (Ed.). *Interacting with Geospatial Technologies*, Wiley-Blackwell, West Sussex, UK.

Jabbour, C., H. Rey-Valette, P. Maurel and J.M. Salles (2019). Spatial data infrastructure management: A two-sided market approach for strategic reflections, *International Journal of Information Management*, 45: 69-82.

Kotsev, A., M. Minghini, R. Tomas, V. Cetl and M. Lutz. 2020. From spatial data infrastructures to data spaces—A technological perspective on the evolution of European SDIs. *ISPRS International Journal of Geo-Information*, **9**(3): 176.

Maguire, D.J. (2007). GeoWeb 2.0 and volunteered GI. *In*: M.F. Goodchild and R. Gupta (Eds.). *The Proceeding of the Workshop on Volunteered Geographic Information*, Santa Barbara, CA, USA.

Minghini, M., A. Mobasheri, V. Rautenbach and M.A. Brovelli (2020). Geospatial openness: From software to standards and data, *Open Geospatial Data, Software and Standards*, **5**(1).

Mooney, P. and M. Minghini (2017). A review of OpenStreetMap data, pp. 37–59. *In:* G. Foody, L. See, S. Fritz, P. Mooney, A.M. Olteanu-Raimond, C. Costa Fonte and V. Antoniou (Eds.). *Mapping and the Citizen Sensor*, Ubiquity Press, London.

Moore, A., M. Johnson, J. Lord, S. Coutts, M. Pagan, J. Gbolagun and G.B. Hall (2016). Applying spatial analysis to the agroecology-led management of an indigenous farm in New Zealand, *Ecological Informatics*, 31: 49-58.

Neteler, M. and H. Mitasova (2013). *Open-source GIS: A GRASS GIS Approach*, Springer, Cham, CH.

Peng, Z.R. and M.H. Tsou (2003). *Internet GIS*, John Wiley & Sons, Hoboken, NJ.

Putz, S. (1994). Interactive information services using World-Wide Web hypertext, *Computer Networks and ISDN Systems*, **27**(2): 273-280.

Rosset, P.M. and M.A. Altieri (2017). *Agroecology: Science and Politics*, Fernwood Publishing, Nova Scotia, USA.

See, L., P. Mooney, G. Foody, L. Bastin, A. Comber, J. Estima, S. Fritz, N. Kerle, B. Jiang, M. Laakso, H.Y. Liu, G. Milčinski, M. Nikšič, M. Painho, A. Pődör, A.M. Olteanu-Raimond and M. Rutzinger (2016). Crowdsourcing, citizen science or volunteered geographic information? The current state of crowdsourced geographic information, *ISPRS International Journal of Geo-Information*, **5**(5): 55.

Skoulikaris, C., J.G. Ganoulis, N. Karapetsas, F. Katsogiannos and G. Zalidis (2014). Cooperative WebGIS interactive information systems for water resources data management, *Hydrology in a Changing World: Environmental and Human Dimensions*, IAHS Publ., 363: 342-347.

Ueda, K. (2020). *iNaturalist Research-grade Observations*, iNaturalist.org. Occurrence dataset retrieved from https://doi.org/10.15468/ab3s5x; accessed on 2 May, 2020.

UN (United Nations) (2015). Transforming Our World: The 2030 Agenda for Sustainable Development, resolution adopted by the General Assembly on 25 September, 2015.

Veenendaal, B., M.A. Brovelli and S. Li (2017). Review of web mapping: Eras, trends and directions, *ISPRS International Journal of Geo-Information*, **6**(10): 317.

Wu, H., Z. Li, H. Zhang, C. Yang and S. Shen (2011). Monitoring and evaluating the quality of Web Map Service resources for optimizing map composition over the internet to support decision making, *Computers & Geosciences*, **37**(4): 485-494.

CHAPTER

8

Geospatial Support for Agroecological Transition through Geodesign

Antoni B. Moore[1] and Marion Johnson[2]*

[1] School of Surveying, University of Otago, PO Box 56, Dunedin, New Zealand
[2] Future Farming Centre, BHU, Lincoln, New Zealand

8.1. Introduction

Geographical information science (GIS) and spatial data-collection technologies contain powerful knowledge and tools to capture, analyze, visualize and crucially, plan for agroecological projects. As such, GIS is well-established as an agricultural decision-support tool worldwide, including in precision agriculture (Cassel, 2007). Specific benefits include increase in production, more efficient land management, cost reduction (Pierce and Clay, 2007) as well as mitigating the impact of agricultural environmental incidents (Wilson, 1999). In agroecology, GIS takes on a necessary community-based aspect and associated public participation (Weiner *et al.*, 2002). Examples, such as, a Maori *iwi* (tribe)-led GIS project (Harmsworth *et al.*, 2005) combines indigenous knowledge and GIS to capture and store cultural and/or traditional values and concepts, along with associated GI data (Landcare Research, 2013). This chapter is an overview of the geospatial support for the geodesign activities in the indigenous agroecology project, He Ahuwhenua Taketake in Aotearoa New Zealand. Aspects of agroecology supported by GIS as part of a geodesign process include mapping to plan plantings and management on two traditional NZ farms. Furthermore, a geodesign approach informed the application of spatial analysis to define locations of plant species used for livestock to self-medicate, along with on-farm tourist facilities, and access for one of those farms.

*Corresponding author: marion.johnson@actrix.co.nz

8.2. Agroecology

8.2.1. Definition and Principles

According to Altieri (1995), agroecology is 'a discipline that defines, classifies, and studies agricultural systems from an ecological and socioeconomic perspective and applies ecological concepts and principles to the design and management of sustainable ecosystems.' The term 'agroecology' was coined by Hanson (1939) with an emphasis on using knowledge of man's relationship with the environment as a major contribution to 'balance and stabilization' of agriculture through ecology. In practice, this has meant the utilization of many species and a chemical-free approach by traditional farmers, backed up by an innate local environmental knowledge. Although much of that accumulated experience has been lost, what remains has led to highly effective and stable farming systems.

It is important to really understand how traditional farmers perceive their environment (Altieri, 2002) and how they translate these perceptions into local agricultural systems. Not all traditional techniques are applicable in the twenty-first century, but they can, after discussion, be adapted and spread from farmer to farmer (Altieri *et al.*, 2012). Knowledgeable extension workers have an important role but farmers should be viewed as equally expert, their local knowledge complementing academic and scientific knowledge. Often decisions about agriculture and agricultural policy occur in cities (Tomich *et al.*, 2011), made by officials who have no visceral connection to the land or agroecosystems. Agroecology requires the engagement and knowledge of farmers and should be adapted to local environmental, social, and economic conditions, and to individual farm situations (Altieri *et al.*, 2012).

The traditional agroecosystems that still survive today have commonalities. Exhibiting high levels of biodiversity and resilience, they are managed carefully, often using traditional technologies, for example, to prevent erosion or manage water supplies. Frequently, social institutions still govern their management and use (Altieri and Toledo, 2011).

8.2.2. He Ahuwhenua Taketake: Aotearoa New Zealand Indigenous Agroecology

In Aotearoa New Zealand, agroecology has a rich history to draw upon. Indigenous agroecology is an ethic of farm stewardship that is being developed, based on the traditional and contemporary experience of Maori (mainland New Zealand) and Moriori (Chatham Islands) agricultural practitioners, invoking the principle of Kaitiakitanga or guardianship. Indigenous agroecology brings a *ki uta ki tai* (from the mountains to the sea) approach, highlighting the inter-relationship between land and water, acknowledging *Papatūanuku* (earth mother) and *Ranginui* (sky father) and our relationship with all living things.

Diversity is central to the restoration of land and the implementation of agroecological methods. Thus, native plants are central to our concept of indigenous agroecology, benefiting land, water, and communities, as illustrated in Fig. 1.

There is a constant conflict with the short-term expedience of industrial ideas, which may cause long-term harm to principles of community, *kaitiakitanga*, local knowledge, and learnings.

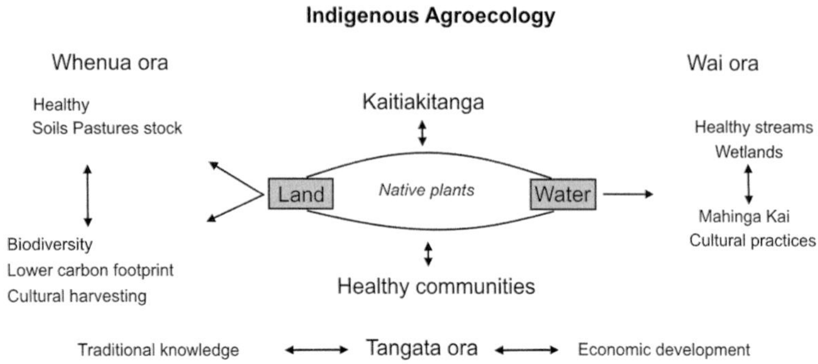

Fig. 1: Native species have a central role in the concept of indigenous agroecology (*Source*: Johnson and Perley, 2015; modified by authors)

Agroecology is based in place. The indigenous agroecology project, He Ahuwhenua Taketake, worked with local peoples and their land to begin to draft an ethic for agroecological land management in Aotearoa New Zealand. By working on farms owned by Maori and Moriori peoples, termed 'research link' farms, the team tried to ground their thoughts and investigations in the land and to develop ideas that grew from the aspirations and knowledge of the local people yet drew on modern techniques to aid decisions around land management.

In other words, although the concepts of indigenous agroecology draw on traditional knowledge, science and technology have much to contribute to the successful agroecological operation. Fundamentally, we have mapped all the research link farms using geographic information systems (GIS) and have examined the meeting of science and traditional knowledge.

8.2.3. Maori and Moriori Farm Trusts: Three Case Study Farms

Much land owned by Maori and Moriori peoples is owned by the community rather than by an individual. The communities frequently manage the land through a trust structure, in which the trustees manage the land for the benefit of all the people, who have an interest in the jointly-owned land. A summary of how Maori land is administered in Aotearoa New Zealand is given in Brady (2004). Three farm trusts (Te Kaio, Taiporutu, and Hokotehi Moriori) in New Zealand (Fig. 2) kindly agreed to work with the indigenous agroecology concept.

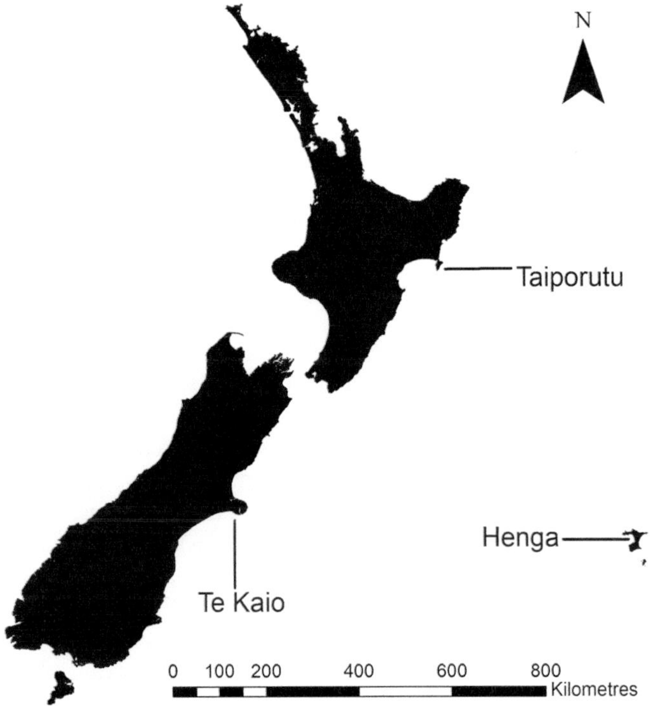

Fig. 2: Location of research link farms
(*Source*: Johnson and Perley, 2015; modified by authors)

Te Kaio Farm Trust manages Te Kaio a 449ha sheep and beef farm on behalf of Wairewa, a people of Horomaka Banks Peninsula in Te Waipounamu South Island. Henga, on Rēkohu (Chatham Island, approximately 500km east of mainland Aotearoa New Zealand) is managed by Hokotehi Moriori Trust on behalf of the Moriori people. The 400ha farm is currently viewed as being uneconomic in its own right and is run in conjunction with the much larger Kaingaroa Station. Taiporutu, on the Mahia Peninsula in Te Ika-a-Maui North Island is managed by the Taiporutu Trust. The farm comprising 99ha of coastal country contains a number of significant cultural sites and is currently leased to a neighbor.

The indigenous agroecology project, He Ahuwhenua Taketake, developed concept agroecological plans for Te Kaio and Henga and addressed the cultural significance of Taiporutu through bioremediation of the spring sites.

Work was focused on these three farms across New Zealand, with global navigation satellite system (GNSS) surveys producing spatial data layers for Te Kaio and Taiporutu, including terrain data, while processed stereo drone photography provided terrain and orthophoto coverage of Henga. In addition, data on farm infrastructure, topographic features, and sacred Maori pa sites were collected. This resource was the foundation for geodesign activities on the Te Kaio and Henga farms, initially forming a GIS base map for recommended, planned, and plotted expert-based use of the farm, as part of a concept agroecological plan. Expertise was also coded as input into spatial analytical procedures for geodesign applied to Te Kaio to: a) devise a mapped scheme of plantings for *rongoa* (traditional medicine) that could be utilized by livestock, through multi-criteria evaluation, and, b) to plan facilities (paths, information platforms) for agritourism visitor farm access through cost-path analysis-calculated maps.

8.3. GIS Support for Agroecology

To support the agroecology research and community initiatives for the indigenous agroecology project it was essential to obtain a robust spatial inventory of the three case-study farms at Banks Peninsula (Te Kaio), Mahia Peninsula (Taiporutu), and Chatham Islands (Henga). Authoritative maps therefore need to exist, as a baseline for the scientific and community work to follow. These maps would contain the on-farm infrastructure (buildings, roads, tracks, fences, etc.) as well as other features of cultural significance (e.g. pa sites). These features should be situated on an elevation surface of high quality. The maps would be compiled and designed, using a geographical information system (GIS).

This section reports on mapping work at all three farms. An initial survey of the geographic data existing for these farms revealed availability of freely available resources (i.e. The NZ national mapping agency, LINZ Topo50 data which provides 20m contours, coastline, rivers and streams, roads and tracks) but at too coarse a scale to adequately map at the farm-scale. Therefore, topographic surveys were undertaken and photogrammetric mapping commissioned to source high resolution, high quality spatial data.

The data management and mapping were carried out by using Esri ArcGIS 10.x and Manifold GIS software.

8.3.1. Te Kaio Farm, Banks Peninsula

A GNSS-based topographic survey of Te Kaio (at the time, named Te Putahi) was undertaken by an experienced surveyor and two supervised senior surveying students over a number of days in 2011.

8.3.1.1. Description of Data

The collected data were in the form of point, polyline, and polygon computer aided design (CAD) file format (dwg and dxf). The CAD data were collected in a local Transverse Mercator projection system, geographic coordinate system GCS NZGD2000/Mount Pleasant Circuit. These CAD files were converted into GIS format files. These are feature classes, stored in an ArcGIS geodatabase.

The main reason for using a geodatabase is to enable logical storage, manipulation, and query, and facilitate analysis and visualization (mapping) of the spatial data collected. This data forms a baseline for subsequent biological and chemical measurements for agroecology research as well as spatial analysis projects (for determining the planting of rongoa species and visitor-access infrastructure).

The feature classes extracted from the CAD data into the geodatabase consist of vector data (point, line, polygon). Once collected and in the geodatabase, these data did not require any further processing, with one exception – the contours were interpolated to produce a continuous elevation surface, a spatial dataset of raster format (1m). This subsequently enabled the production of derived outputs for analysis, such as slope, aspect and hill-shade layers. This dataset was supplemented by an orthophotograph of the farm and layers from Environment Canterbury's database. Finally, there are other data associated with the project that would be desirable but proved difficult to source, such as the soil condition, soil classification, and vegetation. The data used is summarized in Table 1.

Table 1: Te Kaio Spatial Data

Name	Spatial Data Type	Description	Source
Buildings	Vector Polygon	Buildings of dwelling and farm function	GNSS Survey
Contours	Vector Line	Lines of equal height or isolines at 2 m height intervals	GNSS Survey
Spot Heights	Vector Point	Point estimates of height	GNSS Survey
Roads	Vector Line	Roads crossing or adjoining the farm	GNSS Survey
Powerlines	Vector Line	Cables or wires to transmit power to the farm	GNSS Survey
Poles	Vector Point	Overhead powerlines located on the farm	GNSS Survey
Tracks	Vector line	Minor footpaths located within the farm boundaries	GNSS Survey
Banks	Vector line	Any breaks in slope across the farm	GNSS Survey
Breaklines	Vector line	Any significant change in the slope of the ground	GNSS Survey
Waterways	Vector line	Little narrow streams normally leading to the sea	GNSS Survey
Tree Shelter Belt	Vector Line	Demarcated trees being used on the farm as wind shelter	GNSS Survey

(*Contd.*)

Table 1: (*Contd.*)

Name	Spatial Data Type	Description	Source
Fences	Vector Line	Barrier used to enclose and demarcate the farm area	GNSS Survey
Digital Elevation Model	Raster	Elevation model of 1 m resolution	Interpolation from GNSS Survey
Orthophotograph	Raster	Orthophotograph of 0.4 m resolution	New Zealand Aerial Mapping (NZAM)
Slope	Vector Polygon	New Zealand Land Resource Inventory (NZLRI) polygons with the slope attribute GIS dissolved	Land Resource Information Systems (LRIS) Portal, Landcare Research
Land Use Capability	Vector Polygon	New Zealand Land Resource Inventory (NZLRI) polygons with a land use capability (LUC) rating of agricultural production	Land Resource Information Systems (LRIS) Portal, Landcare Research
Soil PH	Vector Polygon	From the New Zealand Fundamental Soils Layer, a relational join of New Zealand Land Resource Inventory (NZLRI) and the national soils database (NSD)	Land Resource Information Systems (LRIS) Portal, Landcare Research
Chemical limitation to plant growth	Raster	Land Environments of New Zealand (LENZ) classification, 25 m resolution	Environment Canterbury database
Erosion type and severity	Vector Polygon	New Zealand Land Resource Inventory (NZLRI) polygons with the erosion attribute GIS dissolved	Land Resource Information Systems (LRIS) Portal, Landcare Research

8.3.1.2. Overview Maps

A subset of the data described above is featured in the following overview maps:

- a 2D topographic map (Fig. 3) featuring collected data, both processed (the raster relief layer of the ground surface) and as captured (a selection of contours, roads and tracks, breaks of slope, buildings, fences, powerlines, and tree shelter belts);

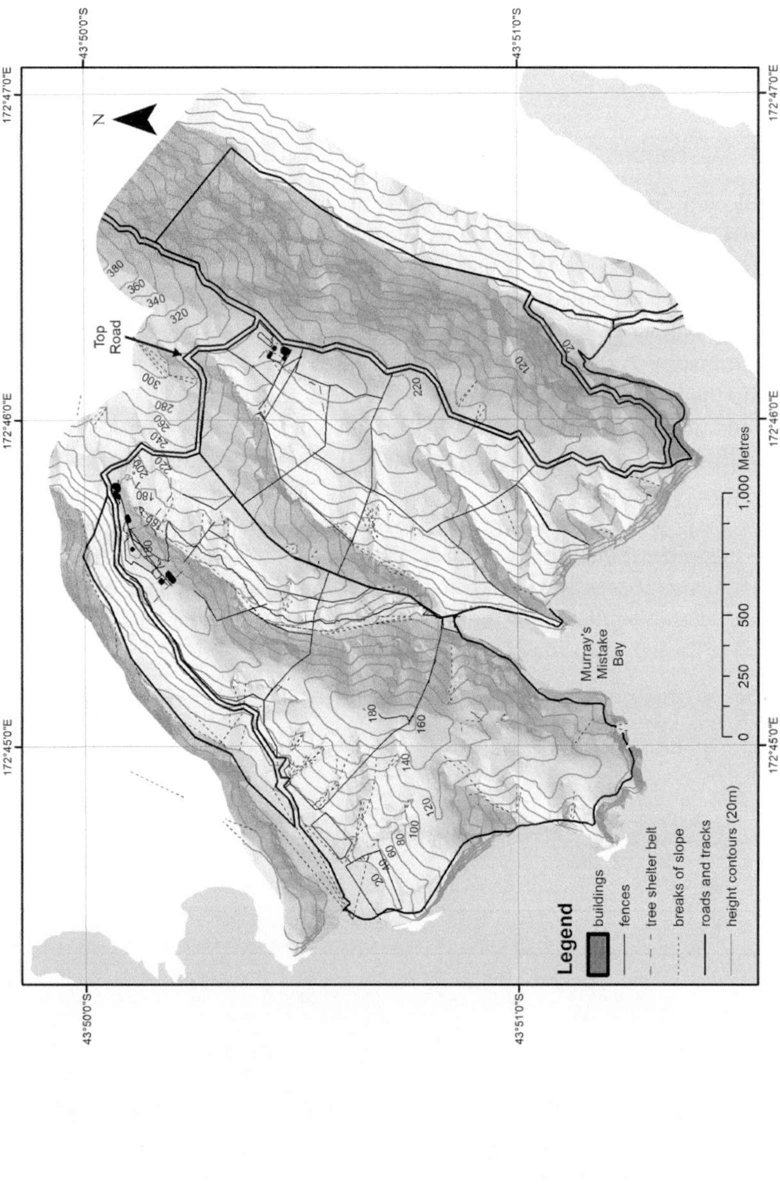

Fig. 3: Context Map of Te Kaio farm (*Source*: Johnson and Perley,2015; modified by authors)

- 3D view of the farm from the south east, draped with an orthophoto (*Source*: NZAM) (Fig. 4).

8.3.2. Taiporutu Farm, Mahia Peninsula

A topographic survey of Taiporutu Farm on the NZ North Island was undertaken over three days in 2014, using GNSS, operated by an experienced surveyor and two supervised senior surveying students. A point of clarification for certain features measured: the bush line and erosion lines on the farm were surveyed with an offset for safety and GNSS reception purposes. Otherwise, no offsets were used.

8.3.2.1. Description of Data

The collected data were received in the form of points in a comma separated variable (CSV) format. The data were collected using the national map projection and coordinate system: New Zealand Transverse Mercator 2000. A single session with the surveyors clarified further what the points signified, whether they should remain as point features, or be turned into line or polygon features. The CSV files were reformatted to reflect the real-world features they represented. Finally, they were imported into ArcGIS as feature classes in a geodatabase.

In addition to the kinds of infrastructural features, such as those collected at Te Kaio (boundaries, buildings, fences, etc.), the surveyors were directed by the farm managers to capture spatial data of on-farm features of cultural significance (e.g. pa sites, potential midden sites). This formed part of a complementary cultural mapping of the farm.

The feature classes extracted from the CSV data into the geodatabase consist of infrastructural, natural, and cultural data. These are all vector data (point, line, polygon) collected, and did not require any further processing.

Fig. 4: 3D Map of Te Kaio farm, draped with an orthophoto, viewed from the southeast (*Source*: Johnson and Perley, 2015; modified by authors)

This dataset was supplemented by an orthophotograph of the farm bought from NZAM, who were also commissioned to digitize, through photogrammetry, contours, and other features from photography. The derived datasets were received as Esri ArcGIS shapefiles (a common GIS exchange format) and were imported into a geodatabase as feature classes.

These are all vector data (point, line, polygon) derived and once in the geodatabase, they did not require any further processing, with one exception – the contours were interpolated to produce a continuous elevation surface, a spatial dataset of raster format (1 m resolution). This subsequently enabled the production of derived outputs for analysis, such as slope, aspect, and hill-shade layers. The Taiporutu dataset is summarized in Table 2.

8.3.2.2. Overview Maps

A subset of the data described above is featured in the following overview maps:

- a 2D topographic map (Fig. 5) featuring collected data, both processed (the raster relief layer of the ground surface and derived contours) and as captured (boundary, roads, and tracks and other infrastructural features; bush, trees, streams, wetlands, and other natural features; pa sites, pits, midden sites, and other cultural features). The data was augmented by LINZ Topo 50 data (surface derived from 20m contours, roads, streams) outside the farm area;
- a 3D view of the farm from the northeast with boundary and cultural features, draped with an orthophoto (*Source*: NZAM) (Fig. 6).

8.3.3. Henga Farm, Rēkohu (Chatham Islands)

A remotely-piloted aircraft system (RPAS) survey backed up by a ground control and ground feature land survey was completed in 2015 on Henga Farm in the Chatham Islands, 500km to the east of the South Island.

8.3.3.1. Description of Data

The main products of the aerial survey of the farm were a high-resolution stitched orthophotograph and a dense cloud of 3D terrain points.

The photography was performed from a Phantom II Vision Plus drone with an integrated camera (14 megapixels; 140-degree field of view). The drone was flying at 112-122 m, generating orthoimagery of 5-7 cm resolution.

Overall, twenty-three to twenty-six blocks out of thirty planned blocks were flown (as wind conditions allowed), with each block or flight having 150-300 photos. The sorties were subject to below 15 km/h wind velocity for optimal flying and photo stability and 30 km/h was the absolute upper limit under which the drone could fly.

AgiSoft photogrammetric software was used to stitch the photos within a block together. To enable this, there was approximately 80 per cent aerial overlap between photos in a flight line or block and 50 per cent overlap between flight lines. The software automatically identified tie points for the stitching process.

As well as a stitched photoset, a digital elevation model (DEM) in the form of a dense 3D point cloud was also produced. From this, a 3D mesh was created with an overlain texture and from this, a cellular DEM was generated with resolution to match the imagery. Contours were then created and used to orthorectify each photo block, before stitching the blocks together.

The final stage within AgiSoft was to export the data and then import it into ArcGIS, for compatibility with the data of the other two farms.

Table 2: Taiporutu Spatial Data

Name	Spatial Data Type	Description	Source
Boundary Pegs	Vector point	Pegs of field boundaries	GNSS Survey
Culverts	Vector point	Man-made drainage feature	GNSS Survey
Old Posts	Vector Point	Old fence posts	GNSS Survey
Power Poles	Vector Point	Overhead powerlines located on the farm	GNSS Survey
3 metres from bush line	Vector Line	3 metre offset from bush line	GNSS Survey
3 metres from erosion line	Vector line	3 metre offset from any breaks in slope across the farm (equivalent to banks and breaklines in the Te Kaio dataset)	GNSS Survey
1 metre from erosion line	Vector line	1 metre offset from any breaks in slope across the farm (equivalent to banks and breaklines in the Te Kaio dataset)	GNSS Survey
Wetlands	Vector polygon	Wetland areas	GNSS Survey
Springs	Vector point	Ground-based water source	GNSS Survey
Streams	Vector line	Little narrow streams normally leading to the sea	GNSS Survey
Trees	Vector Point	Mostly Ti Kouka – cabbage trees, but including karaka and nikau.	GNSS Survey
Fences	Vector Line	Barrier used to enclose and demarcate the farm area	GNSS Survey
Pa Sites	Vector polygon	Old Maori defensive settlement (cultural)	GNSS Survey
Potential midden sites	Vector point	Old Maori household food remains (cultural)	GNSS Survey
Pits	Vector polygon	Old Maori, usually within a pa site (cultural)	GNSS Survey

Pit bottoms	Vector line	Old Maori, usually within a pa site (cultural)	GNSS Survey
Pit drains	Vector line	Old Maori, usually within a pa site (cultural)	GNSS Survey
Boulders	Vector point	Of cultural significance (Maori)	GNSS Survey
Orthophotograph	Raster	Orthophotograph of 0.4 m resolution	New Zealand Aerial Mapping (NZAM)
Buildings	Vector polygon	Farm buildings	Digitised from orthophoto (NZAM)
Contours	Vector line	Lines of equal height or isolines, of 2 m height interval	Digitised from orthophoto (NZAM)
Farm Boundary	Vector line	Boundary of farm	Digitised from orthophoto (NZAM)
Fences	Vector line	Fences on farm	Digitised from orthophoto (NZAM)
Roads	Vector line	Roads crossing or adjoining farm	Digitised from orthophoto (NZAM)
Tracks	Vector line	Tracks crossing farm	Digitised from orthophoto (NZAM)
Streams	Vector line	Streams on and around farm	Digitised from orthophoto (NZAM)
Ponds	Vector polygon	Ponds on and around farm	Digitised from orthophoto (NZAM)
Exotic Bush	Vector polygon	Vegetation of exotic origin	Digitised from orthophoto (NZAM)
Native Bush	Vector polygon	Native (New Zealand) vegetation	Digitised from orthophoto (NZAM)
Digital Elevation Model	Raster	Elevation model of 1 metre resolution	Interpolation from contours

On the ground, a Trimble R8 was used to provide ground control points as well as vector data for a number of on-ground features, to be stored in geodatabase feature class format in the local NZGD 2000 Chatham Islands Transverse Mercator

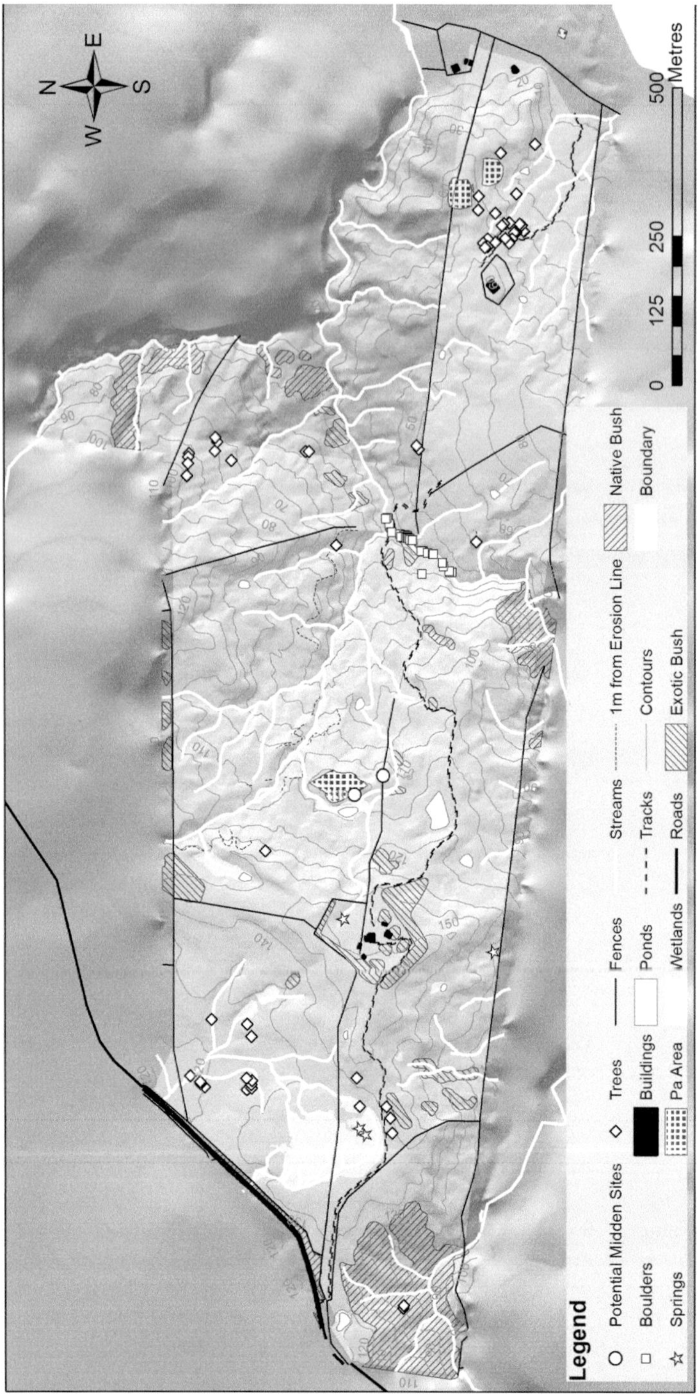

Fig. 5: Context map of Taiporutu farm
(*Source*: Johnson and Perley, 2015; modified by authors)

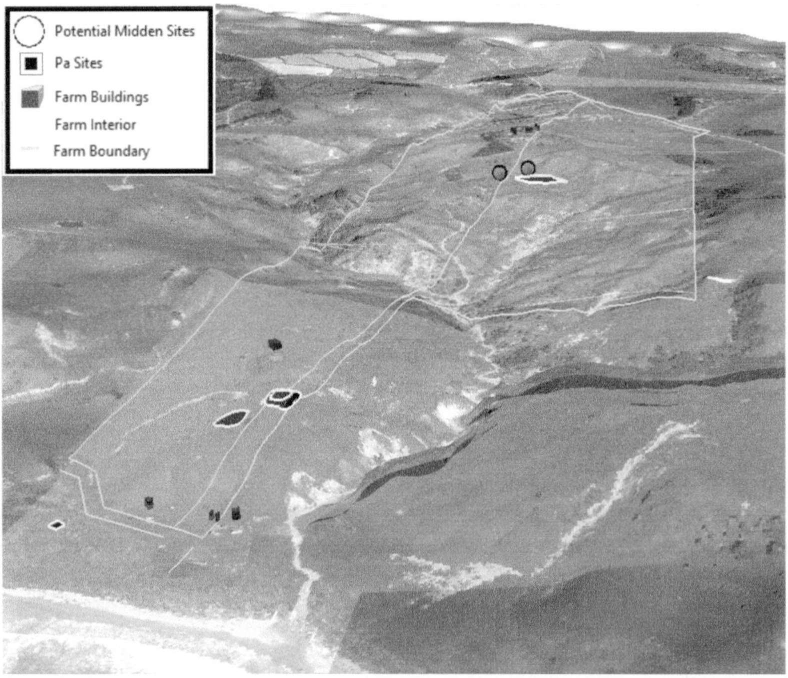

Fig. 6: 3D view of Taiporutu from the northeast, created from elevation (interpolated from NZAM and LINZ Topo 50 contours), orthophoto, boundary, building, pa and potential midden site spatial data (*Source*: Johnson and Perley, 2015; modified by authors)

2000 projection. These are all vector data (point, line, polygon) collected, and did not require any further processing.

This dataset was supplemented by a 30 cm orthophotograph of the farm to fill in areas not covered by the drone. An overview of the Henga spatial dataset can be found in Table 3.

8.3.3.2. Overview Maps

A subset of the data described above is featured in the following overview maps:

- a 2D topographic map (Fig. 7) featuring collected data, both processed (the raster relief layer of the ground surface and derived contours) and as captured (fence lines, buildings, roads, swamp areas, and water trough features). The data was augmented by LINZ Topo 50 data (lakes, drainage, roads);
- a 3D view of the farm from the east with fence boundary and building features, draped with an orthophoto (*Source*: LINZ) (Fig. 8).

Table 3: Henga Spatial Data

Name	Spatial Data Type	Description	Source
Contours	Vector Line	Lines of equal height or isolines, of 3 metre height interval	GNSS Survey
Roads	Vector Line	Local roads	GNSS Survey
Buildings	Vector Polygon	On-farm buildings	GNSS Survey
Power Poles	Vector Point	Overhead powerlines located on the farm	GNSS Survey
Hazards	Vector Polygon	Hazards on the farm	GNSS Survey
Cattle Yards	Vector Polygon	Areas for cattle	GNSS Survey
Swamps	Vector Polygon	Swamp areas on the farm	GNSS Survey
Water Reticulation	Vector Polygon and Vector Line	Network for movement of water	GNSS Survey
Water Storage	Vector Polygon	Fixed water storage (e.g. tanks)	GNSS Survey
Water Troughs	Vector Point	Water source for animals to drink	GNSS Survey
Gates	Vector Point	Gates on the farm	GNSS Survey
Fences	Vector Line	Barrier used to enclose and demarcate the farm area	GNSS Survey
Fence Posts	Vector Point	Isolated fence posts	GNSS Survey
Orthophotograph	Raster	Orthophotograph of 5-7 cm resolution	Drone camera
Digital Elevation Model (DEM)	Vector Point Cloud and Raster	Cloud of located elevation points and derived raster DEM of 5-7 cm resolution	Photogrammetry from orthophoto
Orthophotograph	Raster	Orthophotograph of 30 cm resolution to supplement drone ortho	Geo & Spatial Information Sys. Ltd.

8.4. Using GIS to Plan Agroecological Management

8.4.1. GIS and Geodesign

Geodesign is a planning process that designs changes to the geography of a particular area (Goodchild, 2010; Steinitz, 2012), often using GIS to facilitate this. This marrying of design and geographic approaches has been applied long before

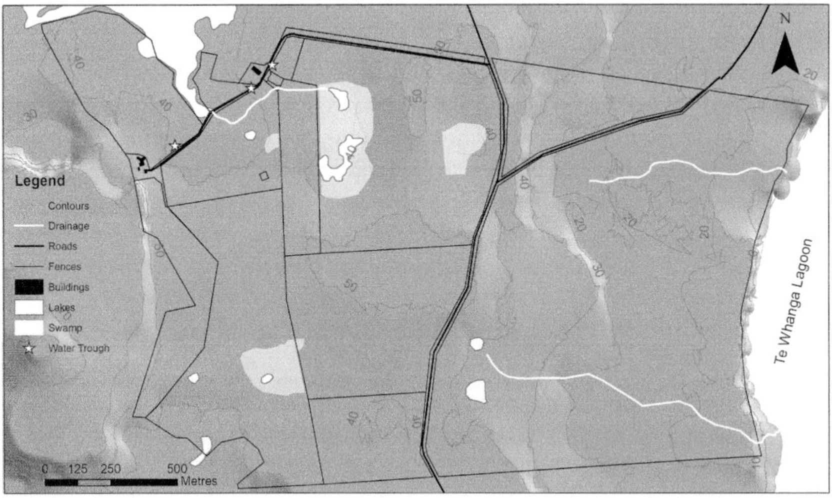

Fig. 7: Context Map of Henga Farm (*Source*: Authors' elaboration)

Fig. 8: 3D View of Henga from the east, created from elevation, orthophoto, fence boundaries, and buildings spatial data (Data source: Neill Glover, Geo & Spatial Information Systems Ltd., except for orthophoto, drainage, lake, and external road data – Land Information New Zealand) (*Source*: Authors' elaboration)

the computer age, but has been greatly facilitated by the coming of GIS, with the 'geodesign' term coming to the fore in the last ten years. Geodesign is based on collaboration within a team consisting of designer(s), geographical scientist(s), technologist(s), and 'the people of the place'. The latter have a key role in needing and providing input into the geodesign process, and making decisions based on the results of geodesign (Steinitz, 2012). The process is iterative; in that if a plan for a particular area is not effective, alternatives are presented to see what other outcomes are possible. Any outcome is judged relative to a group of criteria set by decision-makers. This feedback cycle is implemented, using the constituent groups within the geodesign team. An example of a geodesign project was the proposed wildlife corridor in the Sonoran Desert (Arizona and California), the presentation being augmented by wildlife models and rendered in a 3D environment (Perkl, 2012; Perkl, 2016).

A GIS-aided geodesign process was applied to agroecology planning on Te Kaio and Henga Farms (geodesign was not applied to Taiporutu as the Trust were primarily interested in protection measures and bioremediation for sites of cultural significance rather than an overall agroecological management). In this collaboration, the agroecologists take on the design role, with the geographic and technology roles being assumed by the GIS specialists. The respective Te Kaio and Henga Farm Trusts comprise the community or 'people of the place'.

8.4.2. Agroecology on Te Kaio

Agroecology, integrating ecological, social and agricultural aspects of land management provides an excellent framework for the management of Te Kaio with a particular focus on *mahinga kai*, the Maori term for food which is collected, or other resources that are utilized, for survival. *Mahinga kai* is a cornerstone of culture, and an abundance of food is important for the survival of the people and for *manaakitanga* or hospitality. To help plan the conversion of the farm to an agroecosystem reflecting indigenous agroecology, the farmland has been divided into eight classes, chosen to reflect the land use capability, shown in a GIS-derived map supporting the geodesign process (Fig. 9). The divisions were decided upon in consultation with team members and people associated with the farm and reflected the soils, current state of the land, climate, and the wishes of the community as to how they might restore their land, and what they might like to produce.

8.4.2.1. Intensive Horticultural

Two areas of excellent soils are critical to the provision of *kai* (food) for the community. On these soils, vegetables can be grown to feed local families and traditional crops, such as *kumara* (*Ipomoea batatas*), *kamokamo* (*Cucurbita pepo*), and *taewa* Maori potatoes, such as *tutaekuri* and *kararo* can be grown to keep cultural knowledge alive.

8.4.2.2. Intensive Pasture

Two areas on the farm were identified as being capable of cropping, for example, lucerne or sainfoin. Lucerne is the more traditional crop and in the past was grown very successfully in these areas.

8.4.2.3. Open Grazing

Much of the land on Te Kaio is erosion prone; however, there are areas of stable land, largely along ridge tops. Pastures containing productive deep-rooting grasses (summer dry resistant), clovers, plantain, chicory, and other suitable herbs should be sown. These will provide good feed both for flushing ewes and growing on lambs and calves.

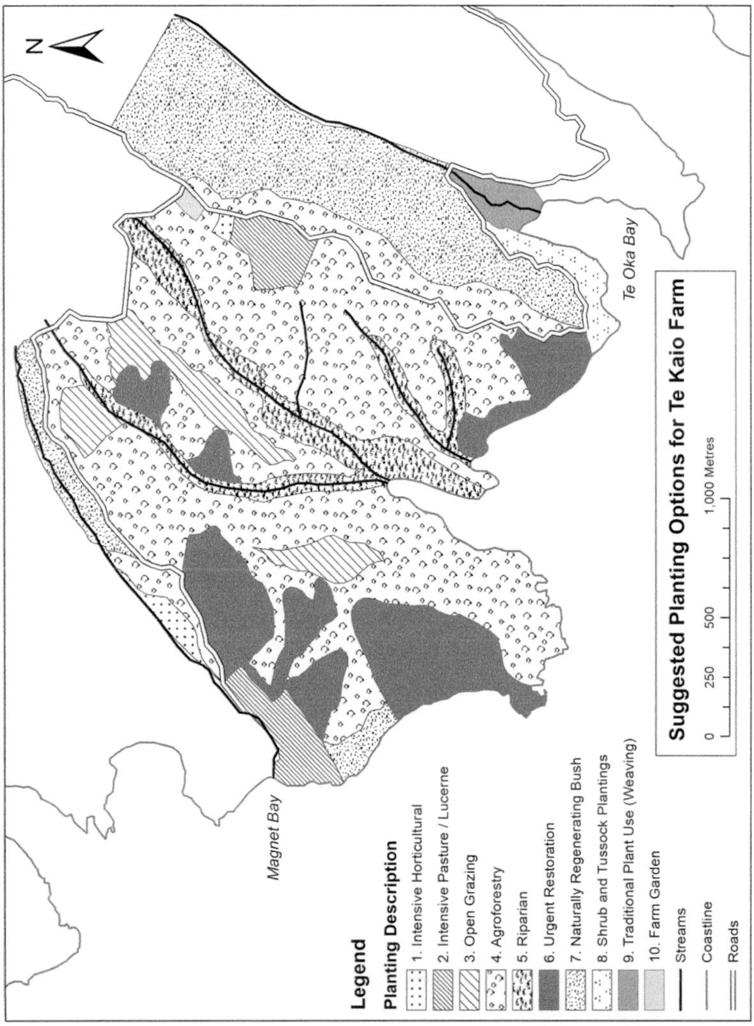

Fig. 9: Map of Te Kaio showing the division of the farm into different management classes (*Source*: Johnson and Moore, 2016; modified by authors)

8.4.2.4. Agroforestry

On the steeper slopes running into gullies the soils require some level of protection. If trees and shrubs are planted, they will help hold the soil; when they have become established, they can be browsed and stock can graze under them. The spacing of the trees should be dictated by species, ground conditions, and their effect on the views from the farm tracks.

Rongoa (traditional medicine) species should be integral to the tree and shrub planting plan, so they can be accessed by the Wairewa community for the treatment of themselves and their livestock and where appropriate animals should be able to access *rongoa* species to self-medicate. The gullies should be thickly planted with shrubs which will help to hold the soil and can be lightly grazed and contribute to animal health, for example, Banks Peninsula *koromiko* (*Hebe strictissima), papapa*, snowberry (*Gaultheria antipoda*) and a range of Coprosma species.

Animals on Te Kaio have been noted to display instinctive healing knowledge, for example, the cattle utilize *maukoro* (*Carmichaelia* sp.) at various times of the year and clearly relish *Harakeke* New Zealand flax (*Phormium tenax*).

In addition to planting *rongoa*, consideration must be given to other species of cultural significance, such as *totara* (*Podocarpus totara*) and also species that will provide habitat for a range of fauna. The management of Te Kaio is committed to protecting *koiorakanorau* (biodiversity) and discussions identified the importance of a number of species to the Wairewa community. To encourage favored birds, such as New Zealand pigeon, *kereru* (*Hemiplegia novae zelandiae*), *tui* (*Prosthemadera novaeseelandiae*) and Fantail *piwakwaka* (*Rhipidura fuliginosa*), fruiting tress and shrubs must be planted to support the invertebrate population. To ensure the breeding success of resident bird populations, food must be available all year round through the provision of a layered range of different plant species, which provide sustenance in sequence, and microfaunal habitats.

8.4.2.5. Riparian Planting

The planting of trees, shrubs, forbs, and grasses on the borders of waterways contributes to the maintenance of water quality and ensures good habitat for aquatic flora and fauna. There are four creek systems associated with Te Kaio – two on the boundary and two internally. The boundary creeks flow through native bush and would only require protection in their lower reaches. The riparian margins of the internal creeks should reflect the surrounding plantings and could include fruiting shrubs, such as fuchsia (*Fuchsia excortica*) and *kaka beak* (*Cliantha puniceus*), providing food sources for native birds (Cunningham, 2012).

8.4.2.6. Temporarily Retired Areas

The badly eroded and actively eroding areas of the farm should be fenced off, stock permanently excluded, and planted in pioneer soil-healing species, such as tree lucerne (*Chamaecytisus palmensis*) or Matagouri (*Discaria tomato*) and in bird attractant species, such as wineberry (*Aristotelia serrata*). The tree lucerne, can be gently harvested for animal fodder if necessary.

8.4.2.7. Naturally Regenerating Bush

The two east-facing ridges on the boundaries of the farm (Magnet Bay Creek and Tumbledown Creek) have large areas of bush remaining on them. These areas should be fenced and left to regenerate, but can act as feed banks for times of drought, and for emergency shelter, and feed in very bad weather.

8.4.2.8. Conservation Areas

Conservation areas should surround public access to Magnet and Tumbledown Bays. There is a small car park for the public on the approach to Magnet Bay, from where surfers walk to the beach. Vegetation-wise, the space could be planted with weaving species, providing a resource for the Wairewa community and a point of interest for tourists. Species could include a range of cultivars of New Zealand flax *harakeke* (*Phormium tenax*), *ti kōuka* cabbage tree (*Cordyline australis*), and perhaps *raupo* (*Typha orientalis*) to stabilize the meanders in the creek. Golden sand sedge *Pingao* (*Ficinia spiralis*) would be a vital component of a planting mix. *Pingao*, a culturally-important sedge was used in ceremonies, for weaving mats, garments, and utensils and also as a food source. In many areas it has been out-competed by the introduced *marram* grass and is now declining.

8.4.3. Agroecology on Henga

See Fig. 10 for the mapped plan supporting GIS-based geodesign of Henga Farm on the Chatham Islands. Hokotehi Moriori Trust invited us to Rekohu, Chatham Island, to visit Henga Farm and the Moriori Ethnobotanical garden. We were introduced to Moriori history and life stories and working with members of the Trust, developed a draft indigenous agroecology plan for Henga. The farm was divided into seven areas reflecting the aspirations of the community, conservation, and production values, and the carrying capacity of the land. As all the produce has to be shipped off the island or air freighted, the key to the plan was to find local or locally adapted plants and breeds that would have a high value in the market and thus offset the costs of export.

8.4.3.1. Indigenous Plantings and Feral Sheep

A proportion of Henga Farm is infested with gorse and many of the paddocks also have dips, hollows, and gullies, so constant vigilance is required for stock safety and there is little suitable feed for the flock of Romney sheep currently run on the property.

Rather than try to eradicate the gorse by costly, imported chemical means, it will be better to plant trees and shrubs amongst it and allow them to grow up and

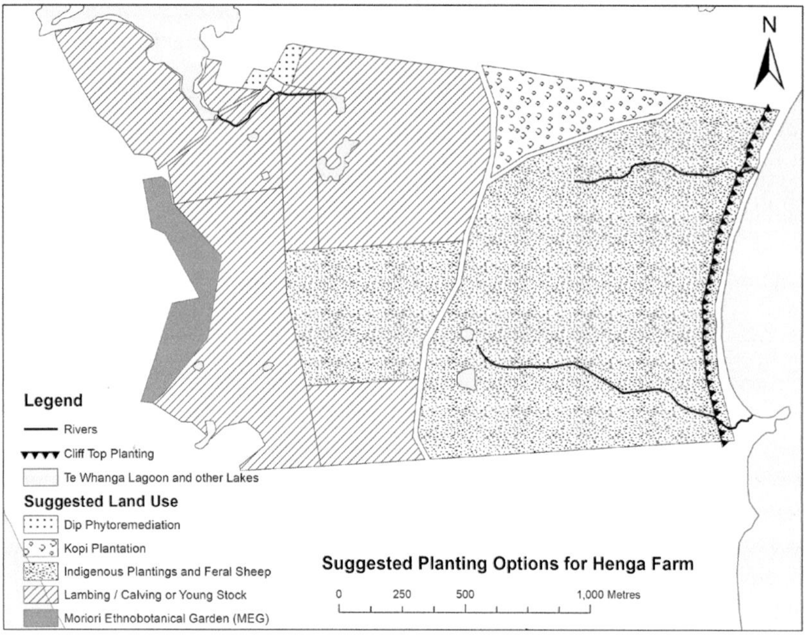

Fig. 10: Map of farm management suggestions for Henga (*Source*: Johnson and Perley, 2015; modified by authors)

shade the gorse out. This process has been successfully demonstrated at Hinewai Reserve on the Banks Peninsula (Wilson, 2002). In a checklist of vascular plants recorded from the Chatham Islands (de Lange *et al*., 2011) the authors provide a record of endemic, indigenous, and introduced plants, which can guide planting. But local knowledge will be the best source of advice on what will grow on the site. In the spirit of trying to conserve as much endemic and indigenous Chatham island flora as possible under planting, the gorse is sensible but some income should be generated from the land. Pitt Island sheep, rugged animals well used to foraging for themselves, would be an alternative to the current Romney breed. There is a good market for feral/wild sheep meat, Pitt island sheep being prized for 'sweet and lean" meat (Rudge, 1983) and they would attract a premium. The Eastern Buff weka (*Gallirallus australis hectori*) was introduced into the islands in 1905 (R&N Beattie Partnership, 2021). They have thrived as there are no predators and many now view them as a pest. It is legal to hunt weka on the islands and numbers are taken each year. However, it is not legal to export the carcass to Aotearoa New Zealand mainland without a permit from the Department of Conservation. Weka are found all through MEG and Henga already and would provide another income stream if they could be exported from the island ready to eat as a 'wild' food.

8.4.3.2. Cliff Top Planting

The lagoon fence on the eastern side of Henga borders cliffs above the lagoon. This would be an ideal site for the regeneration of Lepidium and other herbaceous species, free from stock challenge. *Lepidium rekohuense*, Chatham Islands scurvy grass, found only in the Chathams is now in near terminal decline (de Lange, 2020) and unless steps are taken to establish populations *in situ* (it is hard to cultivate), it will become extinct.

8.4.3.3. The Moriori Ethnobotanical Garden (MEG)

This 12-hectare garden adjacent to Henga scenic reserve serves as a living memorial to the Moriori way of life before the arrival of other peoples and the subsequent changes in society. The garden is a refuge, not only containing plants for food, but also materials for housing, making fire, weaving, fishing, or boat building. The garden nursery can serve as a source for a wide range of cultural plantings for the adjacent farm land.

8.4.3.4. Kopi Plantation

The triangle between the north road and the airport road is also gorse covered but is potentially suitable for *kopi* (*Corynocarpus laevigatus*), a tree revered by Moriori. *Kopi* is deeply intertwined with cultural beliefs and has been used for food, shelter, and gatherings.

Farmers on the island attest to the palatability of *kopi* leaves to stock. Throughout New Zealand in areas where *kopi* (known as *karaka* on the mainland) grows, it is universally acknowledged as a good stock food and a 'medicine' to be given when an animal is unwell. For humans, the *kopi* fruit is a healthy food option, being gluten free with high dietary fiber, a higher energy content than chestnuts, and a fat content similar to walnuts. *Kopi* must be prepared carefully as the kernel is toxic but the fruit can be made into leathers, health bars, liqueurs, and other high-value comestibles.

8.4.3.5. Lambing/Calving or Young Stock

The paddocks along the western boundary of Henga have been fertilized and sown down with improved grasses. The western paddocks might either hold lambing ewes and calving cows or could be used to fatten young stock. Chatham Island wool is sought after (2010) as it is clean and white with a good yield. If the value of wool increases, Romneys could be run on the improved land and Pitt Island sheep on the balance.

8.4.3.6. Phytoremediation

The area around the old sheep dipping tank is likely to be contaminated and should be retired. Many of the early dips contained chemicals harmful to human health, such as arsenic, organochlorides, and organophosphates. These chemicals are persistent in the environment and exposure can harm the health of livestock and humans (Ministry for the Environment, 2006). Exposure usually occurs through ingestion of contaminated soil or produce grown on the site. Additionally, water can be contaminated if run off occurs. After testing for type and level of contaminants, the site can be securely fenced and planted with species, such as willow which are appropriate for the phytoremediation required.

8.4.3.7. Horticulture

A small vegetable unit would provide vegetables for locals and for Moriori returning to the *marae*. It is expensive to purchase imported fruit and vegetables in the local shop and the costs of living on top of the airfare could prove to be a barrier to return for many. The vegetable unit should be sited in a convenient sheltered location with water. There is already a large pear tree in the yards; more fruit trees could be planted in sheltered areas.

8.5. Two Spatial Analysis-based Geodesign Projects on Te Kaio

We have been introduced to three rich examples of Maori and Moriori Trust rural land management. In such operational scenarios, geographic information systems (GIS) as a whole (i.e. spatial analysis as well as mapping) can be a valuable tool to (geo)design and support important decisions that have community- wide implications. Fundamentally, it is easier to visually demonstrate the possible and likely outcomes of spatial change to decision-makers and the public with the help of a GIS than it is to attempt to express the effects of change with words.

8.5.1. Spatial Analysis for Tree Plantings Supporting Rongoa

The managers of the Te Kaio, having viewed the maps and analysis, were keen to plant areas of the farm. Logically the species chosen should be native and ideally of cultural significance. Many species that are used in Maori traditional medicine can be grown on farm to provide a cultural resource and many can be utilized by livestock (Johnson, 2012). In the absence of expertise, where are the best places on the farm to grow the different species? To help identify suitable planting zones, a GIS-based (i.e. spatial) multi-criteria analysis (MCA, Malczewski, 1999) of terrain and proximity was implemented (Moore *et al.*, 2016). This methodological approach uses local botanical knowledge as well as scientific terrain and proximity data inputs to calculate potential optimal areas of growth for traditional medicinal plants. Specifically, spatial MCA is a complex overlay procedure, with competing objectives and their criteria weighted for importance before combination (Malczewski, 1999). Three botanical experts were consulted independently for these weightings, yielding a consistent knowledge-base.

Seven species were modeled in the initial programme, *totara* (*Podocarpus totara*), New Zealand flax *harakeke* (*Phormium tenax*), wine berry *mapou*

(*Myrsine australis*), *kawakawa* (*Piper excelsum*), cabbage tree *Ti kouka* (*Cordyline australis*), *manuka* (*Leptospermum scoparium*), and *kahikatea* (*Dacrycarpus dacrydioides*). The critical conditions for these species were the likelihood of waterlogging, salt wind exposure, wind exposure, frost, drought, and proximity to infrastructure. Each condition had its own set of criteria, for example, waterlogging had aspect and slope criteria (surrogate criteria that could be extracted from terrain data, as no waterlogging data was available). The analysis yielded very useful results (Fig. 11). The resulting map shows where the seven species are likely to grow on Te Kaio. It remains to verify the approach empirically on farm, and there is a need to build a more comprehensive decision-support resource for farm management, with further plant species needing to be modeled, informed by a broader dataset that includes detailed soil data.

8.5.2. Modeling Tourist Access and Use of the Farm

This example of geodesign features the integration of agroecology and tourism on Te Kaio (Moore *et al.*, 2018). An iterative geodesign framework included GIS-implemented spatial analysis to plan a route for visitor access across the farm, with suitable areas identified for information platforms located strategically along the way (Fig. 12). Informed by local agroecological and geospatial expert knowledge, the route was calculated by cost-path analysis (Chang, 2015), with a route devised so as to minimize exposure to steep slopes, damp gullies, and limited landscape views. Terrain data alone is sufficient to facilitate calculation of the required slope, viewshed (i.e. what areas can be seen from any point), and distance to stream line data layers, respectively. A follow-up check of the route ensured that a wide variety of *rongoa* species were travelled through (Figs. 11 and 12). Optimally-sized flat, or near-flat areas (extracted from the slope data layer) were identified as candidates for information platform sites. Visual communication of the results through maps is another GIS-generated component of this geodesign process. This agritourism example promises easy portability to other farm examples, facilitated by increasing availability of pertinent spatial data, free GIS, and local knowledge. The visualization aspect could be enhanced with adaptation into a 3D virtual environment (i.e. a virtual geographic environment – VGE; Gong and Lin, 2006). However, as with the *rongoa* geodesign project, uncertainties need to be addressed, associated with limitations of existing data, the possible vagueness of local knowledge, and missing data (i.e. soils) and knowledge.

8.5.3. Contextualized Description of the Spatial Analysis Outcomes

The farm trustees have had preliminary discussions with other land owners as to the feasibility of a southern bays track, similar to the very popular Banks Peninsula track (Banks Track, 2021). The planning, negotiations, and building of a multi-property track is likely to be a lengthy process. In the interim, a track, based on output from the analytical procedure presented above, could be constructed on

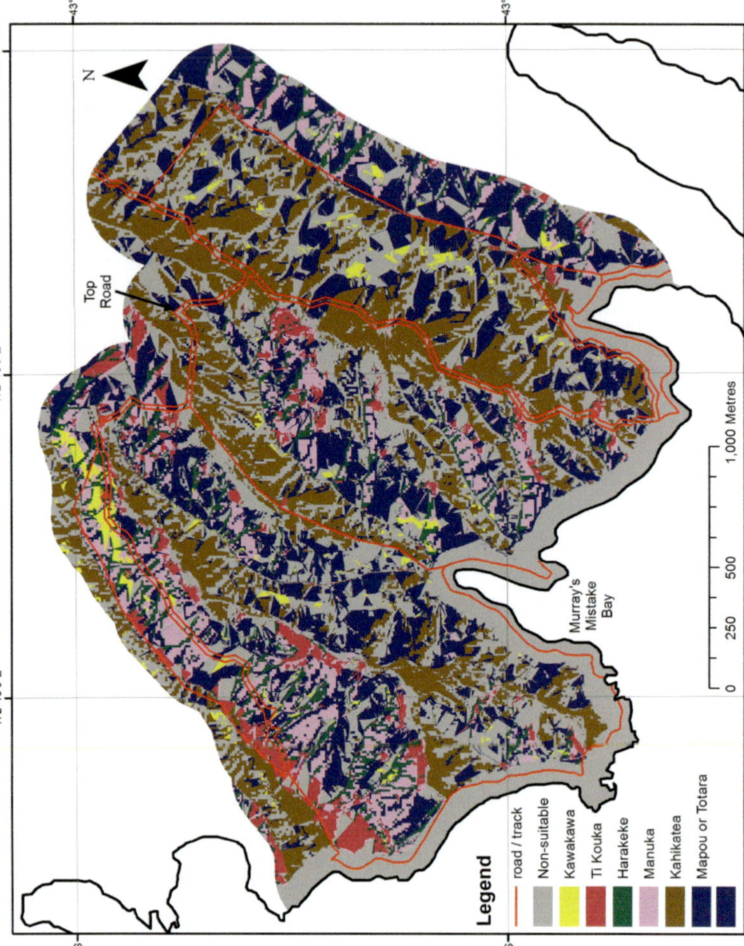

Fig. 11: Map showing where seven rongoa species might be grown on Te Kaio. (*Source*: Moore *et al.*, 2016)

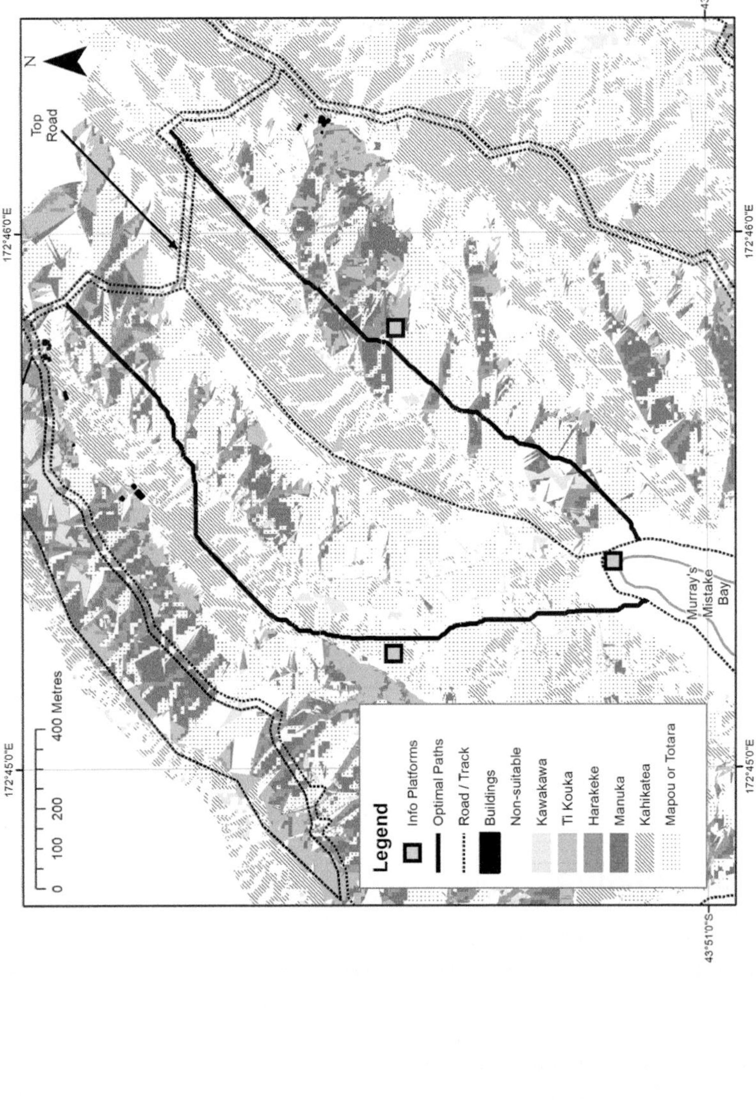

Fig. 12: Map showing the proposed walking tracks (optimal paths) and information platforms, as well as infrastructural features and tree vegetation model results from Fig. 11 (*Source*: Moore *et al.*, 2018; modified by authors)

Te Kaio to cater to walkers, with posts or platforms (giving the visitor local-agroecology information) situated along the track (Fig. 12).

A short walk from the garden brings the visitor to a superb viewpoint, looking out across the farm to Kaitorete spit, a narrow neck of land separating Te Waihora, Lake Ellesmere from the ocean to the southwest of the farm. The spit is a haven for bird life, an increasingly rare wild population of shrubby *tororaro* (*Muehlenbeckia stoinis*) and several species of endemic moth.

Walkers leaving the farm boundary would descend to the private beach by following the eastern arm of the creek. The path passes above regenerating bush and alongside developing agroforestry blocks, which are being planted with trees and shrubs that will provide traditional food, medicine, dyes, oils, and bark. The path descends to the creek and crosses to the eastern bank. The bridge at the crossing point could be crafted locally, reflecting traditional stories. Where the bridge crosses the creek, there is a large open area which could be planted in *totara*, a culturally important species which once clothed much of the peninsula and old *totara* fence-posts can still be found on Te Kaio. The plantings would reflect the natural history of the farm and provide a valuable resource in the future, *totara* for traditional carving, and for sale. Stock will also use this bridge to cross from one side of the farm to the other, so an alternative rope bridge might also be provided for the adventurous. The path now passes through large areas of riparian planting which is protecting the waterway. The riparian margins here are planted with *rongoa rakau* (plants that are used for traditional medicine). Members of the community can harvest these plants, keeping traditional knowledge alive and livestock can access them for self-medication. The path passes above rocky hollows in which a number of native species have survived, including small areas of *kawakawa*. As the path approaches Murrays Mistake, it passes through the restored wetland area overlooked by a number of local artworks. At Murrays Mistake, there is a small memorial cairn to those aboard the ship that was wrecked when the captain put in to the bay many years ago, thinking he was navigating towards the Oashore whale fishery (Jacobsen, 1914). Leaving the beach, the path ascends the ridge through more agroforestry plantings with stunning views out across the farm to the sea. As the walker nears the farm boundary once again, the path drops down to the creek passing through further *rongoa* riparian plantings and passing an old lime (*Tilia* sp.) tree often frequented by *kereru*.

8.6. Conclusion

This is an account of surveying and mapping by using GIS, in support of the indigenous agroecology He Ahuwhenua Taketake project. Three New Zealand farms were featured – Te Kaio on Banks Penninsula, Taiporutu on Mahia Peninnsula, and Henga in the Chatham Islands. A mixture of natural (e.g. terrain), infrastructural (e.g. buildings), and cultural (e.g. pa sites) data was collected, derived, or collated. The purpose of this data was to provide a baseline for the other scientific and community indigenous agroecology activities, and to form a complement to other cultural mapping practices. Specifically, they have supported geodesign processes: using GIS to map concept agroecological plans on Te Kaio and Henga Farms and for Te Kaio, planting locations of *rongoa rakau* (plant species used in Maori traditional medicine and that can facilitate self-medication of livestock), and to define location of tourist infrastructure and access paths. The next stages will involve further exploration of how scientific and cultural mapping can unite for an enhanced capturing of farm geography and history.

Acknowledgments

Thanks to:

1. Nga Pae o te Maramatanga for project funding. Part of this account is derived from the project report (Johnson and Perley, 2015).
2. Brent Hall, Janice Lord, Aubrey Miller.
3. The Te Kaio, Henga and Taiporutu communities, especially Robyn Wybrow, Maui Solomon, and Desna Whaanga-Schollum.
4. Members of the Indigeneous Agroecology Project Group, Hayden Hamilton.
5. Phil Rhodes, Tim Hastings, Riki Cambridge.
6. Mariana Pagan, Sam Coutts, Jeremiah Gbolagun (also for his geodatabase protocol which formed much of the account of Te Kaio data).
7. Anneke Rombouts and Loes van der Ven.
8. Richard Hemi, Sam Mogford.
9. Michael Fletcher (Environment Canterbury), Iain Gover, Alan Mark, Roger May, and Colin Muerk.
10. Neill Glover, Mike McConachie.

Bibliography

Altieri, M. (1995). *Agroecology: The Scientific Basis of Alternative Agriculture*, Westview Press, Boulder, Colorado, USA.

Altieri, M. (2002). Agroecology; The science of natural resource management for poor farmers in marginal environments, *Agriculture, Ecosystems and Environment*, 93: 1-24.

Altieri, M. (2004). Linking ecologists and traditional farmers in the search for sustainable agriculture, *Frontiers in Ecology and the Environment*, 2: 35-42.

Altieri, M. and V. M. Toledo (2011). The agroecological revolution in Latin America: Rescuing nature, ensuring food sovereignty and empowering peasants, *The Journal of Peasant Studies*, 38: 587-601.

Altieri, M., A.K. Bartlett, C. Callenius, C. Campeau, K. Elsasser, P. Hagerman, G. Kenny, K. Lambrechts, W. Miga, J.P. Prado, P. Prove, N. Saracini and K. Ulmer (2012). Nourishing the World Sustainably: Scaling up Agroecology, Ecumenical Advocacy Alliance; Retrieved from: http://www10.iadb.org/intal/intalcdi/PE/2013/10704.pdf; accessed on 16 March, 2021.

Banks Track (2021). Banks Track/Akaroa; Retrieved from: www.bankstrack.co.nz; accessed on 16 March, 2021.
Brady, K.B. (2004). Māori Land Administration: Client Service Performance of the Māori Land Court Unit and the Māori Trustee. Report of the Controller and Auditor General Tumuaki o te Mana Arotake; https://oag.parliament.nz/2004; accessed on 14 September, 2020.
Cassel, S.F. (2007). Soil salinity mapping using ArcGIS, pp.141-162. *In:* F.J. Pierce & D. Clay (Eds.). *GIS Applications in Agriculture*, CRC Press, Boca Raton, USA.
Chang, K.-T. (2015). *Introduction to Geographic Information Systems* (8th edition), McGraw Hill Higher Education, Columbus, USA.
Cunningham, S. (2012). Report on the Establishment and Enhancement of Native Biodiversity on TePutahifarm, Banks Peninsula, Christchurch, University of Otago, NZ.
de Lange, P.J., P.B. Heenan and J. Rolfe (2011). Checklist of Vascular Plants Recorded from Chatham Islands, Wellington, Department of Conservation; Retrieved from: https://www.doc.govt.nz/globalassets/documents/conservation/native-plants/chatham-islands-vascular-plants-checklist.pdf; accessed on 16 March, 2021.
de Lange, P.J. (2020). Lepidium rekohuense Factsheet New Zealand Plant Conservation Network; www.nzpcn.org.nz/flora/species/lepidium-rekohuense; accessed on 27 April, 2020.
Gong, J. and H. Lin (2006). Collaborative virtual geographic environment, pp.186-207. *In*: S. Balram and S. Dragicevic (Eds.). *Collaborative Geographic Information Systems,* IGI Global, Hershey, USA.
Goodchild, M. (2010). Towards geodesign: Repurposing cartography and GIS? *Cartographic Perspectives*, 66: 7-22.
Hanson, H.C. (1939). Ecology in Agriculture, *Ecology*, 20: 111-117.
Harmsworth, G., M. Park and D. Walker (2005). Report on the Development and Use of GIS for iwi and hapu: Motueka Case Study, Aotearoa - New Zealand. Landcare Research NZ Ltd, 33 pp.; Retrieved from: https://wwwuat.landcareresearch.co.nz/__data/assets/pdf_file/0008/39968/Development_use_GIS_Motueka.pdf; accessed on 16 March, 2021.
Jacobsen, H.C. (1914). *Tales of Banks Peninsula Outside the Harbor*, Akaroa Mail Office, Akaroa, NZ.
Johnson, M. (2012). Report on Adapting the Principles of Te Rongoa into Ecologically and Culturally Sustainable Farm Practice; Retrieved from: http://www.maramatanga.co.nz/project/adapting-principles-rongo-ecologically-and-culturally-sustainable-farm-practice; accessed on 16 March, 2021.
Johnson, M. and C. Perley (Eds.) (2015). Report on Indigenous Agroecology TeAhuwhenuaTaketake, University of Auckland NZ, Nga Pae o teMaramatanga; Retrieved from: http://www.maramatanga.co.nz/sites/default/files/project-reports/Indigenous Agroecology FINAL Report 2015-Johnson.pdf; accessed on 16 March, 2021.
Johnson, M. and A. Moore (2016). Parasites, plants and people. *Trends in Parasitology*, 32: 430-432.
Landcare Research, New Zealand, Ltd. (2013). Methods for Recording Maori Values on GIS; Retrieved from, http://www.landcareresearch.co.nz/science/living/indigenous-knowledge/land-use/recording-values.
Malczewski, J. (1999). *GIS and Multi-criteria Decision Analysis*, John Wiley & Sons, New York, USA.

Ministry for the Environment (2006). *Identifying, Investigating and Managing Risks Associated with Former Sheep-dip Sites: A Guide for Local Authorities*, ME 775. Ministry for the Environment, New Zealand.

Moore, A., M. Johnson, J. Lord, S. Coutts, M. Pagan, J. Gbolagun and G.B. Hall (2016). Applying spatial analysis to the agroecology-led management of an indigenous farm in New Zealand, *Ecological Informatics*, 31: 49-58.

Moore, A., M. Johnson, J. Gbolagun, A. Miller, A. Rombouts, L. van der Ven, J. Lord, S. Coutts, M. Pagan and G.B. Hall (2018). Integrating agroecology and sustainable tourism: Applying geodesign to farm management in Aotearoa, New Zealand, *Journal of Sustainable Tourism*, 26: 1543-1561.

Perkl, R. (2012). *Geodesign and Wildlife Corridors*, ArcNews (Summer); Retrieved from: https://www.esri.com/news/arcnews/summer12articles/geodesign-and-wildlife-corridors.html; accessed on 16 March, 2021.

Perkl, R. (2016). Geodesigning landscape linkages: Coupling GIS with wildlife corridor design in conservation planning, *Landscape and Urban Planning*, 156: 44-58.

Pierce, F.J. and D. Clay (2007). *GIS Applications in Agriculture*, CRC Press, Boca Raton, FL.

R&N Beattie Partnership (2021). Pioneering Easy Care Ethical Practice 'Beyond Organics'; Retrieved from: www.rnbeattie.co.nz; accessed on 16 March, 2021.

Rudge, M.R. (1983). A reserve for feral sheep on Pitt Island, Chatham group, New Zealand, *New Zeal. J. Zool.*, 10: 349-363.

Steinitz, C. (2012). *A Framework for Geodesign*, Esri Press, Redlands, USA.

Tomich, T.P., S. Brodt, H. Ferris, R. Galt, W.R. Horwarth, E. Kebreab and L. Yang (2011). Agroecology: A review from a global-change perspective, *The Annual Review of Environment and Resources*, 36: 193-222.

Weiner, D., T.M. Harris and W.J. Craig (2002). Community Participation and Geographic Information Systems, pp. 3-16. *In:* W.J. Craig, T.M. Harris and D. Weiner (Eds.) *Community Participation and Geographic Information Systems*, Taylor & Francis, London, UK.

Wilson, H. (2002). *Hinewai: The Journal of a New Zealand Naturalist*, Shoal Bay Press, Christchurch, NZ.

Wilson, J.P. (1999). Local, national, and global applications of GIS in agriculture, pp. 981-998. *In:* P.A. Longley, M.F. Goodchild, D.J. Maguire and D.W. Rhind (Eds.). *Geographical Information Systems: Management Issues and Applications*, John Wiley & Sons, Hoboken, USA.

CHAPTER

9

Smart Cities and Agroecology: Urban Agriculture, Proximity to Food and Urban Ecosystem Services

Francesca Peroni[1]*, John Choptiany[2] and Samuel Ledermann[3]

[1] PhD Programme in Historical, Geographical, Anthropological Studies, University of Padova, Italy

[2] Farmbetter, United Kingdom

[3] Elliott School of International Affairs, George Washington University, USA

9.1. Introduction: Smart Cities and Urban Agriculture

One of the most inspiring models of urban areas of the future is smart cities (Vanolo, 2016). Smart cities are usually considered efficient ecosystems, where urban processes, resources, and services are optimized and ameliorated (Carvalho, 2015). A unique and clear definition is already not specified; however, the concept generally refers to the integration of technology and data collection into urban context to monitor, manage, and regulate flows (Hollands, 2008; Maye, 2019).

Recently, the concept of smart city has evolved and it is always more frequently related to climate change issues (Angelidou, 2015). Cities are ecosystems of high consumption of energy and resources and they are sources of pollution and contamination due to the high concentration of people (Seto *et al.*, 2012). In addition, it is estimated that by 2050 almost 70 per cent of the world population will live in urban areas, resulting in increased pressure on urban ecosystems (de Amorim *et al.*, 2019). In this regard, the concept of smart cities is at present shifting towards a focus to achieve 'greener' cities, i.e. planned to reduce greenhouse gas emissions, and to promote energy efficiency through adoption of appropriate technologies (Gabrys, 2014).

*Corresponding author: francesca.peroni@dicea.unipd.it

It is worth noting that the components and topics usually developed in a smart-city strategy try to integrate various systems of an urban area; for instance, economy, transportation, and education. According to a literature review on smart cities provided by Albino *et al.* (2015), four aspects come to light: (i) political efficiency and social/cultural development; (ii) urban growth; (iii) social inclusion of citizens; and (iv) environment (Albino *et al.*, 2015). The literature gives significant attention to topics related to the quality of life and environment; nevertheless it does not mention questions like urban food systems and food security (Maye, 2019). Surprisingly, the issue of 'urban food' and how to integrate new technologies to feed a growing population is not a key topic in the smart city debate; moreover, the definition of smart cities does not include any direct references to food production.

On the other hand, academic research as well as the social demands on urban agriculture (UA) is constantly growing. Based on the literature review provided by Artmann and Sartison (2018), UA is a surging topic, especially since 2013. Over the last few years, UA has been considered a key issue in steadily increasing global populations and the ongoing expansion of cities. Indeed, it is estimated that by 2050, the world's population will be 9.7 billion and the supply of food will become a critical issue (Fouilleux *et al.*, 2017; UN Department of Economic and Social Affairs, 2019). Moreover, new urbanization and infrastructure are steadily growing and the pressure on semi-natural soil and peri-UA is drastically increasing. Hence, the lands allotted for food production are decreasing, especially near cities (Eigenbrod *et al.*, 2011). As a consequence, policy makers and urban planners have gradually identified urban areas as a fundamental part of the food system and, especially in Western countries, food policies have been developed to relocate some areas of food production into cities (Sonnino, 2009; Veolia Institute, 2019). Moreover, UA provides other important benefits to citizens, especially in developed countries by increasing social inclusion and equity in the communities involved in the process, as discussed later in the chapter (Poulsen, 2017).

However, a direct link already exists between smart cities, information and communication technologies (ICTs) and agriculture. For nearly twenty years, precision agriculture has been growing in prominence as a farming-management concept. Precision agriculture is related to the use of digital technologies, such as unmanned aerial vehicles (UAV), autonomous machines, global positioning systems (GPS), and geographic information systems (GIS) (Griffin and Lowenberg-Deboer, 2005). The main goals of precision agriculture are to: (i) enhance and optimize agricultural production; (ii) contribute to food security; (iii) minimize the impact of conventional agriculture on the environment and wildlife (Norton and Swinton, 2000; Gebbers and Adamchuk, 2010; Whelan and Taylor, 2013; European Parliament, 2014; Walter *et al.*, 2017; Moreno *et al.*, 2019; de Amorim *et al.*, 2019). This farming system is rather common and has already been applied in different parts of the world to various sectors of agriculture, e. g. viticulture, horticulture, and livestock production (Gebbers and Adamchuk, 2010). Much of the literature has considered precision agriculture more sustainable than conventional agriculture. However, ICT solutions in agriculture could cause some negative consequences (Serbulova *et al.*, 2019). The high cost of these technologies could lead to exclusion of vulnerable people in developing countries or in low-income communities (Zhang *et al.*, 2002; Dobermann *et al.*, 2004). Moreover, despite the high capacity of new technologies to decrease and limit agrochemical input use, such as pesticides and fertilizers, the pressure on the environment continues. As a consequence, many researchers suggest that the turning point would be an agroecological approach, in order to achieve more sustainable ways to produce food (Duru *et al.*, 2015).

However, compared with precision agriculture, agroecology does not usually integrate ICTs. Given this background, it is clear that there is a gap between smart cities and urban food systems, even if smart cities and agriculture already present a connection in the precision farming debate. Hence, the aim of this chapter is to investigate the two research fields (smart city and agriculture) and to analyze which possible links could be achieved between these main topics. By doing so, the purpose is to support an idea of sustainable urban agriculture that aims to reach environmental and social sustainability in cities through the integration of ICTs. In other words, the vision is to combine the debate between smart cities, food systems, and UA by adopting agroecological approaches. To develop these arguments, the chapter begins by reviewing the smart city and urban agriculture concepts respectively. Later, it analyzes some innovative examples of integration of ICTs in urban agriculture and finally, it focuses on a specific example, named 'farmbetter'.

9.2. Are Smart Cities Inclusive Urban Environments?

The debate around smart cities has covered more than twenty years and at present many cities could be labelled as smart cities, due to the integration of ICTs in their urban environment (Hollands, 2008; Allwinkle and Cruickshank, 2011). The most famous forerunners are Singapore, San Diego, San Francisco, and Amsterdam and there are also examples of smart cities built from scratch, like Songdo (South Korea), Masdar City (United Arab Emirates), and PlanIT Valley (Portugal) (Arun and Teng Yap, 2000; Vanolo, 2016; Mullins, 2017). However, over the last few years, many critics have been moved against this urban development model. The critics are concerned with two main aspects: on one hand, there is a strong presence of companies and big societies in smart city development projects, e.g. IBM, Cisco, Google, Sidewalk Labs, and Alphabet (Wu and Lindasay, 2020); on the other hand, there is an undefined role of citizens and citizenship within smart cities (Hollands, 2008).

Companies and entrepreneurs occupy a key role in the development of these technologies and, at the same time, they are producers of ICTs and amplifiers of the concept of smart city. Particularly, they become indispensable actors of smart city debate (Paskaleva, 2011; Söderström *et al.*, 2014). One of the most famous examples is IBM, which, in 2011, officially registered the trademark 'smarter cities' and began a campaign to spread the concept around the world. The campaign highlights the most critical issues that cities are facing, such as the increase in urban populations, aging infrastructure, and pollution of urban environments. The initiative suggested that these problems need more than traditional solutions to be solved. As a result, IBM proposed technological alternatives as the unique medicine of urban problems and the company gained a hegemonic position on other IT societies, animating the debate on smart cities (Söderström *et al.*, 2014). Generally, even the more dominant role of private companies in the public debate about smart cities is causing a lot of concern. The first issue is related to the administration of public services which are managed by private companies (Hollands, 2008). Second, the dependency of cities on technologies provided by companies is increasing in relation to the use of particular technological platforms or devices (Bates, 2012; Hill, 2013). Finally, smart solutions provided by these ICT companies do not account for the uniqueness of places, peoples, and cultures and they are similarly spread to other cities (Townsend, 2014). Moreover, citizens usually occupy a defined role as part of communication systems: they are often described as sensors aimed at collecting data in continuous interactions with sensing technologies used for environmental monitoring and feedback (Gabrys, 2014). ICTs enable city dwellers to track their processes, consumption, and monitoring of their life and, in this way, they feel they are an active and fundamental part of smart cities. However, the language of smart city is not always inclusive and many questions are growing on the role of citizens within these innovative urban systems. Some authors highlight that smart cities are usually not considering social inclusion strategies, participation processes, and empowerment of urban dwellers (Carvalho, 2015; Wiig, 2016). The management and the services of cities are directly regulated by city administrations, often in association with private companies, following a top-down approach (Cardullo and Kitchin, 2019). Indeed, the debate does not include the voices, aspirations, and desires of all citizens and there is a lack of connection with society and people (Vanolo, 2014). Smart cities appeared to be more addressed to solving urban issues by adopting market-led solutions rather than fostering civil, social, and political rights (Swyngedouw, 2016).

9.3. Urban Agriculture: The Living Space of Urban Food Production and Social Inclusion

Worldwide, UA provides different benefits to humankind and their well-being. It is a complex issue that embodies various urban challenges. Indeed, UA aims not only to achieve food security, but is a multifaceted phenomenon that seeks also to enhance social cohesion, equity, education, and mitigation of extreme events (Veolia Institute, 2019). In this regard, there are several definitions of UA.

According to Viraj Puri, CEO of Gotham Greens, UA is defined as 'reconnecting with the community through food, jobs and economic development' (Baltimore Sun, 2018, p. 1) UA is also described as 'growing food in cities' (Taylor and Lovell, 2012, p. 57) and as 'improving the economy, environment, and health of cities' (Food Tank, 2016), while the FAO described UA as the 'crop and livestock production within cities and towns and surrounding areas' (FAO, 2010, p. 1).

Since the early 1990s, food produced in cities has begun to increase worldwide. In just thirteen years, from 1993 to 2005, urban food production doubled from 15 per cent to 30 per cent of all food produces and the trend has been steadily increasing (Martellozzo *et al.*, 2014; Altieri and Nicholls, 2019). Moreover, UA is increasing in parallel to urban population growth. At the global level, in 2007, the population living in urban areas overtook the rural population and, at the same time, concerns related to urban planning and fresh food supply grew (UNFPA, 2007). UA is heterogeneously defined and usually refers to agriculture in areas of limited space, due to the high competition for land in cities. It can involve vertical and rooftop gardens, community and residential gardens, vacant lands or brownfields, containers on balconies, and commercial urban farms (Specht *et al.*, 2014). The scale of urban agricultural activities is related not only to the physical dimension, but also to its categorization. For instance, UA could be practised at the individual level in privately-owned areas, in communities (community gardens or guerrilla gardening[1]), and at commercial scale (Brown *et al.*, 2003; Pearson *et al.*, 2010; Adams and Hardman, 2014; Mok *et al.*, 2014). What is worth noting is that UA is performed in both developed and developing countries, but there are some differences in the role and implementation of UA in both areas. Indeed, in developed countries, the debate is recognized by institutions, policymakers, and citizens for several benefits and services provided and the initiatives related to UA are particularly diffused. The spread of UA in developed countries is mainly located in Europe and the USA, as demonstrated by the majority of academic studies, while in developing countries, research is poor or is composed mainly of grey literature, e.g. technical documents and project reports (Orsini *et al.*, 2013; Artmann and Sartison, 2018). Altogether, in developed countries, the general role identified for UA is to create a more sustainable lifestyle along with social ties within communities. Many initiatives are promoted in European and North American countries, and probably the most famous is the Milan Urban Food Policy Pact in 2015. The Pact highlights the key role of cities to achieve 'sustainable food systems and promoting healthy diets' and it recognizes the important role played by urban farmers and smallholder producers through food production (Milan Urban Food Policy Pact, 2015, p. 1). Hence, cities have to implement and adopt 'food policies', 'programmes and initiatives' that do not have to be the only activities related to departments of agricultural and/or rural development (Milan Urban

[1] Guerilla gardening is an example of grassroot initiatives of UA. The term usually refers to urban dwellers occupying spaces for growing vegetables or plants (Adams and Hardman, 2014)

Food Policy Pact, 2015, p. 2; Kago *et al*., 2019). The papers analyzed in *Artmann and Sartison's Review* (2018) focused on developed countries and they highlight how the scientific literature covers different societal challenges of urbanization, such as climate change, ecosystem services, social cohesion, and food security. On the other hand, UA in developing countries has a different function and it is more related to food and nutrition security (Maxwell *et al*., 1998). However, it is important to highlight that in developed countries, UA also plays an important role in the socio-economic conditions of urban dwellers (Mougeot, 2000). Indeed, it is crucial for self-consumption and it could also become a source of income as well as decreasing the costs of grocery shopping (Moustier and Danso, 2006). Nevertheless, institutions and policymakers do not consider UA in agricultural policies that are usually addressed in rural areas. Hence, many dwellers and urban farmers do not have access to the capital needed to purchase services to improve their production, for instance fertilizers, chemicals, and technical advice (Veolia Institute, 2019).

9.4. How to Grow Food in Cities?

The existing agri-food system causes between 19-29 per cent of global greenhouse gas emissions and agricultural production is responsible for 80-86 per cent of total food system emissions (Vermeulen *et al*., 2012). Conventional agriculture causes pollution in land and water bodies, loss of biodiversity, and degradation of important ecosystems (Goucher *et al*., 2017; Sánchez-Bayo and Wyckhuys, 2019). Furthermore, the current agri-food scheme results in other negative environmental impacts, e. g. soil erosion, loss of nutrients, loss of organic matter, and loss of soil biodiversity (FAO, 2015). The main negative drivers of industrial agriculture are the use of monoculture practices that result in low genetic diversity, intensive tillage, chemical pest control, and the excessive use of inputs, such as fertilizers and pesticides (Woodhouse E, 2010; Holt-Giménez and Altieri, 2013, Horton, 2017). In addition, these highly-mechanised crop systems tend to compromise future yields in favor of high immediate productivity (Gliessman, 2015).

It is important to highlight that the term UA refers to agriculture in general and it includes different agricultural systems, for instance, organic agriculture, agroecology, permaculture, vertical farming, and also industrial agriculture. Some of them could be categorized as alternative methods of agriculture, but they present many more similarities to conventional agriculture practices. Vertical farming could be an example of alternative methods; however, it refers to an intense production model to cultivate plants or animals within skyscrapers (Despommier and Ellingsen, 2008). Compared with conventional agriculture, vertical farming is considered more sustainable, given its emphasis on reduced energy, water, and fossil fuels use. However, sustainability could be called into question, in fact, start-up costs and energy are very high, especially in the beginning (Schmutz, 2017; Al-Kodmany, 2018).

Organic agriculture is another case of a non-traditional form of agriculture, through biodynamic agriculture and permaculture, which are defined as alternative agriculture due to the reduction of synthetic chemical inputs, such as pesticides and fertilizers (Lotter, 2003). However, even if organic farming principles are similar to agroecological ones, market forces are demanding farmers to introduce input substitutions, making their operations dependent, and more intensive, mainly to maximize agricultural production (Rosset and Altieri, 1997). By doing so, environmental impacts are reduced, but farmers remain strictly dependent on companies and too high production costs. The paradox is demonstrated by Californian organic farmers who grow grapes and strawberries, using between twelve and eighteen biological inputs and simultaneously they become trapped in an 'organic treadmill'. It means that while some specific diseases are managed by the use of organic inputs, other plantation aspects may be simultaneously affected by the need of other inputs to be controlled again (Guthman, 2004).

Data clearly show the high production UA could provide to a city (Kennard and Bamford, 2020). In Cleveland (Ohio), a city of 400,000 inhabitants, it is estimated that the urban area should be able to achieve high levels of self-sufficiency in fresh vegetables, fruits, eggs, poultry, and honey, depending on UA and how it is managed (Grewal and Grewal, 2012). Another important example of where UA has been applied is Cuba, starting thirty years ago. In 1990, due to the collapse of the Soviet Union and then of hugely subsidized fuel, the country shifted from an intensive monoculture system to a small-scale system, applying agroecological principles (Vázquez Moreno and Funes Aguilar, 2016). Worldwide, it seems that no other country has achieved these objectives with a form of agriculture that uses the ecological services of biodiversity and reduces food miles, energy use, and effectively closes local production and consumption cycles (Altieri and Funes-Monzote 2012). It is estimated that in Cuba, during 2014, more than 50 per cent of fresh products were produced by urban farmers and UA has been responsible for more than 300,000 jobs (Fernandez, 2017).

9.5. IC Technologies to Spread an Agroecological Approach in Cities

The premise of this chapter is to describe how to develop an agricultural system within smart cities that is sustainable and at the same time can guarantee food security and other social benefits to citizens, at present and in the future. As shown in the previous paragraph, the starting point could not be conventional agriculture or other methods that use high quantities of inputs (energy, water, and agrochemicals). Thus, the origin will be agroecology approaches and principles.

Agroecology can produce higher yields than industrial agriculture without the negative environmental impacts of the latter, even if it requires more labour (Pretty *et al.*, 2006). Indeed, the main objective of agroecology is to improve the efficiency of biological processes and to enhance biological activities above and below the soil, also in the urban environment (Altieri and Nicholls, 2019). Moreover, agroecological principles are usually applied in developing countries in small parcels of land and, therefore, it could be suitable also at the urban scale (Altieri, 1995).

In fact, the urban environment is a concrete place to test a transition from industrial agriculture to agroecology, especially at a small-test scale. However, there are significant bottlenecks to the diffusion of this system, not only in urban contexts but also in rural contexts. It could be possible to define three different crucial issues: (i) practitioners usually need a lot of experience before translating their knowledge into functional agroecological systems in specific contexts; (ii) there are few experts and too few new experts are being trained (Norton 2019); and (iii) although scientific knowledge on this subject is well advanced, it is very difficult to ensure its accessibility to the general public (Raghavan *et al.*, 2016). Knowledge required to manage an agroecological system is multiple and refers to climate information, land topography, water management, local biogeochemical conditions, information about specific plants and animals, and many others. What is more, knowledge is usually site-specific, even though it could be adapted to other sites (Raghavan *et al.*, 2016). In this framework, ICTs could play a key role in disseminating agroecological knowledge in urban environments and smart cities. They could become important in every phase of the agroecological system – from the design of the agroecosystem to the maintenance of the unit – and they could be a great support to experts, new practitioners, and citizens. The introduction of ICTs in agroecosystems could be applied through two different methods: (i) models to disseminate practices and to connect experts and citizens; and (ii) introduction and application of open-source Internet of Things (IoT) technologies. At present, these two methods are an emerging field in the agroecology discourse and only in recent years, researchers and agroecologists have joined forces to develop these applications for serving citizens. Models are very useful methods that could help experts and citizens to plan, to develop, and to maintain their agroecosystems. Moreover, they become fundamental to connect people with different knowledge and in different parts of the world to share their knowledge. Models are developed by researchers and experts and they usually present information about plants, their interactions with other plants, climate and soil models suitable for specific locations, and other information (Raghavan *et al.*, 2016). This information is obtained by the integration of data provided by existing databases that could be satellite data and weather/climate models. Currently, there are two models that are being implemented: (i) the software for agricultural ecosystems community coordinator (SAGE-CC); and (ii) and the smartphone app farmbetter. The first one is developed by the University of California and is specific for the urban environment (Norton, 2019). At the moment, the software is a demonstration and will be further implemented. The main objective of SAGE-CC is to facilitate the design and maintenance of agroecosystems between different owners of a community garden to create community polyculture. More specifically, the SAGE-CC could help neighborhoods to create a suitable sustainable polyculture system and simplify some fundamental processes in agroecology. The software could suggest which plants take advantage of their proximity or whether their relationship could be detrimental. According to the article 'The SAGE Community Coordinator: A Demonstration', a paramount example is reported by the process of pollination: some plants have to be located near other individuals of the same species to ensure wind or animal pollination. The idea of the SAGE-CC is to develop an interface with a map of the gardens where users can design and customize their property, using the most suitable cultivars. Vegetation databases will be developed categorizing plants not only in reference to the type or species but also based on their ecosystem relationship property, e.g. bark protection, fire, and insects, and on ecosystem relationship value. In this way, owners may understand good practices and best strategies to grow their sustainable food production and local food systems. Moreover, the SAGE-CC is developed as free and open-source software. First, users could freely access the system and their plant data; second, other communities could copy and modify it to their needs and environment. The second model is executed by farmbetter and is mainly aimed at developing countries for rural communities. It is implemented for measuring and improving the resilience of farmers and pastoralists in the face of climate change, for example, providing them with the right strategies to adapt to extreme events or specific shocks, such as floods and droughts (Choptiany *et al.*, 2019). As climate change is becoming an urgent issue, as previously mentioned, it is crucial to guarantee food security to citizens and farmers, in part by providing them with the tools to deal with this growing threat (Beddington *et al.*, 2011). New technologies, such as mobile phones, and big data collection, that are available in real-time and are becoming ever cheaper, could certainly help in this task. Hence, the transition of an urban community to one that has long-term local food resilience is complex and faced with many organizational challenges (Norton *et al.*, 2014). The second method presented in this chapter is the introduction of open source IoT technologies to support and implement practices of urban agroecology and to empower citizens. The 'Connected Seeds and Sensors' project was an important case study that took place between 2015 and 2017. It was a joint project between Spitalfields City Farm, a community garden operating since the 1970s in east London, and Queen Mary University of London. More specifically, the aims of the project were to support the practices of food-growing and seed-saving via the use of networked environmental sensors and data visualizations and the creation of an interactive seed library (Heitlinger *et al.*, 2019). The purpose of researchers was to integrate IoT technologies in small-scale urban agriculture, in stark contrast to sensors used by precision farming. The project introduced open source and custom-built IoT devices in the community garden to collect data and to verify how gardeners could respond to devices. The data collected included air temperature, humidity, and pressure, soil moisture and temperature, and ambient light. A second part of the project was to create an interactive library of seeds, where records were gathered and provided by experts, to explain the value and functions of a specific seed. More specifically, at the beginning, citizens were involved in workshops and activities to better understand best practices for food production in cities and the importance of saving seeds. Later, fifteen people of different nationalities were involved as seed guardians to grow one to two crops. In eight of these gardens, IoT devices were set up. These devices were previously designed and customized and they were based on open-source systems. Finally, the data collected from the sensors were presented on an interactive website, where citizens could also view photos and audio from the gardens. By doing so, researchers were able to increase citizens' participation in the integration of IoT technologies, generally managed in a top-down vision of the city. Moreover, the objective of the project was not only to test IoT technologies in urban agriculture, but also to create opportunities for interaction, social cohesion, and the care of common spaces.

9.6. Farmbetter: Building Resilience through Knowledge

In this section, we introduce a vignette of the application farmbetter and review its positioning within the current paradigm of rural development, explaining how it aims to improve farmers' access to knowledge. We subsequently outline the necessary steps to transform it into urban settings and conclude with an outlook on its applicability in contributing toward agroecology in an urban context. Existing approaches to assess the resilience of farmers (FAO, 2018a, FAO, 2018b; UNDP, 2018) fail to provide farmers with actionable recommendations. Instead, they have been designed largely for project staff and monitoring and evaluation officers in order to measure project impact and to design more targeted interventions. Recognizing that climate change presents these farmers with novel challenges, farmbetter was developed in 2018, aiming to build upon the lessons learned from the above tools. This resulted in an Android app (farmbetter) launched in 2019 and provides users with tailored recommendations to empower them to make informed decisions to adapt and withstand shocks and stresses (e.g. droughts). By making the app on a smartphone aimed at farmers (rather than a tablet), it changed the emphasis away from project staff and instead empowered farmers to assess and improve their resilience themselves.

In addition to the above, the dominant approach in integrating ICT in rural areas is focused on capital-intensive solutions that include the improved adoption of high-yielding seed varieties, fertilizers and pesticides. While successful at reducing the yield gap at least in the short term, these intensification strategies are more vulnerable to shocks and stresses and are highly dependent on continuing to produce ever greater amounts of inputs. Integrating more knowledge-intensive, agroecological approaches that have a proven track record can, not only improve agricultural productivity in the long run, but importantly, provide avenues to strengthen smallholder farmers' resilience in rural areas. Farmers in these areas tend to be less well connected with input suppliers and also receive less extension service support. The app 'farmbetter' aims to supplement existing services and value chains by leveraging peer-reviewed sustainable land-management practices from existing databases (e.g. the world overview of conservation approaches and technologies (WOCAT) is the leading database of over 1,200 sustainable land-management practices, hosted by the University of Berne (WOCAT, 2014)). The app starts by asking questions about the farmers' practices, contexts, and interests in improving their livelihood (Fig. 1A). Using farmers' geolocation, global datasets are accessed to understand the local context (soil type, precipitation, altitude, agroecosystem zone, etc.). This information creates a unique profile of the farmer who can link to thousands of best practices from databases, such as WOCAT (Fig. 1B). By using global databases, farmbetter is able to access many more of the best practices (that have the same characteristics as the farmer's context) than if only accessing ones from within a specific country or region.

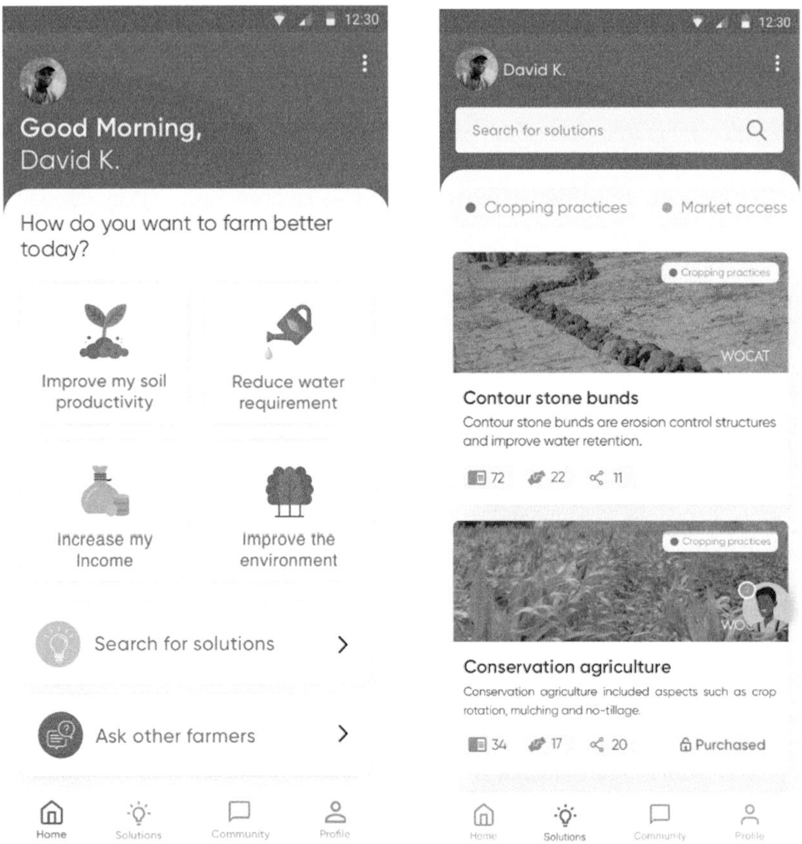

(A) (B)

Fig. 1: Home 2 (A) and solutions 2 (B) (*Source*: Authors' elaboration)

In order to adapt the farmbetter approach from rural farmers to a peri-urban or urban agricultural context, it is expected that the following five main conditions will be needed to change: (i) conditions limiting the reach of ICTs, such as lower literacy and ownership of smartphones in rural areas, will be less prevalent; (ii) as described in the community gardens above, production will need to be intensified given the price premium placed on land in urban settings; (iii) urban production is expected to be less diverse, as high value crops or legumes are grown to offset the costs; (iv) in addition to space, a key limiting capital factor is likely labour, especially in cases where urban agriculture is undertaken as a voluntary or part-time activity to supplement a wage-earning main income; and (v) lastly, with regards to shocks and stresses, urban areas often experience the heat island effect, increasing temperatures between 1-3°C, and hence impacting viabilities of urban food production. Given the varied geography with buildings, wind tunnels, and shades, there are likely to be numerous microclimates that will be more difficult to assess from global datasets based on satellite data. However, one major benefit of using farmbetter in smart cities is that there will be more data available for improving recommendation matching and for more nuanced best practices for urban farmers to implement.

In order to adapt the application to a smart city environment, farmbetter would need the following three main changes: first, the current rural application focuses predominantly on the ability of farmers to receive tailored recommendations. In a context of lower population densities and lack of adequate extension services, the provisioning of new targeted knowledge subsequently can act as a key driver to empower farmers to improve their resilience. In an urban environment, where the barriers to gaining access to services or new knowledge are expected to be lower, the need for tailored recommendations might be lower. With lower expertise in urban agriculture, however, the need for having a forum for farmers to interact with experts and others to get additional advice is heightened. Second, given urban agriculture's positioning outside the field of dominant development actors, there's a need to build upon the relatively recent and limited documentation of urban agricultural practices. For the current application, the existing dataset of solutions from WOCAT will need to be complemented with specific ones that are applicable to urban areas. Examples include databases from the University of Missouri (Hendrickson and Porth, 2012), Interreg Mediterranean (MADRE, 2018), and the United States Department of Agriculture (USDA, 2016).

Third, in order to improve the matching mentioned above, a population density filter would need to complement the existing agroecological filters (e.g. WorldPop, 2021) to ensure that the matching of solutions is adequately weighted based on the dominant characteristics of working in an urban environment. Urban settings would also trigger a reduction in the ranking on select factors, such as soils, as they are usually not a pre-determinant for urban agricultural production, given that many of the sites would either bring in soil or use artificial approaches, such as hydroponics or vertical agriculture. Furthermore, given the effect of the built environment on temperatures or flooding, a higher spatial granularity is likely to improve the matching, and in the event of a lack of such data, adapting the weights would be advisable. Incorporating data from the urban farmers themselves on microclimates, including rooftop farming, could also help identify more opportunities to implement urban agriculture and make developing the profile of the farmer's context more accurate.

Given that farmbetter has been designed as a platform that is based on matching the profiles of farmers to datasets of best practices, the app is relatively agnostic to the content of those databases (as long as the content is of a high quality). As discussed above, some changes would still be recommended to make it more appropriate and effective for urban agriculture. As with developing farmbetter in its current form required many iterations and input from users, adapting farmbetter to the urban context would require similar collaborations. Once databases of best practices are connected to the app, a pilot would be necessary to test its effectiveness and to allow for improvements, using a human-centered design approach. As urban agriculture provides a significant opportunity to reduce environmental impacts and increase food security in urban areas, it is important to realise this potential. The farmbetter app provides one means by which to connect the existing, proven best practices to urban farmers, who want advice tailored to their specific contexts.

9.7. The Benefits of Agroecology in Smart Cities: Ecosystem Services, Food Sovereignty, and Empowerment of Citizens

This chapter aims to investigate and analyze possible links between smart cities and UA by the integration of an agroecological approach. Hence, the chapter provides a critical viewpoint on one of the most inspiring models of the city of the future, namely smart cities, and how that form of urban, social, and political development, with its connected technologies, could be linked with UA and food systems to provide not only food sovereignty, but also, others social and environmental benefits to human well-being.

Particularly, the benefits provided by urban agroecology in cities are multiple. Like other UA systems, it increases food sovereignty in cities thanks to easier access to healthy food in developing countries, but also in underserved communities in developed countries (Kennard and Bamford, 2020). Moreover, urban agroecology could be an important approach due to its educational, cultural, geographic, and economic dimensions (Siegner *et al.*, 2018). However, the spread of this approach, not only in rural areas but also in cities, could be difficult, due to the high competencies required by citizens and farmers. Hence, ICTs could involve innovative devices and resources to support the dissemination and knowledge of this practice in urban environments (Raghavan *et al.*, 2016). Even if the three examples above are tests or at the beginning of their experience, they are examples of how it could be possible and feasible to integrate ICTs into UA and how they could foster the introduction of agroecological practices.

Indeed, all of them show how, through the use of IoT technologies, laypersons could grow their own food wherever they are and support food production in cities through the introduction of agroecological techniques. Although farmbetter is mainly developed for rural areas, it shows how farmers could benefit from ICTs to develop the best agroecological practices to improve their resilience to climate change (Choptiany *et al.*, 2019). At present, SAGE-CC is only a test but it already illustrates how various owners of community gardens could collaborate to implement a sustainable polyculture. Moreover, each example highlights different positive aspects provided by the integration of ICTs and urban agroecology. SAGE-CC highlights how the implementation of sustainable polyculture systems could contrast the dependency of citizens on conventional agriculture and increase plant diversity in urban ecosystems (Norton *et al.*, 2019). Finally, these examples illustrate how ICTs are important tools in the development of cities of the future and, at the same time, how new technologies and data could be applied. In fact, many critics have been moved against the present model of smart cities, due to their inability to guarantee empowerment of citizens and citizenship and also due to the big presence of companies in the governance and administration of cities. The literature on smart cities shows how ICTs are mainly integrating, monitoring, and regulating directly by companies or in a strict partnership between private and public, without the participation of citizens, except for the sensing of data. The examples proposed in this chapter investigate how it could be possible to involve different citizens in the life of the cities of the future. For instance, the Queen Mary University of London study is an example of social inclusion in urban spaces by practicing urban agroecology and, at the same time, providing food for citizens. The project was promoted in a low-income community in the north-east of England and it showed how citizens collaborated to develop their own community garden. The project highlights how IoT technologies could be integrated, not with a top-down approach, but with a bottom-up experience that aims to empower citizens and provide them with knowledge to test and use open source IoT in UA.

In conclusion, the chapter highlights the need for further research and studies to analyze and investigate how to feed a growing population in cities of the future. Moreover, it highlights the potential to capture and integrate agroecological approaches and ICTs into the smart city debate.

Bibliography

Adams, D. and M. Hardman (2014). Observing guerrillas in the wild: Reinterpreting practices of urban guerrilla gardening, *Urban Studies*, 51: 1103-1119.

Al-Kodmany, K. (2018). The vertical farm: A review of developments and implications for the vertical city, *Buildings*, 8: 24.

Albino, V., U. Berardi and R.M. Dangelico (2015). Smart cities: Definitions, dimensions, performance, and initiatives, *Journal of Urban Technology*, 22: 3-21.

Allwinkle, S. and P. Cruickshank (2011). Creating smarter cities: An overview, *Journal of Urban Technology*, 18: 1-16.

Altieri, M.A. (1995). *Agroecology: The Science of Sustainable Agriculture*, CRC Press, Boca Raton, USA.

Altieri, M.A. and F.R.F. Funes-Monzote (2012). The Paradox of Cuban Agriculture, Monthly Review; Retrieved from https://monthlyreview.org/2012/01/01/the-paradox-of-cuban-agriculture/; accessed on 21 February, 2020.

Altieri, M.A. and C.I. Nicholls (2019). Urban agroecology, *AgroSur*, 46: 46-60.

Angelidou, M. (2015). Smart cities: A conjuncture of four forces, *Cities*, 47: 95-106.

Artmann, M. and K. Sartison (2018). The role of urban agriculture as a nature-based solution: A review for developing a systemic assessment framework, *Sustainability*, 10: 1937.

Arun, M. and M. Teng Yap (2000). Singapore: The development of an intelligent island and social dividends of information technology, *Urban Studies*, 37: 1749-1756.

Baltimore Sun (2018). Urban Farm Coming to Former Sparrows Point Steel Mill Site in Baltimore County; Retrieved from http://www.baltimoresun.com/business/bs-md-sparrows-point-farm-20180508-story. html; accessed on 15 March, 2020.

Bates, J. (2012). 'This is what modern deregulation looks like': Co-optation and contestation in the shaping of the UK's Open Government Data Initiative; Retrieved from http://www.ci-journal.net/index.php/ciej/article/view/845/916; accessed on 15 March, 2020.

Beddington, J., M. Asaduzzaman, A. Fernandez, M. Clark, M. Guillou, L. Jahn, M. vand Erda, T. Mamo, N. Van Bo, C.A. Nobre, R. Scholes, R. Sharma and J. Wakhungu (2011). Achieving Food Security in the Face of Climate Change, Copenhagen; Retrieved from www.ccafs.cgiar.org/commission; accessed on 21 February, 2020.

Brown, K.H., M. Bailkey, A. Meares-Choen, J. Nasr, J. Smit, T. Buchanan and Mann Peter (2003). Urban Agriculture and Community Food Security in the United States: Farming from the City Center to Urban Fringe; retrieved from https://community-wealth.org/content/urban-agriculture-and-community-food-security-united-states-farming-city-center-urban-fringe; accessed on 21 February, 2020.

Cardullo, P. and R. Kitchin (2019). Being a 'citizen' in the smart city: Up and down the scaffold of smart citizen participation in Dublin, Ireland, *GeoJournal*, 84: 1-13.

Carvalho, L. (2015). Smart cities from scratch? A socio-technical perspective, *Cambridge Journal of Regions, Economy and Society*, 8: 43-60.

Choptiany, J.M., B.E. Graeub, S. Hatik, D. Conversa and S.T. Ledermann (2019). Participatory assessment and adaptation for resilience to climate change, *Consilience*, 21: 17-31.

de Amorim, W., S.A. Borchardt Deggau, G. do Livramento Gonçalves, S. da Silva Neiva, A.R. Prasath and J.B. Salgueirinho Osório de Andrade Guerra (2019). Urban challenges and opportunities to promote sustainable food security through smart cities and the 4th industrial revolution, *Land Use Policy*, 87: 104065.

Despommier, D. and E. Ellingsen (2008). The vertical farm: The sky-scraper as vehicle for a sustainable urban agriculture. *In:* A. Wood (Ed.). *CTBUH 8th World Congress 2008*, Dubai, United Arab Emirates.

Dobermann, A., B.S. Backmore, S.E. Cook and V. Adamchuk (2004). Precision farming: Challenges and future directions. *In:* T. Fischer, N. Turner, J. Angus, L. McIntyre, M. Robertson, A. Borrell and D. Lloyd (Eds.). *Proceedings of the 4th International Crop Science Congress*, Brisbane, Australia.

Duru, M., O. Therond and M. Fares (2015). Designing agroecological transitions: A review, *Agronomy for Sustainable Development*, 35: 1237-1257.

Eigenbrod, F.V., A. Bell, H.N. Davies, A. Heinemeyer, P.R. Armsworth and K.J. Gaston (2011). The impact of projected increases in urbanization on ecosystem services, 3201-3208. *In:* M.P. Hassell, S.H. Alonzo, G.R. Carvalho, I.C. Cuthill, H.A.P. Heesterbee and M.T. Siva-Jothy (Eds.). *Proceedings of the Royal Society B.*, London, UK.

European Parliament (2014). Precision Agriculture: An Opportunity for EU Farmers – Potential Support with the CAP, 2014-2020; Retrieved from https://www.europarl.europa.eu/thinktank/it/document.html?reference=IPOL-AGRI_NT%282014%29529049; accessed on 21 February, 2020.

FAO (Food and Agriculture Organization of the United Nations). (2010). Fighting Poverty and Hunger: What Role for Urban Agriculture? Retrieved from www.fao.org/fcit; accessed on 15 March, 2021.

FAO (Food and Agriculture Organization of the United Nations). (2015). Status of the World's Soil Resources: Main Report; retrieved from http://www.fao.org/documents/card/en/c/c6814873-efc3-41db-b7d3-2081a10ede50/; accessed on 21 February, 2020.

FAO (Food and Agriculture Organization of the United Nations). (2018a). RIMA|Resilience Index Measurement and Analysis (RIMA); Retrieved from http://www.fao.org/resilience/background/tools/rima/en/; accessed on 15 March, 2020.

FAO (Food and Agriculture Organization of the United Nations). (2018b). Self-evaluation and Holistic Assessment of Climate Resilience of farmers and Pastoralists (SHARP); Retrieved from: http://www.fao.org/in-action/sharp/en/; accessed on 15 March, 2021.

Fernandez, M. (2017). Urban Agriculture in Cuba: 30 years of policy and practice, Urban Agroecology; Retrieved from https://ruaf.org/document/urban-agriculture-magazine-no-33-urban-agroecology/; accessed on 15 March, 2021.

Food Tank (2016). Twelve Organizations Promoting Urban Agriculture around the World; Retrieved from http://www.fao.org/family-farming/detail/en/c/461898/; accessed on 20 February, 2020.

Fouilleux, E., N. Bricas and A. Alpha (2017). 'Feeding 9 billion people': Global food security debates and the productionist trap, *Journal of European Public Policy*, 24: 1658-1677.

Gabrys, J. (2014). Programming environments: Environmentality and citizen sensing in the smart city, *Environment and Planning D: Society and Space*, 32: 30-48.

Gebbers, R. and V.I. Adamchuk (2010). Precision agriculture and food security, *Science*, 327: 828-831.

Gliessman, S.R. (2015). *Agroecology: The Ecology of Sustainable Food Systems*, CRC Press, Boca Raton, USA.

Goucher, L.R., D.D. Bruce, S.C. Cameron, Lenny Koh and P. Horton (2017). The environmental impact of fertilizer embodied in a wheat-to-bread supply chain, *Nature Plants*, 3: 1-5.

Grewal, S.S. and P.S. Grewal (2012). Can cities become self-reliant in food? *Cities*, 29: 1-11.

Griffin, T.W. and J. Lowenberg-Deboer (2005). Worldwide Adoption and Profitability of Precision Agriculture Implications for Brazil; Retrieved from http://www.ers.usda.gov/Briefing/ARMS/; accessed on 21 February, 2020.

Guthman, J. (2004). Agrarian Dreams: The Paradox of Organic Farming in California, *Geographical Review of Japan, Series A*, University of California Press, Oakland, USA.

Heitlinger, S.N., Bryan-Kinns and R. Comber (2019). The right to the sustainable smart city, pp. 1-13, *Proceedings of the 2019 CHI Conference on Human Factors in Computing Systems*, Glasgow, UK.

Hendrickson, M.K. and M. Porth (2012). Urban Agriculture – Best Practices and Possibilities, Retrieved from http://extension.missouri.edu/foodsystems/urbanagriculture.aspx.; accessed on 15 March, 2021.

Hill, D. (2013). On the Smart City or, a 'Manifesto' for Smart Citizens Instead; Retrieved from https://medium.com/butwhatwasthequestion/on-the-smart-city-or-a-manifesto-for-smart-citizens-instead-7e0c6425f909; accessed on 15 March, 2020.

Hollands, R.G. (2008). Will the real smart city please stand up? *City*, 12: 303-320.

Holt-Giménez, E. and M.A. Altieri (2013). Agroecology, food sovereignty, and the new green revolution, *Agroecology and Sustainable Food Systems*, 37: 90-102.

Horton, P. (2017). We need radical change in how we produce and consume food, *Food Security*, 9: 1323-1327.

Kago, J., S. Loose and R. Sietchiping (2019). Implementing the new urban agenda: Urban and territorial integration approaches in support of urban food systems. pp. 271-293. *In:* H. Ginzky, E. Dooley, I.L. Heuser, E. Kasimbazi, T. Markus and T. Qin (Eds.). *International Yearbook of Soil Law and Policy*, Springer, Cham, Switzerland.

Kennard, N.J. and R.H. Bamford (2020). Urban agriculture: Opportunities and challenges for sustainable development, pp. 1-14, *In:* W. Leal Filho, A. Azul, L. Brandli, P. Özuyar and T. Wall (Eds.). *Zero Hunger. Encyclopedia of the UN Sustainable Development Goals*, Springer, Cham, Switzerland.

Lotter, D.W. (2003). Organic agriculture, *J. Sustain. Agric.*, 21: 59-128.

MADRE (2018). Urban and Peri-urban Agriculture, Best Practice Catalogue; Retrieved from https://anima.coop/sites/default/files/madre_catalogue_web_light.pdf; accessed on 15 March, 2021.

Martellozzo, F.J., S. Landry, D. Plouffe, V. Seufert, P. Rowhani and N. Ramankutty (2014). Urban agriculture: A global analysis of the space constraint to meet urban vegetable demand, *Environmental Research Letters*, 9: 064025.

Maxwell, D., C. Levin and J. Csete (1998). Does urban agriculture help prevent malnutrition? Evidence from Kampala, *Food Policy*, 23: 411-424.

Maye, D. (2019). 'Smart food city': Conceptual relations between smart city planning, urban food systems and innovation theory, *City, Culture and Society*, 16: 18-24.

Milan Urban Food Policy Pact (2015). Milan Urban Food Policy Pact; Retrieved from https://www.milanurbanfoodpolicypact.org/the-milan-pact/; accessed on 15 March, 2020.

Mok, H.F., V.G. Williamson, J.R. Grove, K. Burry, S.F. Barker and A.J. Hamilton (2014). Strawberry fields forever? Urban agriculture in developed countries: A review, *Agronomy for Sustainable Development*, 34: 21-43.

Moreno, A., A. Bhattacharyya, L. Jansen, Y. Arkeman, R. Hartanto and M. Kleinke (2019). Environmental engineering and sustainability for smart agriculture: The application of UAV-based remote sensing to detect biodiversity in oil palm plantations, 012008. *In:* A.G. Niam, I. Yusliana and H. Imantho (Eds.). *International Conference on Digital Agriculture from Land to Consumers*, Bogor, Indonesia.

Mougeot, L.J.A. (2000). Urban Agriculture: Definition, Presence, Potentials and Risks, and Policy Challenges Cities Feeding People Series; Retrieved from http://www.idrc.ca/cfp; accessed on 21 February, 2020.

Moustier, P. and G. Danso (2006). Local economic development and marketing of urban produced food. pp. 174-195. *In:* R. van Veenhuizen (Ed.). *Cities Farming for the Future: Urban Agriculture for Green and Productive Cities*, RUAF Foundation, IDRC and IIRR, Philippines.

Mullins, P.D. (2017). The ubiquitous-eco-city of Songdo: An urban systems perspective on South Korea's Green city approach, *Urban Planning*, 2: 4-12.

Norton, G.W. and S.M. Swinton (2000). Precision agriculture: Global prospects and environmental implications, *Proceedings of the 24th International Conference of Agricultural Economist*, Berlin, De.

Norton, J. (2019). Information Systems for Grassroots Sustainable Agriculture, PhD thesis, University of California, Irvine, USA.

Norton, J., B.J. Pan, S. Nayebaziz, B. Tomlinson and S. Burke (2014). Plant guild composer: An interactive online system to support back yard food production, pp. 523-526, *Proceedings of the Conference on Human Factors in Computing Systems*, Association for Computing Machinery, New York, USA.

Norton, J., B. Tomlinson, B. Penzenstadler, S. McDonald, E. Kang, N. Koirala, R. Konishi, G.P. Carmona, J. Shah and S. Troncoso (2019). The SAGE community coordinator: A demonstration, pp. 1-10. *In:* S. Easterbrook (Ed.). *Proceedings of the Fifth Workshop on Computing within Limits*, Association for Computing Machinery, Lappeenranta, Fi.

Orsini, F., R. Kahane, R. Nono-Womdim and G. Gianquinto (2013). Urban agriculture in the developing world: A review, *Agronomy for Sustainable Development*, 33: 695-720.

Paskaleva, K.A. (2011). The smart city: A nexus for open innovation? *Intelligent Buildings International*, 3: 153-171.

Pearson, L., J.L. Pearson and C.J. Pearson (2010). Sustainable urban agriculture: Stocktake and opportunities, *International Journal of Agricultural Sustainability*, 8: 7-19.

Poulsen, M.N. (2017). Cultivating citizenship, equity, and social inclusion? Putting civic agriculture into practice through urban farming, *Agriculture and Human Values*, 34: 135-148.

Pretty, J.N., A.D. Noble, D. Bossio, J. Dixon, R.E. Hine, F.W.T.P. De Vries and J.I.L. Morison (2006). Resource-conserving agriculture increases yields in developing countries, *Environmental Science and Technology*, 40: 1114-1119.

Raghavan, B., B. Nardi, S.T. Lovell, J. Norton, B. Tomlinson and D.J. Patterson (2016). Computational agroecology, pp. 423-435. *In:* J. Kaye and A. Druin (Eds.). *Proceedings of the 2016 CHI Conference Extended Abstracts on Human Factors in Computing Systems*, ACM Press, San Jose, USA.

Rosset, P.M. and M.A. Altieri (1997). Agroecology versus input substitution: A fundamental contradiction of sustainable agriculture, *Society and Natural Resources*, 10: 283-295.

Sánchez-Bayo, F. and K.A.G. Wyckhuys (2019). Worldwide decline of the entomofauna: A review of its drivers, *Biological Conservation*, 232: 8-27.

Schmutz, U. (2017). Urban Agriculture or Urban Agroecology? RUAF; Retrieved from https://ruaf.org/document/urban-agriculture-magazine-no-33-urban-agroecology/; accessed on 21 March, 2021.

Serbulova, N., S. Kanurny, A. Gorodnyanskaya and A. Persiyanova (2019). Sustainable food systems and agriculture: The role of information and communication technologies, 12127. *In:* M. Pasetti and V. Murgul (Eds.). *IOP Conf. Ser.: Earth Environ. Sci.*, Rostov-on-Don, Russia.

Seto, K.C., B. Güneralp and L.R. Hutyra (2012). Global forecasts of urban expansion to 2030 and direct impacts on biodiversity and carbon pools, *Proceedings of the National Academy of Sciences of the United States of America*, 109: 16083-16088.

Siegner, A., J. Sowerwine and C. Acey (2018). Does Urban Agriculture Improve Food Security? Examining the nexus of food access and distribution of urban produced foods in the United States: A systematic review, *Sustainability*, 10: 2988.

Söderström, O., T. Paasche and F. Klauser (2014). Smart cities as corporate storytelling, *City*, 18: 307-320.

Sonnino, R. (2009). Feeding the city: Towards a new research and planning agenda, *International Planning Studies*, 14: 425-435.

Specht, K., R. Siebert, I. Hartmann, U.B. Freisinger, M. Sawicka, A. Werner, S. Thomaier, D. Henckel, H. Walk and A. Dierich (2014). Urban agriculture of the future: An overview of sustainability aspects of food production in and on buildings, *Agriculture and Human Values*, 31: 33-51.

Swyngedouw, E. (2016). The mirage of the sustainable 'smart city': Planetary urbanization and the spectre of combined and uneven apocalypse. pp. 134-143. *In:* O. Nel-lo and R. Mele (Eds.). *Cities in the 21st Century*, Routledge, New York, USA.

Taylor, J.R. and S.T. Lovell (2012). Mapping public and private spaces of urban agriculture in Chicago through the analysis of high-resolution aerial images in Google Earth, *Landscape and Urban Planning*, 108: 57-70.

Townsend, A. (2014). *Smart Cities – Big Data, Civic Hackers and the Question for a New Utopia*, W.W. Norton & Company, Ed., New York, USA.

UN Department of Economic and Social Affairs. (2019). World Population Prospects 2019 Highlights; Retrieved from www.population.un.org; accessed on 21 February, 2020.

UNDP (United Nations Development Programme). (2018). Community-based Resilience Analysis (CoBRA) Conceptual Framework and Methodology; Retrieved from www.undp.org.; accessed 15 March, 2020.

UNFPA (United Nations Population Fund). (2007). State of World Population 2007: Unleashing the Potential of Urban Growth, New York; Retrieved from www.unfpa.org.; accessed on 21 February, 2020.

USDA (United States Department of Agriculture). (2016). Urban Agriculture Toolkit; Retrieved from https://www.usda.gov/sites/default/files/documents/urban-agriculture-toolkit.pdf; accessed on 15 March, 2021.

Vanolo, A. (2014). Smartmentality: The Smart City as Disciplinary Strategy, *Urban Studies*, 51: 883-898.

Vanolo, A. (2016). Is there anybody out there? The place and role of citizens in tomorrow's smart cities, *Futures*, 82: 26-36.

Vázquez Moreno, L.L. and F. Funes Aguilar (2016). *Avances de la agroecología en Cuba Avances de la agroecología en Cuba, Editora Estación Experimental de Pastos y Forrajes Indio Hatuey*, La Habana, Cuba.

Veolia Institute (2019). Urban Agriculture: Another Way to Feed Cities; Retrieved from www.institut.veolia.org.; accessed on 15 March, 2021.

Vermeulen, S.J., B.M. Campbell and J.S.I. Ingram (2012). Climate Change and Food Systems, *Annual Review of Environment and Resources*, 37: 195-222.

Walter, A., R. Finger, R. Huber and N. Buchmann (2017). Smart farming is key to developing sustainable agriculture, *Proceedings of the National Academy of Sciences of the United States of America*, 114: 6148-6150.

Whelan, B. and J. Taylor (2013). Precision agriculture for grain production systems, *International Journal of Digital Earth*, CSIRO Publishing, Collingwood, Australia.

Wiig, A. (2016). The empty rhetoric of the smart city: From digital inclusion to economic promotion in Philadelphia, *Urban Geography*, 37: 535-553.

WOCAT (World Overview of Conservation Approaches and Technologies). (2014). WOCAT; Retrieved from https://www.wocat.net/en/; accessed on 15 March, 2021.

Woodhouse, P. (2010). Beyond Industrial Agriculture? Some questions about farm size, productivity and sustainability, *Journal of Agrarian Change*, 10: 437-453.

WorldPop (2021). Retrieved from: https://www.worldpop.org/focus_areas; accessed on 31 March, 2021.

Wu, D. and G. Lindasay (2020). How to Design a Smart City that's Built on Empowerment – not Corporate Surveillance; Retrieved from https://www.fastcompany.com/90469838/how-to-design-a-smart-city-thats-built-on-empowerment-not-corporate-surveillance; accessed on 21 March, 2020.

Zhang, N., M. Wang and N. Wang (2002). Precision agriculture – A worldwide overview, *Computers and Electronics in Agriculture*, 36: 113-132.

CHAPTER

10

(Free and Open) Satellite Imageries for Land Rights and Climate Justice in Amazon Agroforestry Systems

Daniele Codato[1]*, Guido Ceccherini[2] and Hugh D. Eva[2]

[1] Research Programme Climate Change, Territory, Diversity, Department of Civil Environmental Architectural Engineering, University of Padova, Italy

[2] Bio-Economy Unit, European Commission Joint Research Centre, Ispra, Italy

10.1. The Amazon Region

What is the Amazon? Or even better: what are the 'Amazons'? The Amazon is such a complex system that even its limits are not well defined and vary across the spheres of watershed management, ecology, and geopolitics. The Amazon as a river-basin boundary is the largest watershed in the world that reaches the top of the Andes where the Amazon river and its tributaries originate (Maretti *et al.*, 2014); the ecological Amazon, also called the Amazon biome (*see* Fig. 1), corresponds to the area occupied by the largest moist tropical rain-forest in the world, in combination with other minor vegetation systems, including savannas and grasslands; and the political Amazon is delimited by the eight countries that constitute the Amazon Cooperation Treaty Organization (ACTO) and which are Bolivia, Brazil, Colombia, Ecuador, Guyana, Peru, Suriname, and Venezuela. More than 60 per cent and 11 per cent of Amazonian territory belong to Brazil and Peru, respectively. French Guyana is not part of the OTCA, but its territory is indeed part of the Amazon Biome, though it counts for only about 1 per cent of the Amazon (OTCA, 2020; RAISG, 2012; Maretti *et al.*, 2014). Socio-ecological boundaries for the Amazon also merit mention; the *Red Amazonica de Informacion Socioambiental Georefferenciada*, Amazonian Network of Georeferenced Socio-environmental Information (RAISG), a network of Amazonian organizations and

*Corresponding author: daniele.codato@unipd.it

NGOs, has defined a geographical area that combines biome boundaries with the administrative limits of Brazil and Ecuador, and which are used for socio-ecological research purposes (RAISG, 2012). Consequently, its extension can vary according to the different limits and projects or investigations considered, usually ranging between 6,500.000 and 8,000,000 km^2 (RAISG, 2012; Maretti *et al.*, 2014).

Despite competing definitions of its geographical boundaries, the Amazon region's importance as a reservoir for bio-ecological diversity and ecosystem services, and as a tool for mitigating climate change, is recognized worldwide: it hosts the greatest variety of inland plant-and-animal species and endemism, between 12-20 per cent of global freshwater not stored in ice, much of which is exchanged through atmospheric rivers created by forest evapotranspiration, and almost 10 per cent of global carbon storage (Charity *et al.*, 2016; Maretti *et al.*, 2014).

For 11,000 years, the Amazon has also been the heart of great cultural and agroecological diversity: many indigenous nations live in the Amazon and have learnt to survive and utilize the forest's natural resources. These groups have developed traditional ecological knowledge systems and practices that have allowed for sustainable human subsistence with minimal impacts on forest ecosystems (Charity *et al.*, 2016; Rudel *et al.*, 2002; Kelly *et al.*, 2018; Santos *et al.*, 2018). Besides hunting, fishing, and the use of about 200 different tree species for various purposes, indigenous populations introduced exotic crops and learned to domesticate about eighty-three native plant species, creating a multifunctional agroforestry system based on shifting agriculture which, in the western Amazon (the Amazon of Peru, Colombia, Ecuador, Bolivia), is called a *chackra* or *chagra* (Charity *et al.*, 2016; Vera *et al.*, 2019; Fonseca-Cepeda *et al.*, 2019). Small plots of forest are burned or cut down to plant different layers of crops alongside native and domesticated trees, according to their cultural knowledge system. After some years, people leave the area, enabling the restoration of the tree cover, and they start to utilize another portion of forest (Vera *et al.*, 2019).

Following World War II, rapid globalization and free markets led to increased pressures on Amazonian ecosystem and its capacity to provide ecosystem services, with consequences for bio-cultural diversity. Non-renewable resource extraction (in particular, fossil fuel exploitation and mining), settler invasion, and agriculture, large-scale palm oil and soya cultivation, pastures and cattle farming, road and urban expansion, dams, and wood extraction are causing several direct and indirect socio-ecological impacts at the global and local scale, such as wildfires, deforestation, habitat fragmentation, air, ground, and water contamination, and biological and cultural diversity loss (RAISG, 2012; Finer *et al.*, 2015; Laurance *et al.*, 2009; Codato *et al.*, 2019).

The pervasive Western vision of the Amazon as a rich, empty green space to colonize and exploit, increasingly competes with the Indigenous and conservationist visions of the Amazon, the former of which is broadly known as *sumak kawsay*, which translates to 'good living'. Both visions fight to protect and restore ecosystems and bio-cultural diversity, promote various sustainability projects within the spheres of agroecology, ecotourism, and ecosystem services, and establish forms of socio-environmental legal protection, such as the constitution of natural protected areas (PAs) and the recognition of ancestral indigenous territories (ITs) (Charity *et al.*, 2016; RAISG, 2012). In particular, the indigenous people of the Amazon, with a current estimated population of about 3 million (on a total of more than 35 million Amazonian inhabitants) and over 350 tribes, are considered guardians of the tropical forest, voluntarily or involuntarily supporting global efforts to mitigate climate change and develop sustainably (Charity *et al.*, 2016). At the same time, however, indigenous people are confronted with the negative effects of colonization and exploitation of the Amazon, with some support from NGOs, missionaries, and indigenous organizations, and little or no support from governments. Moreover, in some areas, forest exploitation is also leading to an erosion of indigenous people's cultural and ecological knowledge and practices, changes in their traditional behaviors, and an increase in population density that are forcing them to emigrate to urban centers or use unsustainable agricultural practices at the expense of the forest (Maretti *et al.*, 2014; Charity *et al.*, 2016; Thiede and Gray, 2020; Rudel *et al.*, 2002).

Nowadays, between 21-25 per cent of the Amazon is protected by PAs, with various degrees of success in preventing exploitation (RAISG, 2012, Maretti *et al.*, 2014; Gullison *et al.*, 2018). Indigenous ancestral territories account for between 27-31 per cent of the Amazon and it is a figure that includes formally recognized and non-formally recognized ITs, established and proposed reserves for uncontacted and voluntarily isolated groups (RAISG, 2012). Over 45 per cent of the Amazon is under some kind of direct or indirect protection, when considering both ITs and PAs (*see* Fig. 1).

The complexity and vastness of the Amazon region, which is still largely inaccessible, highlight the need for tools, technologies, and methodologies to improve the capacity for research and monitoring, in order to support decision-making processes and efforts toward protection and sustainable management. Satellite imagery and remote sensing (RS) tools and techniques have proved exceptionally useful for these purposes (Finer *et al.*, 2018; Dos Santos *et al.*, 2014). In this chapter, we briefly explore the principles of remote sensing, the available satellite sensors and data, the techniques and indices used in forestry and agriculture, the available tools and platforms to explore and use remote sensing data, and, lastly, we present an overview of peer review case studies concerning the use of satellite and RS for research in the Amazon.

10.2. Remote Sensing: Principles and Operation

In a broad sense, remote sensing can be understood as the process of 'collect[ing] information about an object without making physical contact with that object. In the context of geospatial analysis, that object is usually the Earth. Remote sensing also includes processing the collected information' (Lawhead, 2019, p. 18).

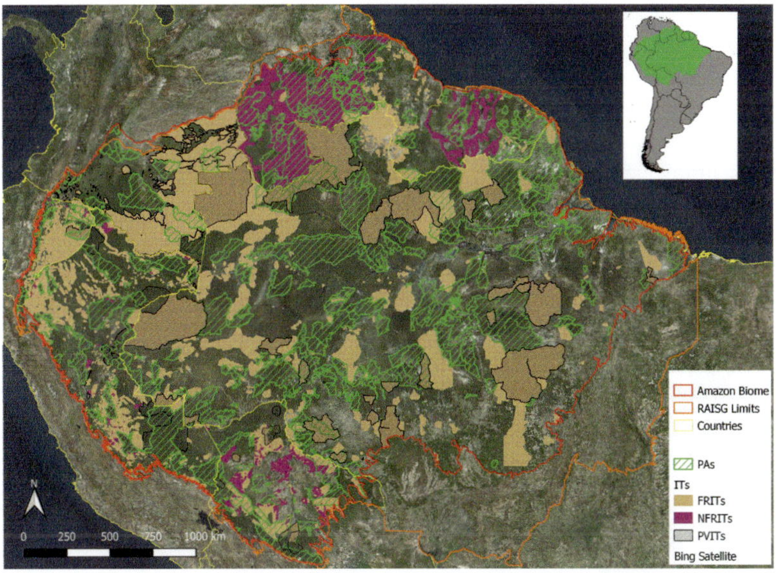

Fig. 1: The Amazon region, with two different boundaries (Biome and RAISG limits) and the geographical distribution of protected areas (PAs) and indigenous territories (ITs), divided in formally recognized ITs (FRITs), not formally recognized ITs (NFRITs) and population in volunteered isolation or uncontacted territories (PVITs) (*Source*: Authors' elaboration. *Data sources*: WWF, 2020; RAISG, 2020; UNEP-WCMC and IUCN, 2020)

The above description is also applicable to a very familiar 'remote sensor' – our eyes. Humans and animals explore the surrounding landscape with their senses and, through the brain, process the data collected and use it for decision making. Human needs and desire for knowledge drove the development of different types of technological sensors, such as cameras and scanners on platforms, including airborne ones (kites, balloons, airplanes, and more recently on unmanned aerial vehicles, better known as drones) and spaceborne ones (satellites) (Adams and Gillespie, 2006; Dainelli, 2011). The challenges presented by a rapidly changing world have made RS essential for Earth monitoring, with a growing level of interest in this sector from both public and private institutions and organizations. RS platforms and technologies have been developed for virtually all sectors and disciplines, and the production of commercial and free and open-source products has increased access to these technologies.

RS expands, supports, and integrates the potential of geographic information system (GIS) technologies and geographical proximity activities (fieldwork with the use of GPS, such as surveys and monitoring) to obtain data on vast territories and record phenomena and changes at a global scale over time (Weng, 2010).

Before presenting the most suitable RS sensors, products, tools, and analysis techniques for the Amazon and indigenous territories, this section answers some propaedeutic questions: What are the physical mechanisms behind RS? How does it work? What are the main characteristics of the available sensors? This section is not exhaustive and so readers should refer to RS literature, available at end of the chapter, for additional resources.

The sun emits energy as electromagnetic radiation that, upon reaching the Earth's atmosphere and surface, is absorbed, reflected, or emitted, depending on the chemical and physical characteristics of the material or substance it hits. The reflected (and in part, the emitted) portion of the energy that originally hits the object is what the sensor can detect and record, and is called reflectance (and emittance for the emitted part) (Fig. 2).

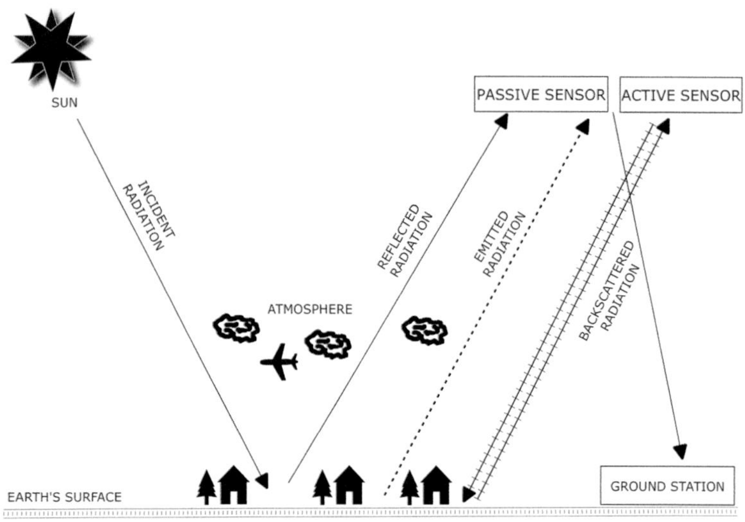

Fig. 2: Key aspects of data collection process by satellite remote sensing (*Source*: Author's elaboration)

The electromagnetic radiation is composed of a continuum of different wavelengths that together constitute the electromagnetic spectrum (ES): the visible light is the small part of the ES that humans can perceive with their eyes, while RS sensors are built and calibrated to detect many more types of radiations, such as those that are part of the infrared or ultraviolet bands.

The distance between two successive wave-crests, ranging from short to long and measured by using the metric system, are conventionally grouped in spectral bands, as shown in Fig. 3.

The human eye sees the ocean as blue and vegetation as green because of the absorption and reflection of the visible wavelengths; RS sensors detect the different wavelengths reflected by different objects: vegetation, for example, besides green, actually reflects more wavelengths in the near-infrared, while water reflectance

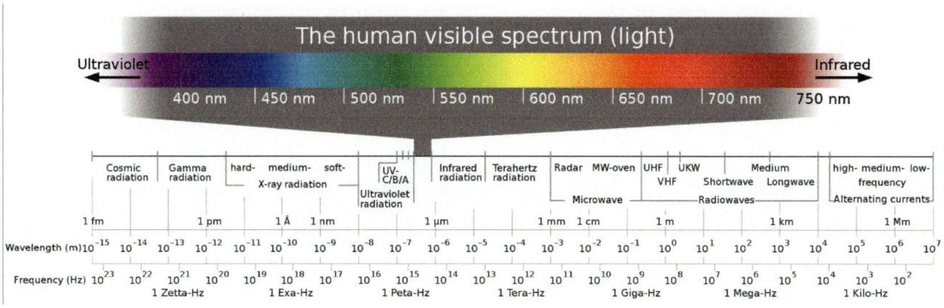

Fig. 3: The electromagnetic spectrum. The visible part ranges from (*left*) violet, blue, green, yellow, to orange and red (*right*) (modified from Horst, 2006)

is primarily in the shorter visible wavelengths (blue), which a calibrated sensor can detect. The way in which an object or substance reflects and emits different quantities of energy within different spectral bands is called its spectral signature (modeled as spectral curves) and is very useful in identifying and categorizing objects and changes in their state over time (Lavender and Lavender, 2016). Many variables can influence the radiation value captured by the sensor. One of these variables is the atmosphere itself, which masks portions of the ES, so that only wavelengths comprising the 'atmospheric windows' can reach the Earth surface (the visible bands and some portions of the infrared and thermal). Moreover, different components of the atmosphere, such as gases, particulates, and clouds, interact with the ES and can be studied through the RS. As a result, the amount of energy that reaches the sensor (radiance) is affected by the atmosphere and needs to be corrected to obtain the real reflectance value on the ground. Satellite sensors, then, usually require that data be corrected or calibrated (Dainelli, 2011).

Sensors can be divided into two typologies in terms of energy detection: passive sensors, which measure the reflected or originally emitted energy from a natural source – usually the sun; and active sensors, which measure the strength, time delay, and changes in the phase of the backscattered radiation emitted from the sensor itself (Lavender and Lavender, 2016) (*see* Fig. 2).

RS resolution, which greatly determines a sensor's usefulness for a project, is divided into spatial, spectral, and temporal categories. Spatial, or geometric resolution, is the minimum ground detail represented in a pixel, where the image is composed of a matrix of pixels, each one recording a unique radiation value. Higher spatial resolution indicates more detail and smaller pixel size. Spectral resolution refers to the number of spectral bands that a sensor can detect and the width of each band. Temporal resolution refers to the revisit time of a satellite, or the time that a satellite needs to record the same area; higher temporal resolution means a faster revisit time, and less time between data captured in the same geographical area. Another important characteristic worth mentioning is the swath – the portion of the Earth's surface collected by every image, which, for satellites, usually varies between tens and hundreds of kilometers wide (Dainelli, 2011; Lavender and Lavender, 2016).

Thanks to advances in technology, there are now many sensors that provide different data types, which have different scopes, sensor types (active or passive), and capture different spectral bands. Sensors are generally grouped according to the following categories: photograph and photogrammetry, which focus on the visible part of the spectrum; multispectral (usually between 3 and 14 spectral bands), and hyperspectral (hundreds or thousands of spectral bands), with a sensor able to detect several spectral bands with different widths, increasing the possibility for detection of objects according to their spectral signature; panchromatic (PAN), with a sensor that records a single broad band (usually in the visible range, for example, between 0.50-0.75 nm) and therefore, presents a limited spectral resolution but with a higher spatial resolution; thermal RS, which focuses on thermal infrared (TIR) bands and is used to measure the temperature (Table 1 shows the most common spectral bands and their interpretation); ultra-violet bands RS, which is used for atmospheric monitoring; active microwave based instruments, such as RADAR (Radio Detection and Ranging) or SAR (Synthetic Aperture Radar) systems, working similar to the sonar system on boats to measure depth, and usually used to analyze topographical features; and LiDAR (light detection and ranging), an active sensor based on a laser scanner that emits a high-frequency microwave pulse for measuring topography (Weng, 2010; Lavender and Lavender, 2016; Dainelli, 2011). A major advantage of active sensors is their 'all-weather' capacity – they are able to collect data, regardless of cloud cover. However, there are some constraints: the SAR backscatter signal

Table 1: Most Common Multispectral Bands and Their Interpretation (Modified from Bevington *et al.*, 2018)

Band Name	Common Interpretations
Panchromatic	Usually samples visible light at a higher resolution
Ultra-blue	Shallow water, suspended sediments, chlorophyll concentrations, algae blooms, and aerosols; also known as the coastal or aerosol band
Blue	Shallow water, land cover, and deciduous/coniferous, sensitive to atmospheric scatter
Green	Emphasizes the true color of vegetation
Red	Discriminates vegetation and chlorophyll absorption for vegetation health
Red edge	Exploits the sharp contrast between red and near infrared
Near Infrared (NIR)	Emphasizes biomass content and shorelines
Short-wave Infrared (SWIR1)	Soil canopy moisture and thin cloud penetration
SWIR2	Soil and canopy moisture and thin cloud penetration
Cirrus	Detection of cirrus clouds
Thermal Infrared (TIR)	Thermal mapping, soil moisture, cloud mapping

is affected by terrain, changes in ground moisture, and a random noise (speckle) effect, which makes consistent land-cover classification at times problematic. However, SAR systems are limited in their number of wavelengths due to the power requirements for sending out a pulse of electromagnetic radiation from the sensor itself (Lavender and Lavender, 2016).

10.3. Satellite Sensors and Data Availability

Satellite and aerial platforms may often carry a number of RS sensors aboard, each with different characteristics for different applications. In general, airborne sensors are used on specific projects, where a customer is willing to pay for guaranteed data over a given area at a given time (e.g. mining, biomass surveys). Satellite-based sensors collect as much data as possible during their orbit, though this is limited by on-board recording capacity and ground station characteristics. Some satellite platforms can change orbit and inclination to carry out 'pay-per-view' programmed acquisitions.

Sensors fall into two main groups, as discussed in the previous section: *passive* sensors, which measure the amount of the sun's electromagnetic radiation reflected back into space (reflectance) from the Earth's surface, and 2) *active* sensors, where the instrument itself sends out a pulse of electromagnetic radiation, and measures the amount reflected back (backscatter). The former are generally called optical systems, often working off the same principle as a digital camera, and the latter are the RADAR, SAR or LIDAR systems (Lavender and Lavender, 2016).

Concerning active sensors, the most common wavelengths employed by the SAR systems is C-band in Sentinel 1 satellite (ESA, 2020a) and L-band in ALOS PALSAR (JAXA, 2020), the former sensitive to target structure such as leaves, the latter to branches and trunks. New systems, which are programmed in the P-band (such as the forthcoming 2022 biomass satellite from the European Space Agency, ESA), will provide more information on biomass (ESA, 2020b). Data from these SAR systems are also used to create digital elevation models (DEM), the most comprehensive being the global 30 m-resolution DEM created from the shuttle radar topography mission (SRTM) (NASA, 2020a).

The active LIDAR systems at present are only airborne and have been used for local-scale estimates of vegetation height (Venier *et al.* (2019).

Passive sensors are designed to detect and record data at different ground resolutions, and much like a digital camera, a fine resolution sensor (as in a digital camera's zoom) images a smaller area than a low-resolution sensor. This is due to the limitations of data downlink to receiving stations. The spatial resolution of the sensor determines its range of applications and spectral resolutions. Hence, satellites designed for weather mapping, such as the National Aeronautics and Space Administration (NOAA) AVHRR (USGS, 2020) and the EUMETSAT Meteosat (ESA, 2020c), have a low spatial resolution (1-3 km) and are able to map entire continents, and record data in the thermal infrared and visible spectral channels to detect cloud formations, and also sea-surface temperature.

Satellites that are designed for urban mapping require a high spatial resolution (< 1 m) but often, few spectral channels – and sometimes only one – are in the visible range.

For monitoring of wide-area land cover and vegetation conditions, the SPOT VGT (*Satellite Pour l'Observation de la Terre Vegetation*) series of satellites, which have been in operation since 1998, provide global daily coverage at 1 km resolution (VITO, 2020a). The system was upgraded to 300 m resolution in 2013 with the SPOT PROBA V sensor (VITO, 2020b). Several other low-resolution sensors provide similar data, notably the moderate resolution imaging spectroradiometer (MODIS) launched by NASA in 1999 on board the Terra satellite, and in 2002 on board the Aqua satellite (NASA, 2020b). Other sensors in this range include ESA's ATSR and MERIS instruments, at 1 km and 300 m respectively, and the NASA/NOAA and Suomi-NPP's visible infrared imaging radiometer suite (VIIRS) sensor launched in 2011 with 375 and 750 m resolution (NOAA, 2020). While data from these satellites are available as individual scenes, covering swaths from 500-2000 km, the data are also available as daily, monthly, and yearly composites at continental to global levels. Commonly, apart from the original individual band data (e.g. visible, NIR, SWIR), data are available as derived products, such as vegetation indices (condition and photosynthesis), land surface temperature, cold cloud cover duration and rainfall estimation, and fire and burnt area detection.

For more detailed monitoring of land use and land cover change, medium spatial resolution satellites (10-30 m) have been deemed suitable. Notable amongst these is NASA's Landsat (LS) satellite series. The Thematic Mapper (LS 4 and 5) sensor has provided images of the Earth's surface since 1982 in the visible and near and shortwave infrared at around 17-day repeat cycles (NASA, 2020c). Prior to this, from 1972 to 1982, NASA had provided data from its Landsat (1 to 3) MSS (multi spectral scanner) at 80 m resolution (NASA, 2020d). The original TM sensor was upgraded to the ETM (enhanced thematic mapper) in 1999 with Landsat 7 which was equipped with an additional 15 m panchromatic band (NASA, 2020e) and again in 2013 with Landsat 8 Operational Land Imager (OLI) (NASA, 2020f). In 2000, NASA began offering free public access to its remotely sensed data, now managing several websites from which these data can be downloaded.

Similar data come from the SPOT (*Satellite Pour l'Observation de la Terre*) HRV (1-4) instrument with VNIR (visible and NIR) data (10 and 20 m) in operation from 1986 to 2013 (CRISP, 2020). Again, the next generation SPOT 5, 6 and 7 saw an upgrade in resolution to 2.5-5 m. However, these data were only available commercially.

The CBERS (China-Brazil Earth resources satellite) data have been available since 1999, with open access since 2004, thanks to cooperation between the Chinese Academy of Space Technology (CAST) and the Brazilian National Institute for Space Research (INPE, *Instituto Nacional de Pesquisas Espaciáis*).

Along times, CBERS satellites have been carrying different sensors at different resolutions, such as the CBERS-4 with a spatial resolution at 5 m (panchromatic) and 10 or 20 m (multispectral VNIR). Having a near-equatorial orbit, much of the captured data is focused on the Amazon, for which the satellite was aimed to monitor (INPE, 2020a).

Under the Joint European Commission/European Space Agency Copernicus Program (EU, 2020), a series of Earth observation satellites – the Sentinel series – has been launched for land, ocean, and atmospheric monitoring. The Sentinel 1-SAR satellites, launched in 2014 and 2016, provide data in the C-band at two polarizations with an effective ground resolution of 10 m in Strip Map mode. The optical Sentinel-2 (2015 and 2017) satellites provide images every five days at the equator at 10 and 20 m resolutions in the SWIR & VNIR wavelengths. Sentinel-3 (2016) is designed to provide daily global data on land and oceans from its OLCI (ocean and land color instrument) 300 m resolution sensor (ESA, 2020d). The Copernicus program provides free and open access to data from all these satellites via internet download through a series of data warehouses. Data products from all these satellites tend to be scene or reference grid-based, with users able to download a specific image from a particular area on a particular date.

In addition, cloud-processing facilities, such as Google Earth Engine or Amazon's Web Services allow remote access to Copernicus and Landsat data, without the need for a download.

Table 2 provides a schematic representation of the main characteristics of the open-access medium spatial resolution satellites, described above.

Very high spatial resolution (VHR) data (< 5 m) are now far more common, but remain for the most part limited to commercial access, though some providers, such as Google and Microsoft, purchase and allow the visualization of some images for free through their services, with limitations. As a result of restricted access to this data, the use of such imagery has generally been limited to the fields of urban planning and oil and extractive mining sectors, with image acquisition facilitated by the buyer. High-resolution data, such as IKONOS, GeoEye and WorldView2 (Satellite Imaging Corporation, 2020) have been used in land monitoring as a surrogate 'ground truth' to validate lower resolution products on an ad hoc basis. More recently, commercial satellite companies have started to put in place constellations of VHR satellites, capable of providing daily near-global data coverage. RapidEye (5 m), PLANET (3 m) are examples of these (PLANET, 2020). The global coverage means that while wall-to-wall mapping with such data remains a challenge due to data volume and costs, statistical sampling schemes can be employed for validation purposes.

10.4. Remote Sensing Techniques and Indexes for Forestry and Agriculture

Techniques for the image analysis and monitoring of agricultural areas and forests have evolved over time from visual interpretation to automatic land-cover

Table 2: Open-access Medium Spatial Resolution Satellites with their Main Characteristics (modified from Bevington *et al.*, 2018)

Satellite	Sensor, Agency	Spectral Bands and Spatial Resolution	Revisit, Swath	Operational Lifetime
Landsat 1-3	Multispectral Scanner (MSS) NASA	VIS (80 m) NIR (80 m)	17 days, 185 km	L1 MSS Jul 1972-Jan 1975 L2 MSS Mar 1978-Jan 1978 L3 MSS Feb 1982-Mar 1983
Landsat 4-5	Thematic Mapper (TM) NASA	VIS (30 m) NIR (30 m) SWIR (30 m) TIR (120 m)	17 days, 185 km	L4 MSS/TM Jul 1982-Mar 1984 L5 MSS/TM Dec 1993-Jun 2013
Landsat 7	Enhanced Thematic Mapper (ETM+) NASA	PAN (15 m) VIS (30 m) NIR (30 m) SWIR (30 m) TIR (120 m)	16 days, 185 km	L7 ETM+ Apr 1999-present
Landsat 8	Operational Land Imager (OLI) NASA	PAN (15 m) VIS (30 m) NIR (30 m) SWIR (30 m) TIR (100 m)	16 days, 185 km	L8 OLI Feb 2013-present
Sentinel 2A and 2B	Multispectral Imager (MSI) ESA	VIS (10 m) NIR (10 m) SWIR (20 m)	5 days, 290 km	S2A MSI Jun 2015-present S2B MSI Mar 2017-present
CBERS	Various Sensors (PAN, MUX, etc.) CAST and INPE	PAN (5 m) VIS (10 m–20 m) NIR (10 m–20 m)	26 days, 60 km–120 km	CBERS-1 Oct 1999-Aug 2003 CBERS-2 Oct 2003-Jan 2007 CBERS-2B Sept 2007-June 2010 CBERS-4 Dec 2014-present

classification. Generally, automatic classification has been used with medium- to high-resolution images to estimate agricultural and forest areas and changes, either with wall-to-wall mapping (i.e. covering the entire spatial domain) or in sampling schemes. Visual interpretation, originally for military purposes, was based on the analysis of aerial analogue photographs, black and white, true color, and near infra-red images. Expert interpreters used color, context, pattern, and texture to identify specified targets. Visual interpretation can be done for thematic maps, resource surveys (using point interpretation) or topographic maps adopting stereoscopic images. The first major use of remotely sensed satellite imagery for forestry and agriculture was based on the same technique: printing out Landsat imagery at a scale of 1:200,000 and visually interpreting deforestation with transparent overlays, which were then digitized into a GIS. Examples of this methodology include the projects, PRODES, by INPE (2020b), AfriCover, and the Food and Agriculture Organization (FAO, 1993), which performed satellite monitoring of deforestation using a point grid over the Landsat printed images.

The use of computer-based image display systems, along with multi-spectral images, enabled the interpreter to improve visual interpretation by changing the display of band combinations, where an image is composed by a combination of three spectral bands into a single red-green-blue (RGB) image (called true-color when the three bands are the real RGB and false-color when other spectral bands are used) (Bevington *et al.*, 2018). Combinations of RED, NIR and SWIR bands are usually used for agriculture and forest applications (*see* Table 3). Image interpretation can be supported by improving the contrast, which is achieved by stretching the input digital values to a large range of display values. The interpreter could then use on-screen digitizing to directly input the data into a GIS – a technique used in projects researching land cover (the CORINE project, Heymann *et al.*, 1994), deforestation (TREES project, Achard *et al.*, 2002) and agriculture (MARS project, Gallego *et al.*, 1993). Several techniques were used as a final measure to further enhance image quality, removing atmospheric effects due to scattering and absorption of sunlight by aerosols, (e.g. radiative transfer models, such as 6S) (Vermote *et al.*, 1997) and Lowtran (Richter, 1990).

Table 3: Common Band Combinations for Forest and Agriculture Applications

Band Combination	**Use/Interpretation**
Red, Green, Blue	True/natural color
NIR, Red, Green	Near Infrared vegetation (false color)
SWIR2, NIR, Red	Shortwave infrared vegetation (false color)
NIR, SWIR1, Blue	Healthy vegetation (false color)
SWIR1, NIR, Red	Vegetation analysis (false color)
SWIR1, NIR, Blue	Agriculture (false color)

To monitor vegetation health, agriculture, and forestry, a number of indices have been developed, making use of satellite band combinations. Initially, the normalized difference vegetation index (NDVI, Tucker, 1979) was adopted, highlighting vegetation condition by using the difference in the NIR and the Red wavelengths (NDVI formula: (NIR-RED)/(NIR+RED). This was followed by a suite of more elaborate indices (e.g. LAI, FAPAR, SAVI, EVI, NBR) (Bannari *et al.*, 1995, Xue and Su, 2017).

These indices and ratios between bands are usually normalized, obtaining values that range between –1 and +1, allowing the use of thresholding values to differentiate between distinct land covers (e.g. a value of NDVI greater than 0.4 usually indicates the presence of forest in that pixel) (Bevington *et al.*, 2018). A comprehensive database of RS indexes, clearly presented with related sensors and applications, is available at the Index Database webpage (Index Database, 2020).

Digital image-classification techniques for land use classification, derived from different mathematical and logic functions, have been developed to support, reduce or substitute the visual interpretation work by humans. Classification techniques falls into two main groups – supervised, where the human supervisor provides information to the system about the classes (e.g. dense evergreen forest, shrubland, urban, etc.), and unsupervised, where classes with common characteristics are identified without prior input of the human supervisor. The most widely used techniques are clustering (k-means, MacQueen 1967; ISODATA, Tou and Gonzalez, 1974), decision trees (Random Forests, Breiman, 2001), Support vector machine algorithms (Vapnik, 1995) and neural networks (Benediktsson *et al.*, 1990). Another promising technique to replace the onerous task of on-screen digitizing has been used, including image segmentation, which divides the image into objects with similar properties, according to color, size, texture or shape – with the user able to set the shape and size of these objects. After an image has been segmented into appropriate image objects (through parameters set by the user), it can then, in turn, be classified by a range of parameters, including mean band values, texture, and geometry (Blaschke, 2010). A major advantage of a pixel-based classification technique, which is generally used for national and international (IPCC, FAO) forest definitions, is the ability to set a minimum mapping unit, such as the size of the smallest feature that could reliably be mapped. For example, the combination of image segmentation and object classification has been used to prepare the FAO's 2010 forest statistics (Raši *et al.*, 2011).

Object-based classification has proven particularly beneficial in agriculture applications: crop classification and monitoring can be achieved by establishing an agricultural parcel as an object or base unit. Thanks to the wide-area coverage of very high resolution data now available, the European Commission has implemented the control with remote sensing (CwRS) programme, which supports the implementation of the common agricultural policy (CAP) using satellite data (Lemajic *et al.*, 2018) to update the cadaster and monitor land use for payments for farmers across Europe.

A higher-level analysis of land cover classifications is undertaken using landscape metrics (Soille and Vogt, 2009). Such landscape metrics are commonly used to compare different landscapes or quantify the ecological response to disturbances. Although developed to better understand ecological processes and the spatial distribution of species and communities, they are also useful in quantifying forest fragmentation. Some recent examples of these applications include FRAGSTATS (McGarigal and Marks, 1995) and GUIDOS ToolBox (JRC, 2020) where algorithms analyze morphological spatial patterns and generate statistics on patch size and connectivity from the input base maps.

10.5. Tools and Platforms to Explore and Use RS Data

In this section, we present a brief description of the most promising tools, geo-portals, geo-platforms, and software to visualize, explore, download, and analyze RS products. To simplify, tools and platforms are grouped according to their typology (such as geo-portals, desktop software, etc.), providers (such as NASA, USGS, ESA, etc.) and available products (high-resolution imageries, Landsat, Sentinel, etc.), ranging from the more user-friendly tools to the more expert-based ones, and focusing on open-source and freeware products. Finally, the principal proprietary software will be discussed.

This 'toolbox' is based on the authors' experience from both didactic and research settings, and each tool is presented in terms of its usefulness for research in the Amazon. However, the list below is not an exhaustive one. In Table 4 a list of the main platforms and tools is provided in order of appearance in this section.

Online map services, provided by commercial organizations, offer free access to very high- (VHR) and mid-resolution satellite imageries. Google Maps, BING Maps and Yandex maps, among others, let users switch between vector and satellite maps, the latter of which provides a collage of imagery with various spatial resolutions and different capture dates, covering the entire globe. Based on the digital earth philosophy (Guo *et al.*, 2020), some providers, such as Google and ESRI, created services and freeware software that let users carry out some GIS analysis in very user-friendly environments: this is the case for Google Earth (Pro), in particular, the desktop version (while at the time of writing (April 2020) the online version is more limited), and ESRI ArcGIS Earth. These virtual 3D globes store an enormous amount of information and can be used to measure lengths and areas, digitalize and share vector data, geotagged photos, videos and GPS tracks in kml/kmz format, explore information uploaded by other users and organizations, print maps, and more. Moreover, Google Earth (Pro) has a historical line option that allows users to explore images taken at different dates, allowing simple diachronically analysis of palm oil expansion or road network deforestation for areas of the Amazon where images are available. However, users must pay attention to the terms of use: in fact, these are not open-source tools, but

Table 4: Main Platforms and Tools with Their Websites (*Source*: Author's elaboration)

Platforms/Tools	**Web Links**
Google Maps	https://www.google.it/maps
BING Maps	https://www.bing.com/maps
Yandex maps	https://yandex.com/maps/
Google Earth Pro	https://www.google.it/earth/download/gep/agree.html
Google Earth (web)	https://earth.google.com/web/
ESRI ArcGIS Earth	https://www.esri.com/en-us/arcgis/products/arcgis-earth/overview
OpenStreetMap	https://www.openstreetmap.org/
USGS EarthExplorer	https://earthexplorer.usgs.gov/
USGS GloVis	https://glovis.usgs.gov/
NASA EOSIS Earthdata	https://earthdata.nasa.gov/
ESA Earth Online	https://earth.esa.int/eogateway/
ESA Copernicus SCIHUB	https://scihub.copernicus.eu/dhus/#/home
ESA EO-CAT	https://eocat.esa.int/sec/#data-services-area
ESA GEOSS	https://www.geoportal.org/?f:dataSource=dab
NOAA NEDIS	https://www.nesdis.noaa.gov/
RAISG Amazonia Socioambiental	https://www.amazoniasocioambiental.org/en/
RAISG MapBiomas	https://amazonia.mapbiomas.org/
MAAP	https://maaproject.org/en/
INPE TerraBrasilis	http://terrabrasilis.dpi.inpe.br/en/home-page/
Global Forest Watch	https://www.globalforestwatch.org/
FAO GeoNetwork	http://www.fao.org/geonetwork/srv/en/main.home
WWF-Sight	https://wwf-sight.org/
Nature Map explorer	https://explorer.naturemap.earth/
Policy Support	http://www.policysupport.org/home
QGIS	https://www.qgis.org/en/site/
gvSIG	http://www.gvsig.com/en/home
GRASS GIS	https://grass.osgeo.org/
SAGA GIS	http://www.saga-gis.org/en/index.html
Orfeo toolbox	https://www.orfeo-toolbox.org/
ESA SNAP Toolbox	https://step.esa.int/main/toolboxes/snap/
Guidos Toolbox	https://forest.jrc.ec.europa.eu/en/activities/lpa/gtb/

(*Contd.*)

IMPACT	https://forobs.jrc.ec.europa.eu/products/software/impact.php
Google Earth Engine	https://earthengine.google.com/
ESRI ArcGIS and ArcGIS Pro	https://www.esri.com/en-us/home
L3Harris Geospatial ENVI	https://www.l3harrisgeospatial.com/Software-Technology/ENVI

freeware ones. Images and derived information cannot be used for commercial purposes without permission, while data credit must be clearly visible for non-commercial, educational or research purposes.

A separate discussion is warranted for the open-source crowd-sourced geographic information project OpenStreetMap (OSM), a global database of geographic information created by citizens, where everyone can contribute, download, share, and freely use the map data and only credits are required. Two aspects of OSM can be useful for collaborative monitoring of the Amazon: editing tools in both the online and desktop versions that allow anyone the ability to digitalize features of interest, such as road networks or new infrastructures, based on available VHR images or GPS tracks; and the presence of a strong community that collaborates to improve access to geographical information, even organizing parties to map areas with limited data for humanitarian or other purposes.

Various free and open online tools have been developed in the wake of new satellite platforms and the vast amount of data they produce. These tools include webGIS/webmap, which allows users to visualize and explore RS imageries and data in combination with vector layers; and geo-portals, which provide a wide array of options for the visualization, combination, exploration, analysis, and download of RS imageries, data, and other information. Usually, geo-portals provide a series of sophisticated filters to find the user's ideal imagery based on the area of interest (using a geolocation service, the ability to upload a vector file of the area, or entering the coordinates); the type of satellite or products (for example, if the imagery is provided with atmospheric correction, etc.); the percentage of cloud cover, the desired period of time, and so on. Once the correct filters are set, the tool provides a preview of the images and the user can select the best ones. Finally, a series of download options enable users to obtain a single image or a set, select the format, or download only the desired bands. Some services allow the direct visualization of the full images online or the ability to carry out some analysis before downloading, such as the adjustment of band combinations or the production of indices (NDVI, NDWI, etc.). Moreover, application programming interfaces (API) are often available, allowing users to access data stored in servers via other services and software. The metadata associated with the data of interest should always be provided by the geo-portal, otherwise the image may be invalid. The most reputable online services are provided by the US Geological Survey (USGS) agency, including the USGS EarthExplorer and GloVis; the US National Aeronautics and Space Administration (NASA) EOSIS Earthdata website; the

European Space Agency (ESA) Earth Online with the Copernicus Open Access Hub (SCIHUB), the EO-CAT and the Global Earth Observation System (GEOSS) portal; and the US National Oceanic and Atmospheric Administration (NOAA) with its National Environmental Satellite, Data and Information Service (NEDIS).

Moreover, many institutions and organizations, both governmental and private, are providing online GIS data and information derived from RS, with temporal data on the Amazon related to its ITs and PAs, deforestation, agriculture, infrastructure expansion, land use change, and climate change impacts. One of the most well-known providers of this type of data is the network RAISG (Amazonian Network of Georeferenced Socio-Environmental Information) with projects that include Amazonia Socioambiental, where users can display and download various socio-environmental and geographical information about the Amazon, including indigenous territories; MapBiomas, which focuses on Amazonian land cover change. Another important source of information is the Monitoring of the Andean Amazon Project (MAAP), which monitors real-time deforestation. The National Institute for Space Research's (INPE) as reference for the Brazilian Legal Amazon with its projects PRODES and DETER, among others, has been providing deforestation data since 1988 and which are freely available in the web portal TerraBrasilis, in addition to other OTCA country institutions that monitor deforestation with the support of international organizations.

Other global-scale projects that provide spatial information pertinent to research in the Amazon include Global Forest Watch, which monitors deforestation; FAO's GeoNetwork; WWF with its WWF-Sight project; Nature Map explorer, concerning biodiversity and ecosystem services; and Policy Support, a web-based decision-support system with various datasets and tools for modeling biodiversity and ecosystem services.

To visualize and carry out in-depth geospatial analysis of downloaded GIS and RS data, different open-source desktop GIS softwares are available and in continuous development: the most well-known are QGIS, gvSIG, GRASS GIS, SAGA GIS, as well as others more focused on RS, such as Orfeo Toolbox and ESA SNAP Toolbox. In particular, QGIS has been improved and diffused rapidly in recent years, thanks to its growing community, its user-friendly multi-language interface, and the integration of third-party algorithms from GRASS, SAGA and R software, which improve its ability to work with raster data, including satellite imagery. Moreover, QGIS allows for the use of a series of plugins for satellite image download, visualization, and analysis, developed by users. Quickmap Services and OpenLayers plugins permit the use of various VHR satellite basemaps from different providers, like Google, Bing, Yandex; and the Semi-Automatic Classification plugin allows for the download, pre- and post-processing, and classification of satellite imagery. Another provider of a very promising open-source desktop GIS is the Joint Research Center (JRC) with its software Guidos Toolbox and IMPACT Toolbox, specifically developed to carry out land cover classification, analyze land cover change and deforestation, and perform landscape metrics analysis.

Developments in the geospatial analysis field allow users to access all the features of a desktop GIS via an online platform, or the so-called cloud GIS, without the need to install software or download data. The most prominent example is Google Earth Engine, a platform for scientific analysis and visualization of geospatial datasets for academic, non-profit, business, and government users. Earth Engine hosts satellite imagery, including from Landsat and Copernicus Sentinel, and stores the data in a public archive that includes historical earth images going back more than forty years. The images, uploaded daily, are made available for global-scale data mining. Earth Engine also provides APIs and other tools to enable the analysis of large datasets. However, the need to use JavaScript or Python languages to carry out geospatial analysis limits its use to GIS users with skills in these programming languages.

To conclude this section, the world of proprietary GIS and RS software merit mention: the most well-known GIS software is owned by ESRI, which, since launching Arcview GIS in the 90s, has created a diverse range of GIS products, ranging from desktop GIS (ArcGIS and the recent ArcGIS Pro) to ArcGIS Earth, to web-based solutions, such as ArcGIS Online and ESRI Storymaps. Concerning proprietary softwares more focused on RS, well-known examples are ENVI (Harris geospatial solution) and ERDAS Image (Hexagon geospatial division).

10.6. Case Studies from the Amazon Region

After this brief exploration of the available satellite image databases and RS tools and techniques, in this section we present case studies where satellite imagery and analysis are used for research and monitoring of the Amazon region, in some cases in combination with other GIS and environmental modeling techniques and field-based data collection.

These case studies are taken from an overview of recent scientific papers, collected from databases of peer-reviewed literature (Scopus and Google Scholar) using different combinations of keywords, such as 'Amazon', 'Remote Sensing', 'Indigenous', and 'Agroforestry', among others.

Finer *et al.* (2018) and Dos Santos *et al.* (2014) highlight the key role of tropical forests on climate regulation, biodiversity, and human well-being, and emphasize the importance and usefulness of satellite imagery and RS, in combination with field data and other GIS information, for research and monitoring of deforestation or forest degradation, forest structure and biomass, as well as the environmental impacts by humans or natural forces. In particular, Finer *et al.* (2018) underline recent developments in available satellites, sensors and RS analytical capabilities that allow almost near real-time monitoring, and their potential in supporting decision-making processes and prioritizing areas of intervention. The potential of satellite imagery, coupled with modern airborne RS and terrestrial RS, can also improve our understanding of pre-Columbian Amazon settlements and allow comparison to current patterns in land use change and associated environmental impacts (Santos *et al.*, 2018).

Multi-temporal satellite imagery associated with spatial data concerning PAs, ITs, and other areas of land conservation, enable users to analyze the effectiveness of different protection categories in reducing deforestation and fire threats, allowing comparison between protected areas and areas without status. Most research confirms that land under some level of protection has fewer environmental pressures (Jusys, 2018; Paiva *et al.*, 2020) and highlights the key role of protected areas as carbon reservoirs that help regulate the climate (Walker *et al.*, 2019; Blackman and Veit, 2018). However, success in preventing environmental stressors can vary widely over time, depending on the protected area's proximity to exploitation activities and the legal, political, and organizational support provided by governments (Lima *et al.*, 2020).

To better understand social dynamics of land-use changes, the effectiveness of sustainability policies and projects, the drivers of unsustainable actions, and the behaviors of different actors (in particular, indigenous people and settlers), researchers often combine social fieldwork (georeferenced surveys, interviews, participatory mapping) and Census data with RS analysis in targeted geographic areas (Rudel *et al.*, 2002; Lu *et al.*, 2010; De Espindola *et al.*, 2012; Ribeiro *et al.*, 2014; Caviglia-Harris and Harris, 2008; Sirén and Brondizio, 2009). This technique is also useful in investigating the role of indigenous traditional ecological knowledge and practices in forest conservation (Olivero *et al.*, 2016; Paneque-Gálvez *et al.*, 2018). Moreover, combination with other spatial data concerning exploitative activities, future projects, and environmental variables (slope, elevation, soil types, etc.) are useful for carrying out environmental and statistical modeling analysis, in order to predict future scenarios, patterns of land-use change, and associated impacts (Lopez and Sierra, 2010; Vijay *et al.*, 2018; Pérez and Smith, 2019; Laue and Arima, 2016).

Satellite imagery and derived data are powerful instruments for monitoring the development of legal or illegal infrastructure and exploitation projects, including oil and gas infrastructure, mining operations, roads, and large-scale palm oil plantations, in remote territories. These tools allow researchers to evaluate the geographic expansion and environmental impacts of these activities, and evaluate whether socio-environmental standards, such as environmental impact assessments, are being met (Rudke *et al.*, 2020; Facchinelli *et al.*, 2020; Finer *et al.*, 2015; Vijay *et al.*, 2016; Bennet *et al.*, 2018; Glinskis and Gutiérrez-vélez, 2019).

The use of RS and high-resolution imagery is essential to do research concerning the geographic distribution and protection of uncontacted or voluntarily-isolated Indigenous people because it precludes direct contact, respecting a group's decision to remain isolated (Kesler and Walker, 2015).

Recently, the Amazon gained worldwide media attention due to the spread of human-caused fires resulting from agricultural and colonization policies, droughts, and climate change. Researchers are testing different multi-temporal RS products from MODIS, VIIRS and Landsat thermal and optical bands to monitor fire events, their spatial distribution, and their relationship with other variables, and establishing fire alert systems (Pivello, 2011; Lima *et al.*, 2020; Santana *et al.*, 2018; Silva Junior *et al.*, 2019).

Multi-temporal RS in combination with landscape metrics is used in landscape ecology and road ecology to investigate habitat loss, fragmentation, and other direct and indirect impacts on Amazon ecosystems due to linear infrastructure, agriculture, and other exploitation activities (Cabral A.I.R. *et al.*, 2018; Godar *et al.*, 2012; Grecchi *et al.*, 2015; Renò and Novo, 2019).

Bibliography

Achard, F., H.D. Eva, H.J. Stibig, P. Mayaux, J. Gallego, T. Richards and J.P. Malingreau (2002). Determination of deforestation rates of the worlds humid tropical forests, *Science*, **297**(5583): 999-1002.

Adams, J. and A. Gillespie (2006). *Remote Sensing of Landscapes with Spectral Images: A Physical Modeling Approach*, Cambridge University Press, Cambridge, UK.

Bannari, A., D. Morin, F. Bonn and A.R. Huete (1995). A review of vegetation indices, *Remote Sensing Reviews*, **13**(1-2): 95-120.

Benediktsson, J.A., P.H. Swain and O.K. Ersoy (1990). Neural network approaches versus statistical methods in classification of multisource remote sensing data, *IEEE Transactions on Geoscience and Remote Sensing*, **28**(4): 540-552.

Bennett, A., A. Ravikumar and H. Paltán (2018). The political ecology of oil palm company – Community partnerships in the Peruvian Amazon: Deforestation consequences of the privatization of rural development, *World Development*, 109: 29-41.

Bevington, A., H. Gleason, X. Giroux-Bougard and T.J. de Jong (2018). A review of free optical satellite imagery for watershed-scale landscape analysis, *Confluence*, **2**(2).

Blackman, A. and P. Veit (2018). Titled Amazon indigenous communities cut forest carbon emissions. *Ecological Economics*, 153: 56-67.

Blaschke, T. (2010). Object-based image analysis for remote sensing, *ISPRS J. Photogramm. Remote Sens.*, 65: 2-16.

Breiman, L. (2001). Random forests, *Mach. Learn.*, 45: 5-32.

Cabral, A.I.R., C. Saito, H. Pereira and A. Elisabeth (2018). Deforestation pattern dynamics in protected areas of the Brazilian Legal Amazon using remote sensing data, *Applied Geography*, 100: 101-115.

Caviglia-Harris, J.L. and D.W. Harris (2008). Integrating survey and remote sensing data to analyze land use at a fine scale: Insights from agricultural households in the Brazilian Amazon, *International Regional Science Review*, **31**(2): 115-137.

Charity, S., N. Dudley, D. Oliveira and S. Stolton (2016). *Living Amazon Report 2016: A Regional Approach to Conservation in the Amazon*, WWF living Amazon initiative, Brasília and Quito, Br and Ec.

Codato, D., S.E. Pappalardo, A. Diantini, F. Ferrarese, F. Gianoli and M. De Marchi (2019). Oil production, biodiversity conservation and indigenous territories: Towards geographical criteria for unburnable carbon areas in the Amazon rainforest, *Applied Geography*, 102: 28-38.

CRISP (Centre for Remote Imaging, Sensing & Processing). (2020). Retrieved from https://crisp.nus.edu.sg/~research/tutorial/spot.htm; accessed on 9 April, 2020.

Dainelli, N. (2011). L'osservazione della terra – Telerilevamento. Manuale teorico-pratico per l'elaborazione delle immagini digitali, *Dario Flaccovio Editore srl*, Palermo, IT.
De Espindola, G.M., A. Paula, D. De Aguiar, E. Pebesma, G. Câmara and L. Fonseca (2012). Agricultural land use dynamics in the Brazilian Amazon based on remote sensing and census data, *Applied Geography*, **32**(2): 240-252.
Dos Santos, J.R., L.S. Galvão and L.E.O.C. Aragão (2014). Remote sensing of Amazonian forests: monitoring structure, phenology and responses to environmental changes, *Revista Brasileira de Cartografia*, 66/7: 1413-1436.
ESA (European Space Agency). (2020a). Retrieved from https://sentinel.esa.int/web/sentinel/missions/sentinel-1; accessed on 6 April, 2020.
ESA (European Space Agency). (2020b). Retrieved from https://earth.esa.int/web/guest/missions/esa-future-missions/biomass; accessed on 6 April, 2020.
ESA (European Space Agency). (2020c). Retrieved from https://www.esa.int/Applications/Observing_the_Earth/Meteosat_satellites; accessed on 21 April, 2020.
ESA (European Space Agency). (2020d). Retrieved from http://sentinel.esa.int; accessed on 10 April, 2020.
EU (European Union). https://www.copernicus.eu/en; accessed on 9 April, 2020.
Facchinelli, F., S.E. Pappalardo, D. Codato, A. Diantini, G. Della Fera, E. Crescini and M. De Marchi (2020). Unburnable and unleakable carbon in western Amazon: Using VIIRS nightfire data to map gas flaring and policy compliance in the Yasuní Biosphere reserve, *Sustainability*, **12**(1): 58.
FAO (Food and Agriculture Organization). (1993). Forest resources assessment, 1990, Tropical Countries, FAO Forestry Paper 112, Rome, It.
Finer, M., B. Babbitt, S. Novoa, F. Ferrarese, S.E. Pappalardo, M. De Marchi and A. Kumar (2015). Future of oil and gas development in the western Amazon, *Environmental Research Letters*, **10**(2): 024003.
Finer, M., S. Novoa, M.J. Weisse, R. Petersen, J. Mascaro, T. Souto and R.G. Martinez (2018). Combating deforestation: From satellite to intervention, *Science*, **360**(6395): 1303-1305.
Fonseca-Cepeda, V., C.J. Idrobo and S. Restrepo (2019). The changing chagras, *Ecology and Society*, **24**(1): 1-16.
Gallego, F.J., J. Delince and C. Rueda (1993). Crop area estimates through remote sensing: Stability of the regression correction, *International Journal of Remote Sensing*, **14**(18): 3433-3445.
Glinskis, E.A. and V.H. Gutiérrez-vélez (2019). Land use policy quantifying and understanding land cover changes by large and small oil palm expansion regimes in the Peruvian Amazon, *Land Use Policy*, 80: 95-106.
Godar, J., E. Jorge and B. Pokorny (2012). Forest ecology and management: Who is responsible for deforestation in the Amazon? A spatially explicit analysis along the Transamazon Highway in Brazil, *Forest Ecology and Management*, 267: 58-73.
Grecchi, R.C., R. Beuchle, Y.E. Shimabukuro and F. Achard (2015). A multidisciplinary approach for assessing forest degradation in the Brazilian Amazon, 1941-1944. *In:* 2015 IEEE International Geoscience and Remote Sensing Symposium (IGARSS), Milan, IT.
Gullison, R.E. and J. Hardner (2018). Progress and challenges in consolidating the management of Amazonian protected areas and indigenous territories, *Conservation Biology*, 32: 1020-1030.
Guo, H., M.F. Goodchild and A. Annoni. (Eds.). (2020). *Manual of Digital Earth*, Springer Nature PP, Singapore, SGP.

Heymann, Y., C. Steenmans, G. Croissille and M. Bossard (1994). Corine Land Cover, Technical Guide; Retrieved from: https://www.eea.europa.eu/publications/COR0-landcover; accessed on 8 April, 2020.

Horst, F. (2006). Elecrtomagnetic Spectrum; Retrieved from: https://commons.wikimedia.org/wiki/File:Electromagnetic_spectrum_-eng.svg; accessed on 12 April, 2020.

Index Database (2020). Retrieved from https://www.indexdatabase.de/; accessed on 10 April, 2020.

INPE (Instituto Nacional de Pesquisas Espaciáis). (2020a). Retrieved from http://www.cbers.inpe.br/; accessed on 9 April, 2020.

INPE (Instituto Nacional de Pesquisas Espaciáis). (2020b). Retrieved from http://www.obt.inpe.br/OBT/assuntos/programas/amazonia/prodes; accessed on 10 April, 2020.

JAXA (Japan Aerospace Exploration Agency). (2020). Retrieved from https://www.eorc.jaxa.jp/ALOS/en/about/palsar.htm; accessed on 6 April, 2020.

JRC (Joint Research Centre). (2020). Retrieved from https://forest.jrc.ec.europa.eu/en/activities/lpa/gtb/; accessed on 10 April, 2020.

Jusys, T. (2018). Changing patterns in deforestation avoidance by different protection types in the Brazilian Amazon, *PLoS One*, **13**(4): e0195900.

Kelly, T.J., I.T. Lawson, K.H. Roucoux, T.R. Baker, E.N. Honorio-Coronado, T.D. Jones and S. Rivas-Panduro (2018). Continuous human presence without extensive reductions in forest cover over the past 2500 years in an seasonal Amazonian rain-forest, *J. Quaternary Sci.*, 33: 369-379.

Kesler, D.C. and R.S. Walker (2015). Geographic distribution of isolated indigenous societies in Amazonia and the efficacy of indigenous territories, *PLoS One*, **10**(5): e0125113.

Laue, J.E. and E.Y. Arima (2016). Spatially explicit models of land abandonment in the Amazon, *Journal of Land Use Science*, **11**(1): 48-75.

Laurance, W.F., M. Goosem and S.G.W. Laurance (2009). Impacts of roads and linear clearings on tropical forests, *Trends in Ecology & Evolution*, **24**(12): 659-669.

Lavender, S. and A. Lavender (2016). *Practical Handbook of Remote Sensing*, CRS Press. Taylor & Francis Group, Boca Raton, FL.

Lawhead, J. (2019). *Learning Geospatial Analysis with Python*, third edition, Packt Publishing, Birmingham, UK.

Lemajic, S., B. Vajsová and P. Aastrand (2018). New Sensors Benchmark Report on PlanetScope: Geometric Benchmarking Test for Common Agricultural Policy (CAP) Purposes; Retrieved from: https://ec.europa.eu/jrc/en/publication/new-sensors-benchmark-report-planetscope-geometric-benchmarking-test-common-agricultural-policy-cap; accessed on 8 April, 2020.

Lima, M., J. Cariele, G.D.M. Costa, R. Carvalho, W. Luiz, F. Correia and C. Antonio (2020). Land Use Policy: The forests in the indigenous lands in Brazil in peril, *Land Use Policy*, 90: 104258.

López, S. and R. Sierra (2010). Agricultural change in the Pastaza River Basin: A spatially explicit model of native Amazonian cultivation, *Applied Geography*, **30**(3): 355-369.

Lu, F., C. Gray, R.E. Bilsborrow, C.F. Mena, C.M. Erlien, J. Barbieri and S.J. Walsh (2010). Contrasting colonist and indigenous impacts on Amazonian forests, *Conservation Biology*, **24**(3): 881-885.

MacQueen, J.B. (1967). Some methods for classification and analysis of multivariate observations, 1: 281-297. *In:* Proceedings of 5th Berkeley Symposium on Mathematical Statistics and Probability, Berkeley, University of California Press, USA.

Maretti, C.C., J.C. Riveros, S.R. Hofstede, D. Oliveira, S. Charity, T. Granizo, C. Alvarez, P. Valdujo and C. Thompson (2014). State of the Amazon: Ecological representation in protected areas and indigenous territories, WWF Living Amazon (Global) Initiative, Brasília and Quito, BR and EC.

McGarigal, K. and B.J. Marks (1995). Fragstats: Spatial Pattern Analysis Program for Quantifying Landscape Structure, US Department of Agriculture, Forest Service, Pacific Northwest Research Station, USA.

NASA (National Aeronautics and Space Administration). (2020a). Retrieved from https://www2.jpl.nasa.gov/srtm/; accessed on 6 April, 2020.

NASA (National Aeronautics and Space Administration). (2020b). Retrieved from http://modis.gsfc.nasa.gov; accessed on 7 April, 2020.

NASA (National Aeronautics and Space Administration). (2020c). Retrieved from https://landsat.gsfc.nasa.gov/the-thematic-mapper/; accessed on 7 April, 2020.

NASA (National Aeronautics and Space Administration) (2020d). Retrieved from https://landsat.gsfc.nasa.gov/the-multispectral-scanner-system/; accessed on 7 April, 2020.

NASA (National Aeronautics and Space Administration). (2020e). Retrieved from https://landsat.gsfc.nasa.gov/landsat-7/; accessed on 9 April, 2020.

NASA (National Aeronautics and Space Administration). (2020f). Retrieved from https://landsat.gsfc.nasa.gov/landsat-data-continuity-mission/; accessed on 9 April, 2020.

NOAA (National Oceanic and Atmospheric Administration). (2020). Retrieved from https://ncc.nesdis.noaa.gov/VIIRS/index.php; accessed on 7 April, 2020.

Olivero, J., F. Ferri, P. Acevedo, J.M. Lobo, J.E. Fa, A. Farfán and D. Romero (2016). Using indigenous knowledge to link hyper-temporal land cover mapping with land use in the Venezuelan Amazon: The Forest Pulse, *Rev. Biol. Trop.*, **64**(4): 1661-1682.

OTCA (2020). Retrieved from http://www.otca-oficial.info/amazon/our_amazon; accessed on 6 April, 2020.

Paiva, P.F.P.R., M. de Lourdes Pinheiro Ruivo, O.M. da Silva Júnior, M. de Nazaré Martins Maciel, T.G.M. Braga, M.M.N. de Andrade and B.M. Ferreira (2020). Deforestation in protect areas in the Amazon: A threat to biodiversity, *Biodiversity and Conservation*, **29**(1): 19-38.

Paneque-Gálvez, J., I. Pérez-Llorente, A.C. Luz, M. Guèze, J.F. Mas, M.J. Macía, M. Orta-Martinez and V. Reyes-Garcia (2018). High overlap between traditional ecological knowledge and forest conservation found in the Bolivian Amazon, *Ambio.*, 47: 908-923.

Pérez, C.J. and C.A. Smith (2019). Indigenous knowledge systems and conservation of settled territories in the Bolivian Amazon, *Sustainability,* **11**(21): 1-41.

Pivello, V.R. (2011). The use of fire in the Cerrado and Amazonian rainforests of Brazil: Past and present, *Fire Ecology*, **7**(1): 24-39.

PLANET (2020). Retrieved from https://www.planet.com/products/monitoring; accessed on 10 April, 2020.

RAISG (Red Amazónica de Información Socioambiental Georeferenciada). (2012). Amazonía bajo presión, Red Amazónica de Información Socioambiental Georeferenciada, Instituto Socioambiental, Sao Paolo, BR.

RAISG (Red Amazónica de Información Socioambiental Georeferenciada). (2020). Retrieved from: https://www.amazoniasocioambiental.org/en/maps/#!/download. Accessed on 2 April 2020.

Raši, R., C. Bodart, H.J. Stibig, H. Eva, R. Beuchle, S. Carboni, D. Simonetti and F. Achard (2010). An automated approach for segmenting and classifying a large sample of multi-date Landsat imagery for pan-tropical forest monitoring, *Remote Sens. Environ.*, 115: 3659-3669.

Renó, V. and E. Novo (2019). Forest depletion gradient along the Amazon floodplain, *Ecological Indicators*, 98: 409-419.

Ribeiro, M.B.N., A. Jerozolimski, P. De Robert and W.E. Magnusson (2014). Forest ecology and management Brazil nut stock and harvesting at different spatial scales in southeastern Amazonia, *Forest Ecology and Management*, 319: 67-74.

Richter, R. (1990). A fast atmospheric correction algorithm applied to Landsat TM images, *IJRS*, **11**(1): 159-166.

Rudel, T., D. Bates and R. Machinguiashi (2002). Ecologically Noble Amerindians? Cattle Ranching and Cash Cropping among Shuar and Colonists in Ecuador, *Latin American Research Review*, **37**(1): 144-159.

Rudke, A.P., V.A. Sikora de Souza, A.M. dos Santos, A.C. Freitas Xavier, O.C. Rotunno Filho and J.A. Martins (2020). Impact of mining activities on areas of environmental protection in the southwest of the Amazon: A GIS- and remote-sensing-based assessment, *Journal of Environmental Management*, 263: 110392.

Santana, N.C., O.A. De Carvalho Júnior, R.A.T Gomes and R.F. Guimarães (2018). Burned-area detection in Amazonian environments using standardized time series per pixel in MODIS data, *Remote Sens.*, 10: 1904.

Santos, M.J., M. Disney and J. Chave (2018). Detecting human presence and influence on neotropical forests with remote sensing, *Remote Sens.*, 10: 1593.

Satellite Imaging Corporation (2020). Retrieved from https://www.satimagingcorp.com/satellite-sensors/; accessed on 10 April, 2020.

Silva Junior, C.H.L., L.O. Anderson, A.L. Silva, C.T. Almeida, R. Dalagnol, M.A.J.S. Pletsch, T.V. Penha, R.A. Paloschi and L.E.O.C. Aragão (2019). Fire Responses to the 2010 and 2015/2016 Amazonian Droughts, *Front. Earth Sci.*, **7**(97).

Sirén, A.H. and E.S. Brondizio (2009). Detecting subtle land use change in tropical forests, *Applied Geography*, **29**(2): 201-211.

Soille, P. and P. Vogt (2009). Morphological segmentation of binary patterns, *Pattern Recognition Letters*, 30: 456-459.

Thiede, B.C. and C. Gray (2020). Characterizing the indigenous forest peoples of Latin America: Results from census data, *World Development*, 125: 104685.

Tou, J.T. and R.C. Gonzalez (1974). *Pattern Recognition Principles*, Addison-Wesley Publishing Company, Reading, Massachusetts, USA.

Tucker, C.J. (1979). Red and photographic infrared linear combinations for monitoring vegetation, *Remote Sens. Environ.*, 8: 127-150.

UNEP-WCMC and IUCN (2020). Protected Planet: The World Database on Protected Areas (WDPA) [On-line], 04/2020, Cambridge, UK: UNEP-WCMC and IUCN. Retrieved from: https://www.protectedplanet.net/en. Accessed on 2 April 2020.

USGS (United States Geological Survey). (2020). Retrieved from https://www.usgs.gov/centers/eros/science/usgs-eros-archive-advanced-very-high-resolution-radiometer-avhrr?qt-science_center_objects=0#qt-science_center_objects; accessed on 6 April, 2020.

Vapnik, V.N. (1995). *The Nature of Statistical Learning Theory*, Springer Inc., New York, USA.

Venier, L.A., T. Swystun, M.J. Mazerolle, D.P. Kreutzweiser, K.L. Wainio-Keizer, K.A. McIlwrick, M.E. Woods and X. Wang (2019). Modelling vegetation understory cover using LiDAR metrics, *PLoS One*, **14**(11): e0220096.

Vera, V.R.R., H.J. Cota-Sánchez and J.E. Grijalva Olmedo (2019). Biodiversity, dynamics, and impact of chakras on the Ecuadorian Amazon, *Journal of Plant Ecology*, **12**(1): 34-44.

Vermote, E.F., D. Tanre, J.L. Deuze, M. Herman and J.J. Morcette (1997). Second simulation of the satellite signal in the solar spectrum, 6S: An overview, *IEEE Transactions on Geoscience and Remote Sensing*, **35**(3): 675-686.

Vijay, V., C.D. Reid, M. Finer, C.N. Jenkins and S.L. Pimm (2018). Deforestation risks posed by oil palm expansion in the Peruvian Deforestation risks posed by oil palm expansion in the Peruvian, *Environmental Research Letters,* **13**(11): 114010.

Vijay, V., S.L. Pimm, C.N. Jenkins and S.J. Smith (2016). The impacts of oil palm on recent deforestation and biodiversity loss, *PLoS One*, **11**(7): e0159668.

VITO (Flemish Institute for Technological Research NV). (2020a). Retrieved from http://www.spot-vegetation.com/pages/VegetationProgramme/generaldescription.htm; accessed on 7 April, 2020.

VITO (Flemish Institute for Technological Research NV) (2020b). http://proba-v.vgt.vito.be/en; accessed on 7 April, 2020.

Walker, W.S., S.R. Gorelik, A. Baccini, J.L. Aragon-Osejo, C. Josse, C. Meyer, M.N. Macedo, C. Augusto, S. Rios, T. Katan, A. Almeida de Souza, S. Cuellar, A. Llanos, I. Zager, G.D. Mirabal, K.K. Solvik, M.K. Farina, P. Moutinho and S. Schwartzman (2020). The role of forest conservation, degradation, and disturbance in the carbon dynamics of Amazon indigenous territories and protected areas, *Proceedings of the National Academy of Sciences*, **117**(6): 3015-3025.

Weng, Q. (2010). *Remote Sensing and GIS Integration: Theories, Methods, and Applications*, McGraw-Hill, New York, USA.

WWF (2020). Amazon rainforest ecoregion boundaries. Retrieved from: https://www.arcgis.com/home/item.html?id=841600854d934d7d8a1656eee32d5847. Accessed on 2 April 2020.

Xue, J. and B. Su (2017). Significant remote sensing vegetation indices: A review of developments and applications, *Journal of Sensors*, 2017: 1353691.

CHAPTER

11

Connecting Farms and Landscapes through Agrobiodiversity: The Use of Drones in Mapping the Main Agroecological Structure

Ingrid Quintero[1,3]*, Yesica Xiomara Daza-Cruz[2,3] and Tomás Enrique León-Sicard[3]

[1] Agroecology Doctoral Program, Faculty of Agrarian Sciences, National University of Colombia, Carrera 30 # 45 – 03 Edificio 500, Bogota

[2] Environment and Development Masters Program, Environmental Studies Institute, National University of Colombia (IDEA/UN), Calle 44 # 45 – 67 Unidad Camilo Torres Bloque B2, Bogota

[3] Environmental Studies Institute, Environmental Agrarian Studies Group, National University of Colombia (IDEA/UN), Calle 44 # 45 – 67, Unidad Camilo Torres Bloque B2, Bogota

11.1. Introduction

Through symbolic, organizational, and technological systems tools, human societies have modified ecosystems. Agriculture has been the oldest and most important transformational relationship between human beings and ecosystems (Ángel, 1995, 1996).

In the seventies, the environmental movement arose along with the first evidence of the negative impact of the use of agrochemicals and synthetic fertilizers on human health and ecosystems, caused by the current model of conventional agriculture (Carson, 1962). This movement criticizes the practices for their effects on nature and their implications for human well-being (Ehrlich and Ehrlich, 1987; Daily *et al.*, 1997).

*Corresponding author: ingridquin@gmail.com

The complex environmental discourse[1], whose purpose is to understand the role of humanity within nature, introduces new philosophical approaches and arguments into the nascent science of agroecology[2]. This is an interdisciplinary and integrative science that deals with environmental problems in the cultural and natural areas where agrarian activity takes place.

Agroecosystems, as basic units of analysis, are represented and evaluated in different ways under the contrasting visions of conventional agriculture and agroecology.

Conventional agriculture, as per the Green Revolution, envisages agroecosystems as those necessary to obtain food, fibers, and energy. The main goal from this perspective is to achieve higher yields, productivity, and efficiency, as required for the accumulation of capital. As such, biodiversity is simplified, and its structures, relationships, and functional processes depend on the farmer, who substitutes these characteristics with inputs like fertilizers, pesticides, GMOs, and fossil energies in order to maintain control and production (Ruttan, 2003; Patel, 2013; Holt-Giménez and Altieri, 2013).

Regarding agroecological concepts, the farmer is part of the agroecosystem which is in turn shaped by the social, economic, political, and technological stamps of his culture (León-Sicard, 2014). In this light, biodiversity plays an important role as it is essential to guarantee the naturally occurring processes and interactions that ultimately provide a wide range of benefits to the farmer.

[1] It recognizes that conservation per se is not the main objective of the environmental discussions as humans cannot conserve without transforming; in this way "... the environmental proposal should be based on well-oriented sciences and technology which allow the establishment of new limits to living systems..." (Ángel, 2003, p. 13). This vision establishes five basic premises with which to model reality: (1) with depth and width; (2) from an aesthetic and ethical perspective; (3) considering its interrelationships and consequences on temporal and spatial scales; (4) including its dynamism; and (5) with respect to subjectivity and different interests (Carrizosa, 2001).

[2] Agroecology is the result of the epistemological contributions of different sciences, such as agronomy, ecology, ethnography, rural sociology, and agrarian economics, and it is also considered to be an agriculture system, a social movement, and a symbolic system that has been permeated by environmental movements. Its aim is to harmonize the ethical and aesthetical existence of different forms of life in decent conditions within agroecosystems. Additionally, agroecology is a response to the politics of rural development imposed in Latin-American countries, whose purpose is to impose capitalist models of production, which are incompatible with the social, economic, and biogeographical local conditions. Finally, this science has, as an action axis, a set of ethical and aesthetic values regarding respect, solidarity, and the inclusion of several sources of knowledge and rituals of different actors in agrarian activity (Arocena, 1995; Wezel and Soldat, 2009; Sevilla-Guzmán and Soler, 2006; Calle-Collado *et al*., 2013; Sevilla-Guzmán and Woodwate, 2013; León-Sicard *et al*., 2018).

In this sense, the farmer should favor agrobiodiversity[3] by combining ecological principles and social contributions (with local knowledge, social fabric, and organizational forms of collective action) in order to achieve ecological and cultural sustainability of the agroecosystem (Altieri, 1999; Altieri and Toledo, 2011; Calle-Collado *et al.*, 2013; Sevilla-Guzmán and Woodgate, 2013; Durú *et al.*, 2015).

León-Sicard (2014) proposes the MAS (main agroecological structure) to characterize agrobiodiversity in the context of farms, where a complex environmental relationship exists. This approach incorporates the possibility of integrating production activities (agriculture, livestock, and forestry) with natural or semi-natural ecosystems. The MAS includes aspects of the regional ecosystem and the internal organization of the farm, as well as management and conservation practices, and other cultural determinants (León-Sicard *et al.*, 2018).

These MAS characteristics assume the need for spatial characterization of both farms and the landscape that surrounds them. For this reason, remote tools that allow the capture of images, virtualization of spatial information, and GIS (geographic information systems), comprise the set of technologies necessary to carry out this kind of research (Lo and Yeung, 2002).

Unmanned aerial vehicles (drones) have also become a viable and relatively low-cost way to obtain such spatial information when satellite images of the areas of interest are either not available, or of a very low resolution or quality (Colomina and Molina, 2014).

The present study has been undertaken within the framework of the agrarian environmental studies line of research at the Environmental Studies Institute (IDEA, from its initials in Spanish) of the National University of Colombia. It presents a methodological approach to the evaluation of the spatial criteria of the MAS with the use of drones and the integration of different sources of spatial information,. such as participatory mapping with the help of local farmers. The application of this methodology on two farms with contrasting styles of management (conventional and ecological) is also presented.

This set of methods allows an integrative approach to the study of agroecosystems and their agrobiodiversity in different ecosystems, even those whose special conditions are complex, like in the Andean mountains of Colombia.

11.2. The Main Agroecological Structure of Agroecosystems (MAS): A brief theoretical framework

The MAS is based on Latin American environmental thinking and the science of

[3] Agrobiodiversity is a concept that includes every component in agroecosystems that play a role in food production. It is defined as "the variability and variety of animals, plants, and microorganisms at genetic, species, and ecosystems levels which are necessary to sustain the key functions in agroecosystems, their structures and processes" (Annexure 1 of Decision III/11 of the Conference of Parties of the Convention on Biological Diversity, 2012).

agroecology, both of which accept the idea of a complex agroecosystem (León-Sicard, 2014).

The environmental dimension has been defined by various authors as complex, constant relationships with different meanings, magnitudes, and intensities that are established between ecosystems and cultures (Ángel, 1993, 1995, 1996, 2000; Noguera, 2006; León-Sicard, 2014) or, in other words, between human societies and the rest of nature. Culture, in this context, refers to all the theoretical positions and actions undertaken by humanity to adapt or transform its ecosystem environments. Culture includes both symbols that originate in human thought, and all forms of societal organization (social, economic, political, and military), which are expressed on various technological platforms (Ángel, 1998; León-Sicard, 2014).

From the meanings and theoretical advances made by humanity from the knowledge and appropriation of the immediate surroundings, societies constructed social relationships, balances of power, hierarchies, and authorities that influenced their modes and forms of production, trade, and accumulation that, in turn, generated specific relationships with their biophysical environments (Noguera, 2000).

Agriculture is a fundamental part of human activity, based on these symbolic processes and strongly rooted in socio-economic, political, and military relationships, and undoubtedly constitutes human beings' biggest intervention in ecosystems (Harari, 2011).

The particularly serious environmental effects produced by the current conventional agriculture model has provoked reactions from different sectors of society, including academics, farmers, and consumers, who propose alternative agricultural systems, some of which are based on the paradigm of agroecology (Altieri, 1989; Altieri and Nicholls, 2000; Acevedo-Osorio and Chohan, 2019).

Agroecological systems propose different approaches, philosophical principles, and methodologies for food production, which involve different options, both at the technological level (not using synthetic chemicals for the phytosanitary management of crops, for example), and in socioeconomic (autonomy, solidarity, fair markets, among other aspects) and symbolic terms (dialogue of knowledge, comprehensive vision of science, and normative regulations for food sovereignty and security). These proposals include knowledge and appropriate use of agrobiodiversity (Calle-Collado *et al.*, 2013; Gliessman, 2013; Sarandón and Flores, 2009).

Agrobiodiversity refers to the set of beings that intervene, in one way or another, in agroecosystems, either as crops or as plants and animals that are introduced – intentionally or not – but that have various functions on the farms. Pastures, trees, temporary and permanent crops, weeds, forage banks, living fences, forest remnants, grasslands, bushes, organisms of different trophic levels and animals for breeding or fattening that are within the boundaries of the farms, constitute the visible components of this agrobiodiversity, spatially arranged in variable ways (SCDB, 2008). In addition, there is edaphic biodiversity, usually invisible to the human eye, but of vital importance to life and the continuity of agroecosystems (fungi, bacteria, actinomycetes, algae, insects, arachnids, mollusks, protozoa). Part of this agrobiodiversity is planned and intentionally introduced by farmers, according to their perceptions, interests, and cultural possibilities (Vásquez, 2013).

It is in this scenario of cultural and ecosystem complexity that the MAS finds its true theoretical and practical justification. Indeed, the MAS includes five criteria of an ecosystem nature in its valuation, and another five of a cultural variety, to express how agrobiodiversity is linked to different social, economic, and technological factors (León-Sicard *et al.*, 2018).

It is also in this sense that the MAS can be considered as a 'dissipative structure' to mitigate disturbances external to farms. A larger agroecosystem or farm can absorb various economic, social or ecosystem disturbances (e.g. market-price variations, food insecurity, droughts, frosts, or floods), depending on its structure, expressed in terms of both the spatial arrangement of its land cover, and of the true capabilities and possibilities each farmer possesses to establish and maintain it (León-Sicard *et al.*, 2018).

As an environmental index that measures cultural and ecosystem aspects of agrobiodiversity, the MAS incorporates a conceptualization in which agroecosystems can be considered a special category of socio-ecosystem. Consequently, within the perimeters of individual plots or farms, the use and management of agrobiodiversity expresses not only the characteristics and requirements of cultivated plants and associated animal diversity, but also all of the cultural intangibles that enhance or limit their establishment (León-Sicard *et al.*, 2018).

But farms or larger agroecosystems are not alone in their physical or geographic space. In territories that have been highly transformed by human activities, farms are generally immersed in agricultural landscapes, sharing their boundaries with other farms that, in many cases, have within them remnants of forests or biological corridors that affect and are affected by the activities carried out in the surrounding agroecosystems (Perfecto *et al.*, 2009).

In this sense, it is possible to designate a different category, on a geographically smaller scale, known as the 'agroecosystems matrix', which groups together all the farms of similar MAS in a given geographic or geomorphological space. In this way, the MAS includes, in its conception and its methodology, the relationships of individual farms with their immediate surroundings (León-Sicard, 2020). It is worth clarifying that in other areas with higher-intensity land use, such as in the industrialized countries of Europe, Asia or North America, the agroecosystems matrix may be immersed in a more complex network, including industrial production units, cities or roadway infrastructure, a situation that poses an additional challenge for future studies of the application of the main agroecological structure of agroecosystems (Wu *et al.*, 2006; Gillespie *et al.*, 2012; Gingrich *et al.*, 2015).

In summary, the MAS is an environmental index that brings together elements of both cultural and ecosystem aspects. It can be used in different types of approaches to measure the effect of agrobiodiversity on diverse variables, such as crop productivity and plant health, soil and water conservation, resilience to climate variability, food security, ecological connectivity or land-use planning.

11.3. The Main Agroecological Structure (MAS): Criteria and Indicators to Measure Agrobiodiversity

León-Sicard (2014, 2021) and León-Sicard *et al.* (2018) proposed ten criteria for the cultural and ecosystem description of agroecosystems, with twenty-seven indicators to evaluate agrobiodiversity as an attribute of farms (Table 1). Six of these criteria need explicit spatial analysis and are related to: a) density of patches of natural vegetation and bodies of water in agroecosystems' areas of influence (with this latter value depending on the farm's dimensions); b) average distance from these patches to the center of the farm; c) length and diversity of the external and internal connectors; and d) land use. These criteria are explained by agricultural and livestock management practices in the production systems, the biodiversity conservation practices, and the farmers' perception, awareness, and knowledge of, as well as their capacity for action regarding agrobiodiversity (for which the maintenance of natural cover is very important).

11.4. Spatial Approach for Evaluation of the MAS

This chapter presents a methodological proposal for spatial analysis of the MAS, including different strategies that allow the study of complex aspects, such as land use and the interpretation of land cover. These strategies include the use of satellites or aerial images, floristic studies, interviews, participatory mapping, and software for spatial analyses, and the use of unmanned aerial vehicles or drones.

This study approaches agrobiodiversity on farms from the inductive perspective, and includes the use of both quantitative and qualitative variables. The research requires fieldwork to gather quantitative spatial and floristic information, as well as the assistance of farm owners and local inhabitants to determine the qualitative variables. The complementary analysis of the cultural and ecosystem aspects of the farms is basic information intended to support the producers in planning and managing them.

The proposed methodology is a five-phase process for the spatial description and explanation of the MAS (Fig. 1).

11.4.1. Phase I: Characterization of the Study Area and the Main Actors

During this phase, an exploration of the region containing the study of

Table 1: Evaluation Criteria of the Main Agroecological Structure (MAS) and Indicators of the Farm's Agrobiodiversity

Criterion		Indicator	Description
1	Connection with the main ecological landscape structure (CMELS)	Fragment distance (FD)	Average distance between the fragments/patches of natural vegetation in the farm's area of influence and its center.
		Distance between water bodies (DW)	Average distance between the water bodies in the farm's area of influence and its center.
		Fragment area percentage (FA)	Percentage of the area covered by fragments of natural vegetation and water bodies in the farm's area of influence.
2	Extension of external connectors (EEC)	EEC	Total length of vegetation connectors as a percentage of the farm's total perimeter.
3	Extension of internal connectors (EIC)	EIC	Percentage of linear extension of connectors with vegetation over the total length of the internal divisions of the farm separating production areas.
4	Diversity of external connectors (DEC)	Richness of external connectors (REC)	Richness or number of plant species in vegetation connectors on the perimeter of the farm.
		Stratification of external connectors (SEC)	Number of vertical strata or diametric classes (grazing, herbaceous, shrub, arboreal, emergent) in vegetation connectors on the perimeter of the farm.
5	Diversity of internal connectors (DIC)	Richness of internal connectors (RIC)	Richness or number of plant species in vegetation connectors in the internal divisions of the farm separating production areas.
		Stratification of internal connectors (SIC)	Number of vertical strata or diametric classes (grazing, herbaceous, shrub, arboreal, emergent) in natural and semi-natural connectors in the internal divisions of the farm separating production areas.

(*Contd.*)

Table 1: (*Contd.*)

Criterion		Indicator	Description
6	Land use (LU)	LU	Percentage of land use within the farm that favors biodiversity.
7a*	Agriculture management practices (aMP)	Seeds (SED)	Origin, type, production, and conservation of the seeds used by the producer.
		Soil preparation (SP)	Tillage type and intensity, complementary conservation practices prior to cultivation
		Fertilization (FZ)	Types of fertilizers, rotations, and complementary practices.
		Phytosanitary management (PM)	Management (ecological or not) of crop health: use of weeds and complementary practices.
		Crop diversification (CD)	Number of species and breeds cultivated.
7b	Livestock management practices (lMP)	Soil preparation (SP)	Tillage type and intensity, fertilizers, and soil conditioners, complementary conservation practices prior to sowing forage grasses.
		System arrangement (SA)	In silvo-pastoral systems, grass and legume diversity, scattered trees, and mixed forage banks.
		Pasture rotation (PR)	Usage and non-usage of rotational grazing and practices to measure the productivity of pastures.
		Water management (WM)	Origin, transport, storage, and control of the quality of water for livestock consumption.
		Sanitary management (SM)	Management (ecological or not) of livestock health: pest control.

8	Conservation practices (CP)	Soil conservation practices (SCP)	Evidence of erosion and practices for its control.
		Water conservation practices (WCP)	Protection of water sources, water harvesting, and recycling, quality analysis and complementary practices.
		Biodiversity conservation practices (BCP)	Maintenance and enrichment of habitats, integration of important animals into the production system.
9	Knowledge - awareness - perception (KAP)	(KAP)	Recognition of the importance and/or benefits and conceptual awareness of agrobiodiversity.
10	Level of capacity for action (CA)	Economic and financial capacity (EFC)	Income, savings, credit capacity, and access to support programs to improve or maintain agrobiodiversity and agroecological practices on the farm.
		Logistic capacity (LC)	Workforce, accessibility, means of transportation, infrastructure to improve or maintain agrobiodiversity and agroecological practices on the farm.
		Management capacity (MC)	Institutional relations, associativity, information management and planning to improve or maintain agrobiodiversity and agroecological practices on the farm.
		Access to technical assistance and training (ATT)	Supply, quality, frequency, and access to training to improve or maintain agrobiodiversity and agroecological practices on the farm.

* The management practice criteria (aMP and lMP) can be selected according to the most representative production system of the farm.

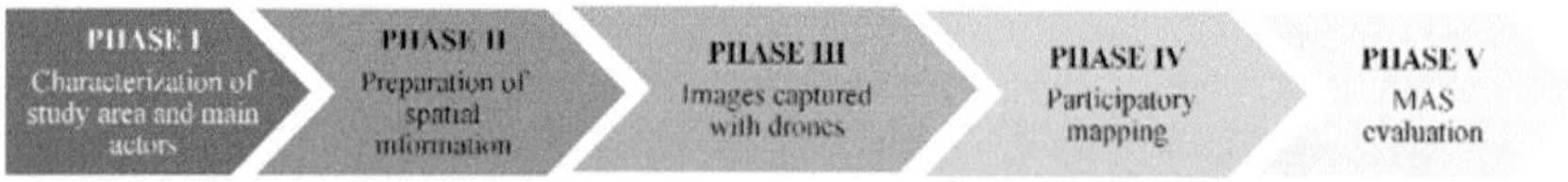

Fig. 1: Phases for determining the spatial components of the MAS (*Source*: Authors' elaboration)

agroecosystems is performed, and their biophysical and cultural characteristics, including socio-economic dynamics, are described. During this reconnaissance, the key actors that function as mediators between farmers and researchers are identified. In general, they are public officials that provide extension services and agricultural or environmental technical assistance in the municipalities.

Initial contacts are established with the community, a fundamental actor for the carrying out of field studies. For this reason, the purposes and scope of the research are communicated by using simple, non-technical language, in order to establish social networks and relationships of trust.

11.4.2. Phase II: Preparation of Spatial Information

This phase includes the gathering of information for spatial characterization of agroecosystems and the surrounding landscape, divided into three stages, as presented in the following subsections:

11.4.2.1. Initial Exploration of Agroecosystems

The researchers, accompanied by the farmers, explore the topographic characteristics and spatial boundaries of the farms, with the help of topographic maps if available. Also, the main areas of interest (natural and semi-natural land cover, pastures, crops, bodies of water, infrastructure, among others) are geolocated with GPS (Garmin, 60CSx).

11.4.2.2. Definition of Polygons and Areas of Influence

Areas of influence refer to zones that are adjacent to the farms and influence them, both via their own production processes and their ecosystem functions. The landscape elements and covers inside them include water bodies, fragments of vegetation, and infrastructure, which in one way or another affect the farms' relationships with their environment and vice versa.

As the farms' boundaries cannot be seen in the images obtained by remote sensors, they need to be viewed in one of two ways: (1) by consulting national cadastral databases, which are preferable in terms of time and resources; and (2) by using a GPS to determine the geolocation of the farms' boundaries. From the polygon that approximates a farm's boundary, the farm's area of influence is defined by the area of a circle whose radius r is twice the longest side of the farm L (equation 1) minus the farm's actual area (equation 2) (León-Sicard, 2014; 2021). Fig. 2 shows an example of the definition of a farm's area of influence.

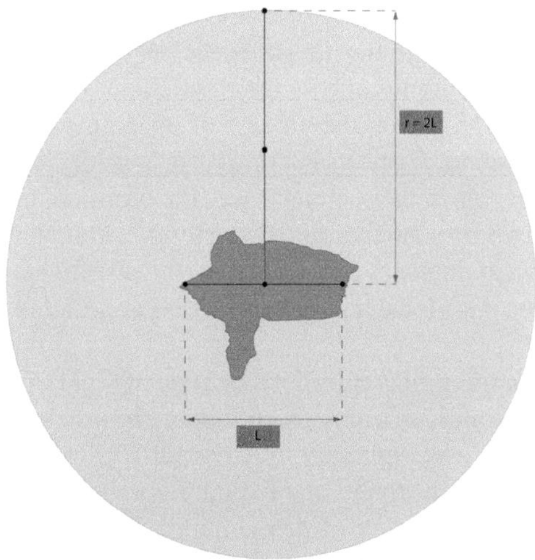

Fig. 2: Determination of a study farm's area of influence
(*Source*: Authors' elaboration)

$$r = 2L \quad (1)$$

$$IA = \pi r^2 - FA \quad (2)$$

Where,
L: Measurement of the longest side of the farm
r: Radius of the area of influence measured from the center of the farm
IA: Area of influence
FA: Farm area

11.4.2.3. Search with Remote Sensing Tools

Once the study area has been explored and spatially defined, it is necessary to obtain higher-resolution satellite images; for example, the 10m resolution Sentinel images available on platforms like Google Earth, Bang and Esri (free to use only for research purposes) or licensed images by private organizations. Another source is the collection of satellite images obtained by state institutions in charge of compiling geographic information of the study areas. Finally, in the case of spatial information not being available, it can be obtained by using aerial platforms, such as airplanes, helicopters or drones[4].

[4] When remote sensing images do not exist or do not have high quality, it is possible to partially describe the spatial dimension of the index with participatory mapping (using the knowledge and experience of the owners and the local community) and up-to-date topographic maps. Some examples are summarized in León-Sicard (2021).

11.4.3 Phase III: Images Captured with Drones

When satellite or aerial images meeting the necessary conditions for the study are not available, the use of drones allows the acquisition of images of high quality and spatial resolution. However, the process of image-capturing in rural areas is not only limited by technical aspects but must also consider topographic, climatic, and cultural complexities in the regions where the farms are located.

Because the MAS does not restrict the analysis within the farms' boundaries and drone flight activity includes external areas, the overflights (or flight mission) must be agreed upon with local authorities as well as with the community in the territory.

Moreover, researchers must consider the safety of flights in which topographic and meteorological conditions and reception of the satellite transmission are key factors. Therefore, it is important to carry out a prior reconnaissance of the study area, to find safe flight zones, and to plan rigorously. In Fig. 3 the four-step sequence for the ideal execution of drone flights is presented. The constituent steps and substeps are discussed in the following sections:

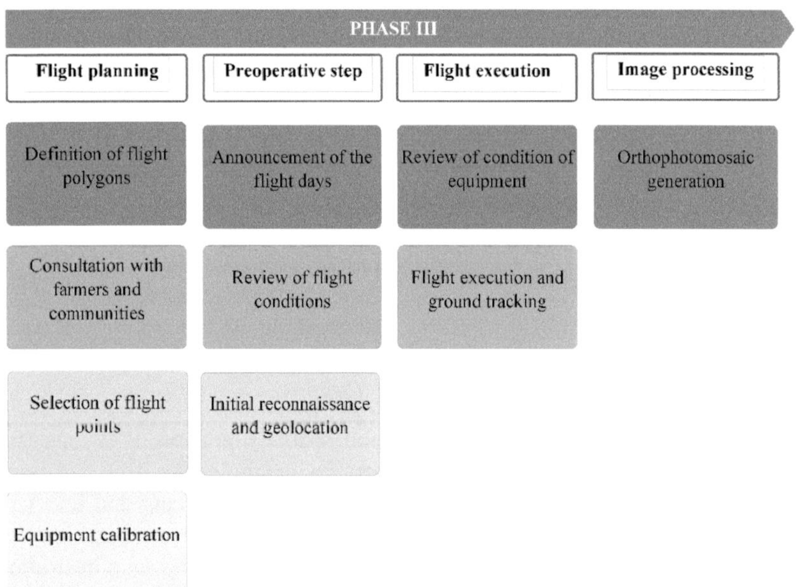

Fig. 3: Stages for the execution of effective and safe drone flights (*Source*: Authors' elaboration)

11.4.3.1. Flight Planning

Because of the long distances involved, the best option is to perform automatic flights whose areas are defined as rectangular polygons that circumscribe the areas of analysis (or areas of influence). This study methodology allows more efficient flights in terms of battery consumption and area overflown. Then, the polygons are uploaded to flight applications (e.g. Drone deploy, DJI ground station, or Pix4D Capture) (Fig. 4).

In many cases, when the photos require high precision, it is essential to use ground control points. Since the MAS indicators aim to determine the boundaries between land covers and measure their areas and distances, the automatic corrections made by the orthophotomosaic models are often sufficient to ensure that required information is of adequate quality.

Fig. 4: Example of flight polygons on the drone deploy platform

As the drone flight may overfly forest cover, electric antennas or even mountain peaks, it is necessary to geolocate safe take-off points, enabling the drone to travel long distances of up to 1,500 m from the point of origin. The use of tools that indicate contour lines, such as Google Maps and Google Earth (Fig. 5), is recommended.

Next, the protocols for updating the software must be followed, both for the verification of the technical and satellite conditions for take-off and flight performance.

As previously mentioned, it is important to consult local communities and leaders about the planned drone flights, so that the process can be carried out in a context of trust and mutual respect with a consensus having been reached about the dates, places, and conditions of the fieldwork zones (Fals-Borda, 2015; Vásquez-Fernández *et al.*, 2017). This can be done with personal interviews and group meetings to exchange knowledge about the study area and the project's goals and benefits. By involving the local inhabitants, these activities can lead to better management of sensitive steps, such as the choice of adequate take-off points and flight paths.

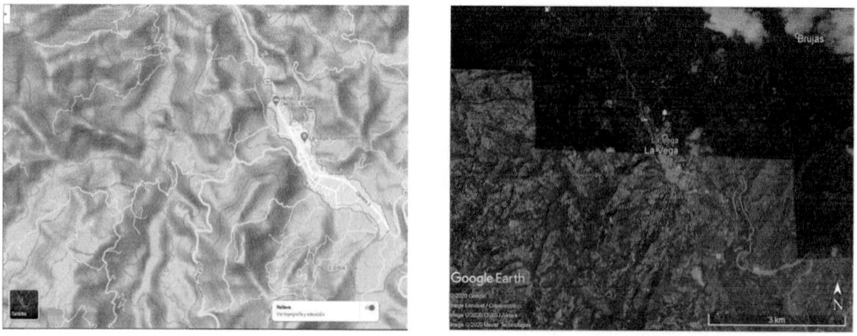

(A) (B)

Fig. 5: Examples of applications, e.g. Google Maps (A) and Google Earth (B), used for height verification to select flight points (*Source*: Google platforms images)

Additionally, it is important to follow regulations set by the national authority regarding the areas and conditions for overflying the territory. This allows for effective flight planning and the establishment of a relationship with the local authorities in the image-capturing exercises. In Colombia's case, Resolution No. 4201 of 2018 on civil aeronautics pertains.

11.4.3.2. Preoperative Step

After the technical aspects have been defined and before the flight image capture step is performed, the researchers must verify that the field conditions permit a safe flight.

To begin the process, the pre-selected take-off points must be verified by an on-site survey. By doing this, the researchers can confirm access possibilities, such as tertiary roads, trails, or routes on private farms, and means of transport (e.g. public transport, private vehicle, on foot).

Once the flight points have been pre-selected, the atmospheric conditions and satellite connection must be verified to guarantee safer flights. Applications like UAV Forecast can be used for this purpose.

11.4.3.3. Flight Execution

When the take-off point is reached, the drone must be monitored, reviewed, and calibrated. In the case of the various versions of DJI Phantom equipment, DJI Go software is used for this operation. Once in flight, it must be constantly monitored with both the control equipment and visual tracking by a second observer (Fig. 6).

11.4.3.4. Orthophotomosaic Generation

The set of photographs must be spliced to generate orthophotomosaics, which are the base files on which the spatial analysis is performed. Among the software available for this procedure is Pix4D Mapper, which was used in this study.

(A) (B)

Fig. 6: Flight day in typical mountainous landscape and monitoring with ground control (A) and visual tracking (B) (*Source*: Authors′ archive)

11.4.4. Phase IV: Participatory Mapping

Since the MAS evaluation is an exercise that introduces cultural aspects as conditioner agents of the configuration of the agroecosystems, participatory mapping is a spatial-information collection process that is closely linked with the experience and perceptions of the farm owners and/or managers (Fig. 7).

Once the orthomosaics containing the area of each farm have been generated and printed in color on a scale from 1:4000 to 1:5000, the owners or administrators can use a marker to delimit the perimeter, based on their recognition of physical points. Afterwards, they can classify and highlight the different uses of land by making their geometric shapes in detail, allowing for the creation of a very precise interpretative map of the covers present on the farm.

Fig. 7: Participatory mapping exercises to consolidate spatial information about the farms (*Source*: Authors′ archive)

Carrying out this exercise with the owners allows for comparison and precise definition of the perimeters of each farm and facilitates the subsequent interpretation of the land covers present in its area of influence.

These activities are also opportunities for the researchers to conduct interviews about the agroecosystems' conditions and investigate the reasons behind the spatial structure of the farms and their agrobiodiversity, which are evaluated by using the non-spatial indicators of the MAS index. By doing this, the owners and/or administrators become active agents in the research.

Finally, the overall methodology and sources used for the determination of the spatial aspects of the farms allow a holistic vision of the realities that define the agroecosystems and their relationship with the landscape.

11.4.5. Phase V: MAS Evaluation

After the spatial images have been obtained, the MAS analysis process begins, using the spatial criteria and indicators. Spatial analysis platforms, such as QGIS and ArcGIS, are used for this purpose.

This phase begins with the creation of a GIS project for each farm with all the spatial information generated in order to monitor, analyze, and define the MAS. The first step starts with the definition of the farms' polygons, in accordance with the information given by the owners (coordinates of key points, and orthomosaics) for the processing and generation of polygons associated with the types of land covers and uses present on the farms, and in their respective areas of influence. This image-interpretation process is supported by the in-field corroboration: the floristic surveys, the verification made by the authors in some sites where the research was conducted, and the participatory mapping. Finally, the covers are classified, following the Corine land cover methodology for Colombia (IDEAM, 2011).

The polygons representing the land covers and uses are the key to generating the landscape metrics selected within the MAS: density of fragments and water bodies in the area of influence, distances from the centers of the farms, length of vegetation corridors (external and internal), and areas taken up by different land uses. Naturally, the MAS analysis includes the phytogeographic and floristic characterization of the connectors in order to evaluate their biodiversity, as well as the participatory mapping process and semi-structured interviews to describe the cultural characteristics of the agrobiodiversity present on the farms.

11.5. Application Case of the Proposed Methodology

11.5.1. The Study Area

The studies that underpinned this methodological approach were carried out in the municipalities of La Vega and Nocaima, in the department of Cundinamarca,

Colombia (Fig. 8). This region is located in the watery province of Gualivá, and contains a set of forest relicts that contribute to the north-south biological connectivity on the western slopes of Cundinamarca Department towards the inter-Andean valley of the Magdalena river (Drews *et al*., 2019).

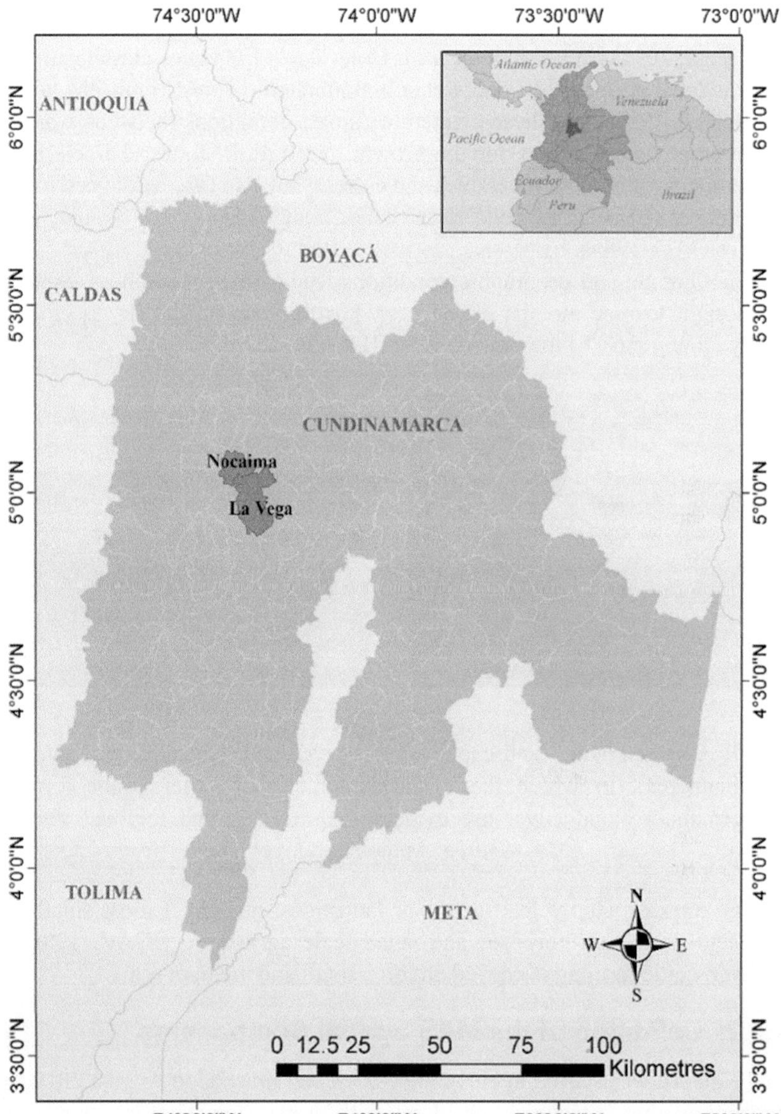

Fig. 8: Map of the study region located in the municipalities of La Vega and Nocaima (Cundinamarca, Colombia) (*Source*: Authors´ elaboration)

The region is mainly mountainous, with slopes ranging between 25-95 per cent, with few flat sectors. Consequently, the physiography is made up of mountains, hills, valleys, fan-shaped terrain, and terraces, constituting a highly heterogeneous territory (CAR, 2014; Municipio de Nocaima, 2012; Municipio de La Vega, 2018). The mean temperature varies between 8-24°C and the mean annual rainfall is between 1,400-1,700 mm, with two rainier periods from April to May and September to November (Municipio de Nocaima, 2012; Municipio de La Vega, 2018). In accordance with Holdridge's life zones classification, the region is located in a moist and wet sub-mountainous forest (bmh-PM and mh-PM) (IGAC, 1985a, b). These temperate climate conditions (between warm and cold thermal floors) allows for the growth, not only of foods characteristic of cold-temperate climates (vegetables and coffee), but also those produced in warm temperatures (banana, cassava, sugar cane, mango, tangerine, orange, among others) (IGAC, 1985a, b).

The climatic and orographic conditions lead to high cloudiness, especially during eight to nine months of the year. For this reason, the sky is cloudy to heavily cloudy most of the time (Fig. 9) (IDEAM, 2020).

(A) (B)

Fig. 9: Characteristic landscape of La Vega and Nocaima municipalities (Cundinamarca), in which the local phenomena of water vapor formation and permanent cloud cover are evident in local (A) and regional view (B) (*Source*: Authors' archive)

The terrain's slopes lead to major landslides, making it most suitable for ecosystem protection purposes and small-scale agriculture (IGAC, 1985a, b). However, cattle farming is carried out on a small and medium scale.

11.5.2. Definition of the MAS Spatial Components

The fieldwork was started in November 2018 and finished in August 2019, with interruptions due to technical problems with drones, logistical issues related to flying over farms with over 300m of altitude gradient in a total area covering 6km^2, as well as weather conditions that did not allow for safe execution of the flights on all the field trips.

The first step was to contact the farmers through the municipal technical assistance units (UMATA, from its initials in Spanish) in the municipalities of La Vega and Nocaima. There, the officials led the pre-selection of farms based on criteria given by the researchers: agricultural or livestock farms, managed in a conventional or ecological way, among other characteristics. In doing this, officials acted as a bridge between the researchers and owners.

Once the farms had been explored and georeferenced, the spatial information was cross-referenced with the country's cadastral system. The digital polygons of each farm were obtained from the Agustín Codazzi Geographical Institute (IGAC, from its initials in Spanish).

With these data, the spatial information phase started, and free-access satellite images were consulted on the United States Geological Survey platform (USGS, 2020) and Google Earth (for the years 2017 and 2018). In the first case, the images did not have a high enough spatial resolution, with a maximum of 15 mper pixel in Landsat images (in panchromatic), and 30min. multispectral. In the second case, the satellite images were of poor quality and contained a great deal of cloud cover.

Furthermore, the Colombian institutions in charge of spatial generation and/or compilation, IGAC, and the Regional Autonomous Corporation of Cundinamarca (CAR), did not have detailed information on the study areas. Finally, we consulted the private satellites Ikonos, OrbView, and SPOT (Apollo mapping, 2021; Harris Geospatial, 2021) with spatial resolution up to 1 m, but no information was available. These difficulties led to the decision to obtain spatial information from drones.

The image-capturing phase started with flight planning. The flight areas were first mapped on the Google Earth platform to cover farms and their areas of influence. This tool was also used to find and geolocate the highest topographic points, both inside and outside the areas of influence, to ensure safe flights. Then, as the flight areas were defined, the new polygons were uploaded to the flight software drone deploy.

For every field operation, the community, local leaders, and officials of the UMATA were informed by face-to-face visits or through radio messages at La Vega's radio station. In some cases, the national police accompanied the drone fieldwork.

When optimal take-off points had been located, the drone was calibrated for the flight conditions with the support of the software DJI Go v. 4.0. Additionally, the ideal flight conditions were checked, and the flights were performed by using the drone deploy software. 60 per cent picture overlapping and a speed of 10 m/s generated a precision of 5 m per pixel by using a camera with a resolution of 12 megapixels.

After the image capturing had been performed, the images captured in each flight plan were spliced into Pix4D Mapper software (v.4.3.31). The process started from the generation of cloud points and a digital spatial model that finally generated an orthophotomosaic in GeoTIF format.

Subsequently, in participatory mapping activities, the owners or administrators delimited the farm perimeters and identified the present land covers inside them on the physical map (the orthomosaic).

11.5.3. The MAS Analysis Process

Based on the spatial information, the MAS analysis process was carried out for 14 farms in the ongoing research. However, for the purposes of this chapter, only the information regarding two contrasting (in terms of production management) agroecosystems is presented; the two cases refer to the Palma de Cera (ecological agroecosystem) and La Aldea (conventional agroecosystem) farms (Table 2).

The first farm, Palma de Cera Nature Reserve, focuses primarily on the ecological production and sale of coffee; this is complemented by other production activities, such as agritourism, ecological tourism, and the development of the orchard, which is very important in their search for food autonomy. Palma de Cera is run by a family whose heads are a doctor and a forest engineer, the latter also being a university teacher. They call themselves 'neo-rural' because they have left the city to live in the country and be in close contact with nature. Accordingly, they set up the production activities so they could be in harmony with the ecosystem.

The second farm is La Aldea, whose purpose is the conventional production of livestock. Here, the owners' relationships with nature are mainly economic, and pastures constitute the main land use as the best way to take advantage of the space. This land cover is only interrupted by gallery forest. The owners are retirees who were born in the region. Their children migrated to Bogotá and they do not intend to take charge of the farm, which could lead to the sale of the land in the short or medium term.

The MAS and its indicators (*see* Table 1) were evaluated through a set of methods that incorporate the use of drone and satellite images, participatory mapping, floristic and phytogeographic surveys, and semi-structured interviews (Table 3 and Figs. 10 to 13). The spatial information made particularly important contributions to the assessment of four of the MAS criteria (CMELS, EEC, EIC, and LU).

According to Criterion 1, both agroecosystems had the same valuation of three (3), representing similar landscape connectivity conditions. In general, the regional landscape has a similar pattern: the productive farms are part of a relatively complex matrix of recreational, crop and livestock farms, interspersed with small patches of natural vegetation in different degrees of succession, and gallery forests. Similar values were reported by León-Sicard (2021) in six horticultural farms with ecological management in the Bogota savanna (located at an altitude of 2,600 masl). They found that even in the cases of farms with ecological management, the surrounding landscape was in general very different from their interiors: the farms were immersed in a matrix of paddocks for cattle, with very low-density patches of vegetation.

Table 2: General Information on Major Agroecosystems in the Study

Type	Name	Area (ha)	Primary Use	Municipality	Type of Property	Coordinates and Altitude (Mamsl)
Ecological	Palma de Cera	6,17	Conservation and coffee	La Vega	Private	4°56'33.66" N 74°21'47.87" W 1,778
Conventional	La Aldea	24,38	Cattle raising	Nocaima	Private	5° 3'16.11" N 74°21'44.45" W 1,281

Table 3: Assessment of MAS for the Two Contrasting Major Agroecosystems. CMELS = Connection with the Main Ecological Landscape Structure; EEC = Extension of External Connectors; EIC = Extension of Internal Connectors; DEC = Diversity of External Connectors; DIC = Diversity of Internal Connectors; LU = Land Use; MP= Management Practices; CP = Conservation Practices; KAP= Knowledge-Awareness-Perception

Farm	CMELS	EEC	EIC	DEC	DIC	LU	MP	CP	KAP	CA	V/r	MAS Description
Palma de Cera	3	10	10	10	6	10	6,25	10	10	8	83,25	Very strongly developed
La Aldea	3	6	6	10	4	6	3	3	3	3,75	47,75	Moderate to Slightly developed

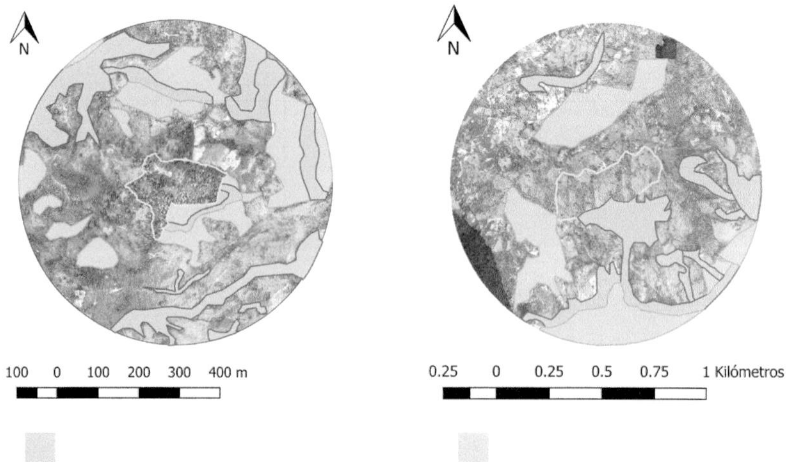

Fig. 10: Criterion 1: Connection with the main ecological landscape structure (CMELS) in La Aldea (A) and Palma de Cera (B) (*Source*: Authors´ elaboration)

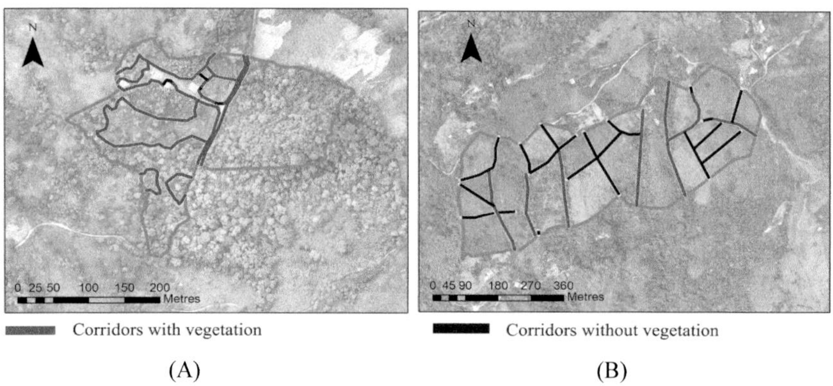

Fig. 11: Criterion 2: Extension of external connectors (EIC) in Palma de Cera (A) and La Aldea (B) (*Source*: Authors´ elaboration)

The evaluations of the MAS's other ecosystem criteria showed better conditions for agrobiodiversity in the ecological system. This is because the extension of internal and external connectors, as well as their diversity (evaluated in criteria 2 to 5), are closely related to the type of production system. Coffee farms in La Vega and the surrounding regions were originally implemented as shading systems, in which they have preserved trees as associated agrobiodiversity (similar results were found in León-Sicard, 2020). On the other hand, in livestock systems, most of the tree vegetation has been eliminated, with only a little conserved for water sources, protection, or to delimit the boundaries of the farm or paddocks.

The greatest contrast was found in the cultural criteria. In the conventional livestock production system, management and conservation practices were very

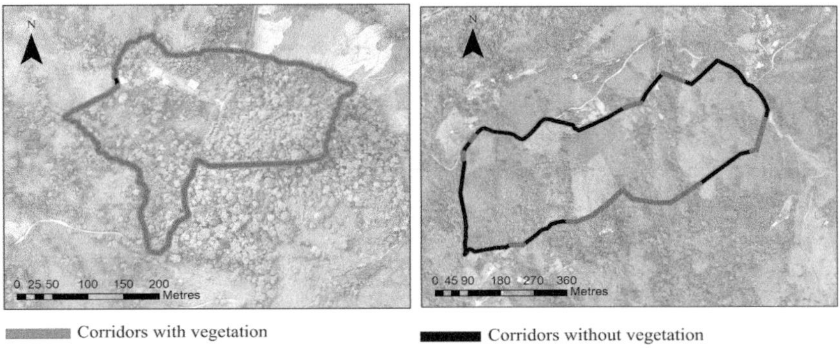

(A) (B)

Fig. 12: Criterion 3: Extension of internal connectors (ECI) in Palma de Cera (A) and La Aldea (B) (*Source*: Authors´ elaboration)

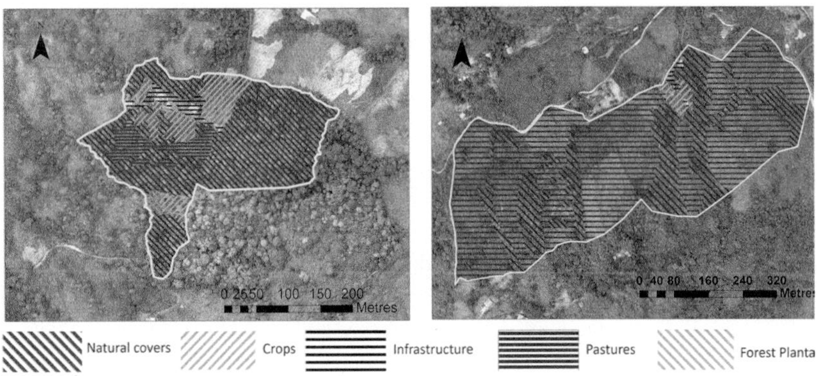

(A) (B)

Fig. 13: Criterion 6: Land use (LU) in Palma de Cera (A) and La Aldea (B) (*Source*: Authors´ elaboration)

weak and there was evidence of the prevalence of economic interest and personal benefit, and limited awareness of the importance of agrobiodiversity.

The better evaluations for the ecological agroecosystem denote a close relationship with ecosystems, influenced by the cultural characteristics of farmers: academic training in natural sciences, high awareness of the need for conservation, and a persistent interest in the search for ecological production processes.

11.6. Final Reflection and Lessons Learned

The main agroecological structure of Agroecosystems is an environmental index that measures different aspects of agrobiodiversity. It does not replace existing studies on the function of that biodiversity, but it does provide various criteria and attributes that measure the spatial connectivity and the richness in species of plant cover present on the farms, information that facilitates analysis of the functional relationships of distinct biotic components of the larger agroecosystems.

Its applications include the description of an emerging quality of agroecosystems that can be used in the future for taxonomic studies and to facilitate the comparison of thematic studies. The MAS can also be applied to research on various subjects, ranging from environmental history and ecological economics to relationships with plant health, climate change resilience, soil conservation, ecosystem services, land-use planning, food productivity, and food security, among others.

The MAS is calculated, based on spatial valuation metrics whose center and purpose is to understand the farm within its perimeter and in relation with the landscape. However, to complement these metrics, the MAS includes intangible variables that are not yet recognized in many scientific disciplines. Among these variables, it directly estimates farmers' perceptions, knowledge, and environmental awareness, while drawing attention to their economic and financing opportunities and their access to agroecological training to establish and maintain functional vegetation connectors over time.

The MAS's measurements require scaling, from the individual farm to the landscape. This requires technological instruments that allow the viewing of individual farms in relation to their surroundings; consequently, remote sensing technology, from conventional photographs to satellite images or those obtained by drones, become indispensable.

Based on the research we conducted, the use of drones to characterize agroecosystems' agrobiodiversity is especially appropriate when it is not possible to access satellite images of sufficient quality and spatial resolution for detailed analysis. In addition, the high-quality images obtained and the participatory mapping process are useful tools for farmers to appropriately manage the agrobiodiversity of their farms and territory.

Furthermore, in the image-capturing exercise, it is possible to establish a fluid interaction with the farm owners and neighboring communities, facilitating a better understanding of the dynamics behind the production processes in agroecosystems and the agro-landscape – aspects that can explain the differences in locally-observed agrobiodiversity.

Additionally, the MAS serves to partially characterize agroecosystems or farms. This characterization is only partial because a complete description should include other qualities or characteristics: size, shape, geographical conditions, climate, relief, types of soils, and, perhaps most importantly, types of producers. This is the challenge for agroecological science, as regards addressing or categorizing farmers' environmental actions. Future research may consider this work as a starting point to construct a taxonomy that helps to understand the particularities of agroecosystems that exist in Africa or Asia versus those of the Americas or Europe. This taxonomy, not yet developed, will most likely present enough challenges and rewards to eventually become an autonomous branch of agroecology.

But beyond its usefulness as an identifier, descriptor, and characterizer of agroecosystems, the MAS has the potential to become an effective tool in land-use planning and in predicting the behavior of agroecosystems in relation to their quality, plant health, nature conservation, or economic efficiency.

Any farmer can plan the use of his farm, improving its agrobiodiversity, and therefore its main agroecological structure, to ensure the greatest possible number of interrelations and, at the same time, the lowest possible number of interventions. Complex agrobiodiversity in itself provides the farmers, not only with materials to maintain soil fertility, but also to control herbivorous pests or disease agents. Of course, this will depend on the application of the general principles of agroecology and the individual capacity to establish a functionally-adequate MAS.

The MAS allows awareness to be raised of both the farms and the farmers, whose decisions are fundamental in organizing the territory on those smaller scales that are used in land-use planning.

If farmers understand the importance of managing their farm by planning the use of its agrobiodiversity, this can lead to replacing monocultures with polycultures, incorporating patches of forest into the farm, and using hedges as fences, or more generally, encouraging practices that increase plant cover and biodiversity in all its forms. In this way, farmers can contribute to the formation of an ecological landscape structure, which will favor community support for the protection of watersheds, soils, waters, and biodiversity, improving community living conditions, together with the quality of products for external consumers.

The MAS's contribution to this process is to raise awareness on various scales when applied to planning and landscapes. Indeed, the MAS is a useful tool for the land-use planners' analysis to move from a top-down perspective to a bottom-up approach, where planners work with the local inhabitants on the scale of the farm.

These capacities of farmers and/or planners can be improved with the help of monitoring systems based on unmanned aerial vehicles, such as drones, when cartographic information is not available or is of low quality. Much of the information obtained by these means must be validated with the help of local people, who have the necessary knowledge to identify the agrobiodiversity components on their farms, as well as the advantages and disadvantages of its use on different farm and landscape scales.

Acknowledgments

The financial resources required to purchase the drone and related software, as well as to carry out the field work, were granted by the bilateral cooperation agreement to support doctoral training signed between the Institute of Environmental Studies of the National University of Colombia (IDEA/UN) and the Institute of Development Research (ZEF) of the University of Bonn in Germany, with funding from the German Academic Exchange Service (DAAD, from the initials in German). This project also had the support of the Agricultural Sciences Faculty of the National University of Colombia and the Administrative Department of Science, Technology and Innovation of Colombia (Colciencias).

Bibliography

Acevedo-Osorio, A. and J. Chohan (2019). Agroecology as social movement and practice in Cabrera's peasant reserve zone, Colombia, *Agroecol. Sust. Food Sys.*, 4: 331-351.

Altieri, M. (1989). Agroecology: A new research and development paradigm for world agriculture, *Agric. Ecos. Environ.*, 27: 37-46.

Altieri, M. and V. Toledo (2011). The agroecological revolution of Latin America: Rescuing nature, securing food sovereignty and empowering peasants, *J. Peas. Stud.*, 38: 587-612.

Altieri, M. and C.I. Nicholls (2000). Agroecology and the search for a truly sustainable agriculture, *United Nations Environment Programme – Environmental Training Network for Latin America and the Caribbean*, PNUMA, Mexico City, Mx.

Altieri, M., S. Liebman, M. Magdoff, F. Norgaard, R. Sikor and O. Thomas (1999). *Agroecología, Bases científicas para una agriculturasustentable, Nordan-Comunidad*, Montevideo. Uy.

Ángel, A. (1993). *La trama de la vida. Bases ecológicas del pensamientoambiental*, Serie Ecosistema y Cultura No 1, Universidad Nacional de Colombia, Instituto de EstudiosAmbientales, Ministerio de Educación Nacional, Bogotá, Co.

Ángel, A. (1995). *La tierra herida, Las transformacionestecnológicas del ecosistema*, Serie documentosespeciales No 2. Universidad Nacional de Colombia, Instituto de EstudiosAmbientales, Ministerio de Educación Nacional, Bogotá, Co.

Ángel, A. (1996). *El reto de la vida. Ecosistema y cultura, Una introducción al estudio del medio ambiente*, Serie: Construyendo el futuro No. 4. Ecofondo, Bogotá, Co.

Ángel, A. (2000). *La aventura de los símbolos, Una visiónambiental de la historia del pensamiento*, Ecofondo, Bogotá, Co.

Ángel, A. (2003). *La diosa Némesis: desarrollosostenible y cambio cultural*, Corporación Universitaria Autónoma de Occidente, Cartographics S.A., Cali, Co.

Apollo mapping (2021). Retrieved from: http//apollomapping.com; accessed on 25 March, 2021.

Arocena, J. (1995). *El análisis de lo social desde la perspectiva del desarrollo local*, pp. 147-171. *In:* J. Arocena (Ed.). *El desarrollo local: un desafíocontemporáneo, Nueva Sociedad*, Caracas, Venezuela. not in text

Calle-Collado, A., D. Gallar and. J. Candón (2013). *Agroecologíapolítica: la transición social haciasistemasagroalimentariossustentables, R. Econ. Crít.*, 16: 244-277.

CAR (Corporación Autónoma Regional de Cundinamarca). (2014). Plan de Manejo Integral de la Cuchilla El Chuscal (La Vega, Cundinamarca); Retrieved from: https://datosgeograficos.car.gov.co/datasets/4c3938bbcf8e4ed299ced294acc04504_0?geometry=-74.457%2C4.882%2C-74.196%2C4.942; accessed on 18 November, 2019.

Carrizosa, J. (2001). *Qué es el ambientalismo? La visiónambientalcompleja*, Centro de Estudios de la RealidadColombiana (CEREC), Instituto de EstudiosAmbientales, Universidad Nacional de Colombia, Bogotá, Co.

Carson, R. (1962). *Silent Spring*, Crest Books, Fawcet World Library, New York, USA.

Colomina, I. and P. Molina (2014). Unmanned aerial systems for photogrammetry and remote sensing: A review, *ISPRS J. Photo. Rem. Sens.*, 92: 79-97.

Conference of Parties of the Convention on Biological Diversity (2012).

Daily, G.C., S. Alexander, P.R. Ehrlich, L. Goulder, J. Lubchenco, P.A. Matson, P.A. Mooney, S. Postel, S.H. Schneider, D. Tilman and G.M. Woodwell (1997). Ecosystems services: Benefits supplied to human societies by natural ecosystems, *Iss. Ecol.*, 4: 1-16.

Drews, D., M. Santamaría, C. Durana and M. Hernández (2019). La Laja, un núcleo de conservación entre las nubes, pp. 230-237. *In:* Matallana, C., Areiza, A., Silva, A., Galán, S., Solano, C. and A.M. Rueda (Eds.). *Voces de la gestión territorial*: *Estrategiascomplementarias de conservación de la biodiversidad,* Instituto de Investigación de Recursos Biológicos Alexander von Humboldt y Fundación Natura, Bogotá, Co.

Durú, M., O. Therond and M. Fares (2015). Designing agroecological transitions: A review, *Agron. Sustain. Develop.*, 35: 1237-1257.

Ehrlich, P.R. and A.H. Ehrlich (1987). *Extinción I, Salvat Editores*, Barcelona, Es.

Fals-Borda, A. (2015). *Una sociologíasentipensante para América Latina*, CLACSO, Siglo XXI Editores, Mexico City, Mx.

Gillespie, T., S. Pincetl, S. Brossard, J. Smith, S. Saatchi, D. Patakim and J.D. Saphores (2012). A time series of urban forestry in Los Angeles, *Urban Ecosyst.*, 15: 233-246.

Gleissman, S.R. (2013). *Agroecología: Plantando las raíces de la resistencia,,.Agroecol.*, 8: 19-20.

Gingrich, S., M. Niedertscheider, T. Kastner, H. Haberl, G. Cosor, F. Krausmann, A. Tobias, T. Kuemmerle, D. Müller, A. Reith-Musel, M. Rudbeck-Jepse, A. Vadineanun and K.H. Erb (2015). Exploring long-term trends in land use change and aboveground human appropriation of net primary production in nine European countries, *Land Use Pol.*, 47: 426-438.

Harari, Y.N. (2011). *Sapiens: A Brief History of Humankind*, Penguin Random House, London, UK.

Harris Geospatial (2021). Retrieved from: http//3harrisgeospatial.com; accessed on 25 March, 2021.

Holt-Giménez, E. and M. Altieri (2013). Agroecología, soberanía y la nuevarevoluciónverde, *Agroecol.*, 8: 65-72.

IDEAM (Instituto de Hidrología, Meteorología y Estudios Ambientales). (2010). Leyendanacional de coberturas de la tierra. *Metodología CORINE Land Cover adaptadapara Colombiaenescala*, 1:100.000. IDEAM, Bogotá, Co.

IDEAM (Instituto de Hidrología, Meteorología y EstudiosAmbientales). (2020). Retrieved from: http//ideam.gov.co.; accessed on 20 September, 2020.

IGAC (Instituto Geográfico Agustín Codazzi). (1985a). Estudio general de suelos y zonificación de Tierras, Tomo I, Departamento de Cundinamarca, Bogotá, Co.

IGAC (Instituto Geográfico Agustín Codazzi). (1985b). Estudio general de suelos y zonificación de tierras, Tomo II, Departamento de Cundinamarca, Bogotá, Co.

Kuper, A. (1999). *Cultura, La versión de los antropólogos*, Editorial Paidós, Barcelona, Es.

León-Sicard, T. (2009). *Agroecología: desafíos de una cienciaambientalenconstrucción*, pp. 45-68. *In*: Altieri M. (Ed.). *Vertientes delpensamientoagroecológico*: *Fundamentos y aplicaciones*, *Sociedad Latinoamericana de Agroecología*, Medellín, Co.

León-Sicard, T. (2014). *Perspectivaambiental de la agroecología, La ciencia de los agroecosistemas*, Universidad Nacional de Colombia, Instituto de Estudios Ambientales, Bogotá, Co.

León-Sicard, T. (2021). *La Estructura Agroecológica Principal de los agroecosistemas, Perspectivasteóricoprácticas del pensamientoambientalagrario*, Universidad Nacional de Colombia, Instituto de EstudiosAmbientales (in press), Bogotá, Co.

León-Sicard, T., J. Toro, L.F. Martínez-Bernal and J.A. Cleves-Legízamo (2018). The main agroecological structure (MAS) of the agroecosystems: Concept, methodology and applications, *Sustain.*, 10: 3131.

Lo, C.P. and A.K.W. Yeung (2002). *Concepts and Techniques of Geographic Information Systems*, Prentice Hall Inc., New Jersey, USA.

Municipio de La Vega (2018). *Plan municipal para la gestión del riesgo PMGRD*, Technical Report of the Consejo Municipal para la gestión del riesgo de desastres, La Vega (Cundinamarca); retrieved from: https://repositorio.gestiondelriesgo.gov.co/bitstream/handle/20.500.11762/28924/PMGRD_LaVegaCmarca_2018.pdf?sequence=1&isAllowed=y; accessed on 20 September, 2020.

Municipio de Nocaima (2012). *Plan municipal de gestión de riesgo de desastres CMGRD*, Technical Report of the Consejo Municipal para la gestión del riesgo de desastres, Nocaima (Cundinamarca); retrieved from: https://repositorio.gestiondelriesgo.gov.co/bitstream/handle/20.500.11762/432/PMGR%20Nocaima.pdf?sequence=1&isAllowed=y; accessed on 20 September, 2020.

Noguera, P. (2004). *El reencantamiento del mundo*, Universidad Nacional de Colombia, PNUMA, Manizales, Co.

Patel, R. (2013). The long Green Revolution, *J. Peas. Stud.*, 40: 1-63.

Perfecto, I., J. Vandermeer and A. Wright (2009). Nature's matrix: Linking agriculture, conservation and food sovereignty, *Earthscan*, London, UK.

Ruttan, V.W. (2002). *Controversy about Agricultural Technology: Lessons from the Green Revolution*, Department of Applied Economics of University of Minnesota, St Paul, USA.

Sarandón, S.J. and C. Flores (2009). Evaluación de la sustentabilidadenagroecosistemas: Una propuestametodológica, *Agroecol.*, 4: 19-28.

SCDB (Secretaría del Conveniosobre Diversidad Biológica). (2008). *La biodiversidad y la agricultura, Salvaguardando la biodiversidad y asegurando la alimentación del mundo*, PNUMA, Roma, It; retrieved from: https://www.cbd.int/doc/bioday/2008/ibd-2008-booklet-es.pdf; accessed on 13 May, 2019.

Sevilla-Guzmán, E. and M. Soler (2006). Del desarrollo rural a la agroecología, Hacia un cambio de paradigma, *Doc. Soc.*, 155: 25-41.

Sevilla-Guzmán, E. and G. Woodgate (2013). Agroecología: Fundamentos del pensamientoagrario y teoríasociológica, *Agroecol.*, 8: 27-34.

USGS (US Geological Survey). (2020). Retrieved from: http//usgs.gov; accessed on 20 September, 2021.

Vásquez, L. (2013). Diagnóstico de la complejidad de los diseños y manejos de la biodiversidadensistemas de producciónagropecuariaentransiciónhacia la sostenibilidad y la resiliencia, *Agroecol.*, 8: 33-42.

Vásquez-Fernández, A.M., R. Hajjar, M.I. Shuñaqui Sangama, R.S. Lizardo, M. Pérez Piñedo, J.L. Innes and R.A. Kozak (2017). Co-creating and decolonizing a methodology using indigenist approaches: Alliance with the Asheninka and Yine-Yami peoples of the Peruvian Amazon, *ACME An. Int. J. Crit. Geogr.*, 17: 720-749.

Wezel, A. and V. Soldat (2009). A quantitative and qualitative historical analysis of the scientific discipline of agroecology, *Int. J. Agr. Sustain.*, 7: 3-18.

Wu, Q., H. Li, R. Wang, J. Paulussen, Y. He, B. Wang and Z. Wang (2006). Monitoring and predicting lands use change in Beijing using remote sensing and GIS, *Lands Urban Plann.*, 78: 322-333.

Part IV
Conclusions and Perspectives

CHAPTER

12

Agroecological Transitions in the Era of Pandemics: Combining Local Knowledge and the Appropriation of New Technologies

Miguel Angel Altieri[1], Alberto Diantini[2]*, Salvatore Eugenio Pappalardo[3] and Massimo De Marchi[4]

[1] University of California, Berkeley, CA, USA
[2] Research Programme Climate Change, Territory, Diversity – Department of Civil Environmental Architectural Engineering – Postdoc Researcher at the Department of Historical and Geographic Sciences and the Ancient World, University of Padova
[3] Laboratory GIScience and Drones 4 Good, University of Padova
[4] Director Advanced Master on GIScience and Unmanned Systems for Integrated Management of Territory and Natural Resources, Department of Civil Environmental Architectural Engineering, University of Padova

During the preparation of this book, the editors organized a conference to reflect, in the context of the Covid pandemia, about the relationship between agroecological knowledge and the appropriation of new technologies of geographic information. The conference was coordinated by Massimo De Marchi with the intervention of Miguel Angel Altieri, as keynote speaker, and two discussants: Alberto Diantini and Salvatore Eugenio Pappalardo. The event was part of the annual kick-off seminar of the International Joint Master Degree on Sustainable Territorial Development, Climate Change Diversity Cooperation (STeDe - CCD). Considering knowledge and academic work as a common good, the conference was not only part of an academic activity but was opened and shared online to a wide public.

We collect in these pages, maintaining the structure of the dialogue, the interaction among the speakers and the debate with the participants.

*Corresponding author: alberto.diantini@unipd.it

12.1. Introduction

De Marchi M.: In the acme of the pandemia (April, 2020), Boaventura De Sousa Santos published the book, *The Cruel Pedagogy of Virus*. It is an account of the role of the virus in opening the eyes of people to the critical conditions of normality. Despite the narration on the unity of humanity and of a virus making people equally at risk, de Sousa Santos highlights how we are living in a world where colonialism and patriarchy are still well alive.

The tragic transparency of the virus demonstrates how there is a 'south of the quarantine', a group of people paying a higher tribute to the pandemic: women, informal and autonomous workers, peddlers, homeless, poor, refugees, immigrants, displaced people, elders, prisoners, disabled persons.

However, the pandemic can be an opportunity for change, and a new future can start now. So, what can we learn from agroecology in the current pandemic context? How can agroecological knowledge provide the basis for a path to technological sovereignty?

12.2. Agroecological Transitions Towards a Sustainable Food System

Altieri M.A.: In the world, industrial agriculture dominates the landscape. Globally, about 80 per cent of the 1.5 billion hectares of arable land are devoted to industrial monocultures, reshaping the landscape, and impacting the biosphere by promoting deforestation and with a yearly injection of about 5 billion pounds of pesticides (Altieri and Nicholls, 2020). Way before the pandemic, agroecologists started warning that industrial agriculture had become too narrow ecologically, highly dependent on off-farm inputs, and extremely vulnerable to insect pests and climate change (Altieri *et al.*, 2015). And now, as demonstrated by the Covid-19 pandemic, it is evident that the conventional food system is very prone to a complete shutdown by this unforeseen crisis. Certainly, one thing that the Covid-19 is revealing is how closely linked human, animal, and ecological health are.

When we practice agriculture, we manipulate nature by simplifying ecosystems. This simplification has substantially reduced the biodiversity of agroecosystems overriding ecological principles, which in turn trigger ecological disasters and affect human health. Even though industrial agriculture occupies about 70-80 per cent of the world's arable land, it uses about 5.2 billion pounds of pesticides, consumes 70 per cent of the water, and 80 per cent of the fossil fuels, and emits 30 per cent of the greenhouse gases, but it only produces 30 per cent of the food that we eat. So, it is a myth that the food we eat in the world is produced by industrial agriculture. It is actually mostly produced by smallholder farmers in small plots, using almost no modern agricultural technologies.

The effects of climate disruptions are already visible. For example, in May 2012, in the midwest of the United States, there was the worst drought in fifty years, affecting transgenic soybean and corn production with a yield reduction of about 30 per cent. So the latest technology of genetic engineering was demonstrated to be extremely vulnerable to climate change. In the last ten years, California has suffered prolonged droughts that put out of production about 200 thousand hectares of monocultures, with a loss of about 1.5 five billion dollars. Another example comes from the recent hurricanes that have been affecting the Carribean: in the 2017 hurricane, Maria decimated the monocultures of bananas and other plantations in Puerto Rico, showing the lack of resilience of this kind of monocultural production systems.

Large-scale monocultures have advanced, causing wide deforestation and natural habitat loss, and migration of wild animals which coexist with hundreds of virus species towards human settlements. This, in combination with the way we raise animals for human consumption, thousands of genetically homogeneous animals confined in small spaces, created the conditions for the evolution and spread of new deadly viruses and pathogens. In South America, soybean production now covers about 57 million ha, mostly transgenic, being produced at the expense of natural forests (Oliveira and Hecht, 2017). In these ecosystems, different animals coexist with different viruses, but they normally remain within the forests. When the forest is destroyed, these pathogens spillover into livestock and then into human populations – a common pathway for zoonotic diseases. This is exactly what seemed to have happened with Covid-19 and previous epidemics, like avian flu and swine fever.

What is happening now is that Covid-19 is revealing the socio-ecological fragility of the current industrial globalized food system. The effects of the pandemic on the food supply chains are already being felt in terms of widespread food shortages, price spikes, and diet changes. Because of the pandemic, a lot of people do not have access to fresh food anymore. Many migrant workers have lost their work or they are more exposed to the Covid-19 because they are not guaranteed safe working conditions. Another problem is children's access to school lunches. For example, in Latin America and the Caribbean, over 10 million children rely on school lunches, which is perhaps the only meal that they have during the day. Considering that often schools are closed due to the pandemic, they do not have access to that food anymore (Altieri and Nicholls, 2020). Moreover, small farmers are being highly affected because in many countries, restrictions on travel, trade, and lockdown of entire cities restrict access to markets. This is remarkably problematic, especially in cities where millions of people live, requiring thousand tons of food per day, which mostly comes from areas on average about 1000 km far from cities. The decline of transportation has reduced the possibility to move fresh food for long distances. This has undoubtedly increased levels of food loss and waste, reducing access to fresh food especially for the poor (Purdy, 2020).

So, we are already feeling the effects of the pandemic on the food supply. Therefore, what we need is a huge transition to a more socially, just, and ecologically resilient and localized food system.

Diantini, A.: What role can agroecology play in building a more sustainable post-Covid-19 agriculture?

Altieri, M.A.: Well, what we need to do is to move from industrial agriculture, which causes high environmental degradation, depends on fertilizers, pesticides, and petroleum, to a more sustainable and diversified agriculture, based on natural biological interactions and ecological processes that emerge from complex cropping systems. The way to implement and guide this transition is a science called 'agroecology'. Basically, agroecology is a science that is composed on one side by Western sciences, such as ecology, agronomic sciences, and sociology, and on the other, by the knowledge of traditional people who have been farming the land for thousands of years (Francis *et al.*, 2003). In Latin America, we are blessed to have traditional agriculture systems that have stood the test of time (more than five thousand years in the Andes) and still exist. Therefore, it is from this dialogue of wisdoms that the principles of agroecology emerge as the potential basis to guide the much-needed agrarian transition. So what we are looking for is an agriculture that is decoupled from fossil fuel dependence, characterized by diversified agroecosystems that replace monocultures, which have high environmental impacts, and reduce diversity. This new agriculture should be resilient to climate change and multifunctional; producing ecological services as well as providing social and economic services to the communities, thereby enabling the foundation of local food systems. Such systems reduce the distance between producers and consumers and ensure the maintenance of the local culture, such as the traditional culinary traditions and sustainable ways of natural resource management.

Thus, agroecology is a science that shows a different way forward, by providing the principles on how to restore and re-design agricultural systems that can withstand future crises, such as pest outbreaks, diseases, pandemics, climate disruptions, and eventual financial meltdowns (Altieri and Nicholls, 2020). Agroecological systems are resistant because they have a high level of diversity and resilience – both emergent properties increasingly recognized for their potential to reduce risk from climate change and other threats (Nicholls, Altieri and Vazquez, 2016).

So, let me explain how I see agroecology as providing the basis for the reconstruction of new agriculture after Covid-19. First of all, it is important to say that a return to normality would be a disaster: it is just this 'normality' that caused the crisis we are facing today. We need to come up with alternatives and actions to restore the environments weakened and impoverished by conventional agriculture and farming, rethinking how to redesign the agroecosystem matrix and the landscapes that surround agricultural systems. We need to promote rural agriculture as well as urban agriculture based on agroecology, and we also need to promote more ecologically-sound management of pests and diseases without pesticides. This change will certainly lead to healthier conditions for wild and domestic animals and humans too, keeping pathogens in their habitat, as diverse vegetation in the borders of crop fields act as ecological firebreaks. Thus biodiversity, which is worldwide alarmingly declining, is better conserved in mosaics of small farms inserted in complex landscapes. Moreover, the agroecological transition of agriculture will provide nutritional and food security for people since they will eat more fresh fruits and vegetables produced in the proximity. Food will be free of pesticides as a result of a sustainable agroecological system. All this will lead to better livelihoods, local food sovereignty, greater ecological integrity, and in the end, environmental and human health.

To establish an agroecological-based system, one of the first steps is to overcome the pesticide treadmill. Industrial agriculture in the world injects into the biosphere about 2.3 billion kgs of pesticides (Pimentel *et al.*, 1980). Some of these pesticides are endocrine disruptors, while many are immunosuppressive (Repetto and Baliga, 1996). This issue represents a potentially serious risk especially in case of a pandemic, such as this we are living in. To go beyond the pesticide treadmill, we have to replace monocultures with safe agriculture systems, such as polycultures and agroforestry. This gives farmers greater autonomy, as they need not depend on inputs from corporations for pesticides or fertilizers, but rather, rely on the ecological interactions within the agroecosystems. For example, if we break the monocultures into polycultures, we create ecological conditions for richer biodiversity of natural predators and parasites, enhancing biological control (Altieri and Nicholls, 2014). A diversified system has also more favorable conditions for pollinators, which are essential in agriculture, especially considering that they are experiencing a critical decline due to the massive use of pesticides in industrial systems (Constanza *et al.*, 2014). In California, experiments have been done in vineyards where different species of flowering plants were seeded to promote the presence of beneficial insects. Enhancing plant diversity in agroecosystems is a mechanism to support soil fertility, attract pollinators and predators, reduce the use of external inputs and ensure higher productivity (Altieri and Nicholls, 2014).

Diantini A.: *As you said, for post-Covid-19 agriculture, we also need to restore the environments compromised by conventional agricultural practices. How can agroecology and ecological restoration be combined in this light?*

Altieri M.A.: Another important approach for the reconstruction of post-Covid-19 agriculture is to restore the environment. For this, we need to combine agroecology with ecological restoration to create sustainable and resilient agro-landscapes. In agroecology, what we prefer in terms of landscape pattern, is a complex matrix of farms surrounded by forests linked with ecological corridors. In such environments, ecosystems are rich in biodiversity that perform services for agriculture. The forest also acts as an ecological barrier, preventing wildlife with potential pathogens to move into agricultural systems.

The key point is to increase diversity and complexity. In Asia, some studies highlighted that in rice fields surrounded by complex landscapes, compared to simple conventional agroecosystems, there are more beneficial insects and predators, including specific fish species that consume insect pests. These systems show a reduced incidence of plant diseases and insect pest presence (Koohafkan and Altieri, 2016; Zheng and Deng, 1998).

Moreover, a complex and diverse matrix surrounding cultivated fields also provides food for the people. Many small farmers not only depend on the crops that they produce in the fields, but also depend on the wild fruits and weeds that grow in the borders, plants which in industrial agriculture are usually eliminated with herbicides, destroying an important source of food. There are cases in Mexico where after ecological restoration guided by agroecological principles, farmers created agroecosystems that enabled extra-economic income from the sale of fruits and vegetables harvested from the field borders.

There are other examples of degraded landscapes on which, using different techniques, like windbreaks, terraced agroforestry systems, silverbush, corridors, etc., the environment can be fully restored. A specific example comes from Sierra Mixteca, in the highlands of Mexico. Here the agroecosystem was completely degraded by deforestation and overgrazing, but the small farmers did not want to migrate; so they started an ecological restoration project, reforesting the top of the mountains with autochthonous plants, creating terraces and using traditional water-harvesting techniques. In this way, the community was able to stay in the territory and revitalize the production systems. Another example is from Colombia, where a community that did not have any water because the watershed was deforested, started to work to restore the environment. Now they have enough water for themselves, the animals, and the crops. What they have done is not only to restore the watershed, but have also modified their agriculture, which depended on monocultures of mostly tuber crops. Today that system has changed, becoming a highly diversified agroecosystem resilient to insect pests, diseases, and droughts. So the results are that they restored 75 per cent of the forest cover and now they are producing 90 per cent of what they consume, including fruits, coffee, and vegetables. This new agroecosystem undoubtedly enhanced the community's food security. Additionally, the restoration of the landscape implemented in harmony with nature resulted also in a higher level of social cohesion, as the entire community, including children and women, was involved in the process.

Another important property inherent to complex systems is that they are far more resilient to climate change as demonstrated by studies conducted in Central America and the Caribbean. For example, after Hurricane Mitch in 1998, Eric Holt found that the farmers who had monocultures suffered more mudslides than those who had polycultures. In Cuba, studies show that in many cases, monocultures were completely destroyed by hurricanes, whereas the more diversified systems, such as agroforestry systems and farms surrounded by complex borders, were more protected against the strong winds.

In the end, ecological restoration of agricultural landscapes characterized by diversified agroecosystems represents an essential adaptive strategy. Indeed, the resilience of agricultural systems is deeply linked to their diversity: the more complex and diverse the territory inside and around the farm, the better it is in coping with climate change and pest pressure.

Question: How can the principles of agroecology also support a more ecological animal production system?

Let us consider the large list of deadly pathogens linked to large-scale conventional animal production systems: from 5N1-Asian Avian Influenza (H5N2) to multiple Swine Fluvariants (H1N1, H1N2), a variety of influenzas (Weiss, 2013) and, lastly, at least for the moment, Covid-19. The agroecological perspective can facilitate the development of alternative sustainable and effective livestock production systems, such as the sylvopastoral systems (SPS), which combine the production of forage grasses and leguminous herbs with shrubs and trees for animal feeding and complementary uses. It is like a building with different layers of plants that are going to provide different services. These agro-landscapes promote biodiversity and create complex habitats that support animals, plants, and a richer soil biota. Trees and other plants can provide farmers, cattle, and wild fauna with food. In these systems, since the animals live in very complex environments and eat plants grown organically, antibiotics are rarely used, given that the animals' immune system is high (Altieri and Nicholls, 2020). In such systems, the health of the animals is better, leading to higher milk and meat production and a reduced risk to human health as well. For example, in SPS, milk production is confirmed to be sensibly higher than in conventional systems (Murgueitio *et al.*, 2015). SPS grant healthy animal conditions together with increasing the resilience of the agroecosystem.

12.3. Revitalizing Traditional Peasant Agriculture and Urban Agriculture

Diantini, A.: *What is the link between agroecology and traditional peasant farms?*

Altieri, M.A.: Many effective agroecological practices are part of traditional agriculture and farming, thus representing a co-evolution of nature and culture, where farmers developed systems that did not depend on modern technology, such as pesticides and other external inputs (Francis *et al.*, 2003). Evidence shows that agroecology can restore the production capacity of small traditional peasants and farmers by increasing biodiversity which usually leads to less pests and improved soil fertility (Altieri, 1999). Studies on several agroecology projects realized in Africa, Asia, and Latin America highlighted that productivity of traditional farming can be significantly increased if they strictly follow the principles of agroecology

(Rosset and Altieri, 2017). Even in Italy, where there are many rural traditions where farmers possess a very intimate knowledge of their agricultural systems (for example, vines intercropped with olives), the adoption of agroecological practices could lead to successful results in terms of production.

Agroecology can optimize traditional agricultural systems, but traditional farms can also be an important resource for agroecology as well. Indeed, about 7 thousand crop species and 2 million local genetic varieties are in the hands of small peasants, representing the genetic basis for the agriculture of the future. Given this rich agrobiodiversity, it is ironic that the diet of most of the people in the world is composed of three major crops: wheat, rice, and corn (UNSCN, 2020). Crop diversity is essential for agricultural climate adaptation. The Green Revolution has simplified this variety, moving from a traditional diverse and rich agricultural production to an ecologically poor and homogenized agroecosystem and leading to major consequences for the provision of ecosystem services, as well as crop sustainability, and food sovereignty (Jackson, Pascual, and Hodgkin, 2007). Therefore, traditional peasants and farmers have an essential role in maintaining a high crop species diversity in the agroecosystems, which is one of the pillars of agroecology.

Despite the fact that small farmers only control 25-30 per cent of the world's arable land, use 30 per cent of the water and 20 per cent of the fossil fuels, they produce 50-70 per cent of the food that we eat (ETC, 2017). So every time we eat, we need to thank a small farmer, not big corporations, because those industrial agricultural systems do not produce the food that we eat.

There are many examples of traditional farming in different parts of the world. One case, from Chile, for example, is related to half-a-hectare farm. Here a family of two adults and three children divided their land into six plots in a rotational system. Production levels reached about 1.12 tons of vegetables per year, with more than 2,500 eggs, which a family of five would not even eat in one year. So they can produce what they need for themselves, except salt, pasta, and rice. The surplus is sold, bringing income to help economically sustain the family. They do not have to use pesticides or fertilizer, so the cost of production is low and they also have extra time, since they didn't need to invest time in the application of external inputs. Another example comes from Cuba, where up to 72 per cent of the small farmers adopted agroecological practices (Rosset *et al.*, 2011). An illustrative case in Cuba is that of a family which obtained the land from the government and originally was used to growing tobacco and corn in the conventional way. But after training in agroecology, they transformed their farm into a very diverse system where you have a combination of vegetable crops and agroforestry systems with pastures for animals, producing eggs, milk, meat, fruit, vegetables, wood, and water. Many Cuban small farmers adopting agroecology produce food per hectare, sufficient to feed about fifteen to twenty people per year, showing an energy efficiency of around 10:1 (Funes and Vasquez, 2016). This means that for every kilocalorie invested in the management of the farm, they obtain ten back. This is highly efficient, considering that industrial agriculture has an efficiency of 1.5:1.

Pappalardo S.: We live on an urbanised planet, as most of the people live in cities. Can the agroecological principles also be applied to urban agriculture?

If we consider that 60 per cent of the world's population and 56 per cent of the world's poor live in urbanised areas (de Bon, Parrot, and Moustier, 2009), it turns out that today, more than ever, we need to promote localized food systems within the cities to overcome the difficulties posed by the current pandemic in terms of food access. In effect, in many cities, there is a lot of abandoned land that could be put into production. In 2005, the UNDP (United Nations Development Programme) did a study and found out that 30 per cent of the food consumed in the world's major cities came from urban agriculture, and the global urban production ranges between 20-180 million tons per year. One of the benefits of urban agriculture is that it ensures access to fresh vegetables and fruits, improving local food security and nutrition, particularly in not well-served communities.

The same agroecological principles adopted in rural areas can effectively work also in urban areas, designing biodiversified home, school and community gardens with increased soil fertility, crop protection, and production with very few external inputs (Altieri and Nichols. 2020).

One example of urban agriculture that has been very successful comes from Cuba. In the 1990s, after the collapse of the Soviet Union, Cuba, which was highly dependent on pesticides, fertilizers, and fossil fuels, had no more access to these external resources. Practically, there was no way to bring food from rural areas to urban areas because there was no fuel for trucks and cars. Therefore, urban agriculture started flourishing on the island to the point that 50 per cent of the vegetables that are consumed in the major cities come from urban agriculture. The agroecological production in urban areas is very high, reaching an average 15-20 kg/m^2/year (Funes and Vazquez, 2016). In Cuba, one square meter of an agroecological well-designed urban garden can yield ten cabbages every ninety days, thirty-six heads of lettuce every sixty days and a hundred onions every 120 days. If we consider that each person eats 72 kg of vegetables per year, in one year a 10 m^2 garden produces 200 kg of food, potentially satisfying 55 per cent of the annual vegetable needs of a family of five (Clouse, 2014). More than 26 thousand urban gardens in Cuba are producing about 25 thousand tons of food per year, generating jobs particularly for elderly people, women and young people. Today, in times of pandemic, generating jobs is critical. For example, the unemployment rate in Colombia is currently 42 per cent among young people. There is no job in the cities, so a solution would be for the government to promote rural enterprises run by young people.

Overall, the potential of urban agriculture is enormous and its development is not just possible but strategic to enhance access to locally produced food.

12.4. Combining Local Knowledge and Technologies of Geographical Information in Agroecology

Diantini A.: This Covid-19 pandemic showed us there is a strong link between human beings, domestic and wild animals, and plants within the ecosystems. This is the basis of ecology, which can be summarized as the relationship between living beings and the environment. In this light, we can consider what Charles Darwin said about the struggle for existence. If you think about this pandemic, it is due to a virus which is simply a particle that is ten times smaller than a million parts of a metre, made of RNA, and covered by proteins. Viruses have been here on the Earth for billions of years. They initially started to fight against bacteria, then against plants, then against animals; finally, around 2-3 hundred thousand years, against humans. We surely are a young species compared to viruses and maybe we are not in an advantageous position against them. But in this struggle we have a plus, which is that we can think, we can plan, we can learn from the past and build a more sustainable future, for example, through agroecology. Among the pillars of this discipline, some important steps are, for example, shifting from monocultures to polycultures and combining agroecology and ecological landscape restoration. As explained above, agroecology is already used in traditional agriculture in many parts of the world.

My question comes from the fieldwork I did in the oil extraction context in the Amazon forest. I spent some time in indigenous communities and I was very surprised on going into their forest gardens, which in their local language are called *chakras*. There they cultivate, for example, around or even more than twenty species of plants, creating complex ecosystems. This is a pure example of agroecology. But talking with them, some told me they want to deforest their areas to implement their 'big projects', like monocultures since this seems to be the only alternative to oil activities in the area. Have you ever experienced this kind of situation, that maybe can be called 'globalization of industrial agriculture', also spreading inside indigenous populations?

Pappalardo S.: Agriculture is practised in many different countries of the world. This pandemic undressed the structural issues related to the global development and production models, questioning the way we manage environmental resources. I have to say that here, in Italy, from an institutional point of view, we are a little bit behind the concepts and practices of agroecology. Sometimes even in some academic environments, it is a kind of taboo, although there are some experiences in farming networks that are growing and making more sustainable agriculture possible. I'm interested in the opportunity to strengthen these networks and also increase their knowledge about agroecology.

I would like to try to make some reflection about innovation and new technologies in agriculture. You summarized it very well. There is a traditional knowledge coming from experiences of many centuries or sometimes also millennia of local people using natural resources in harmony. They haven't gone through the environmental suicide committed by Western societies. So what about technology in agriculture? I am especially interested in the application of geographic information systems and the use of drones in agricultural systems. In many parts of the world, in the collective memory of the farmers, the application of such technology is receiving attention but it is far from being widespread. Anyway, at least in Italy, whenever this technology is used in agriculture, the main target is always to increase crop production. Overall, I don't know if there is a kind of cultural gap or perhaps a condition that seems that traditional knowledge and new technologies do not fit together. Maybe it is more of a digital gap, which means there is a lack of access to technology, which is basically a problem of democratization of technology. So what do you think about this problem in agriculture?

Altieri M.A.: On the first question, I think it is important to say that indigenous people, traditional farmers, and peasants are connected to the world, and many of them receive information to change their systems because the dominant discourse is that they need to link into the global economy. Well, it is not our role as scientists to go there and tell them what to do; these are complex decisions that they have to make on themselves. As researchers, our role can be to facilitate the decision-making process. We can become facilitators of a process so that they become aware of the implications of adopting a particular technology, as Freire's pedagogy teaches. Will they become dependent on external sources of knowledge and inputs? So, as agroecologists, one effective way for spreading ideas and practices is to identify communities where farmers are successfully using agroecology and enable an exchange of information with farmers from other communities. This works very well because as soon as farmers see other farms that are operating in a much more ecological and sustainable way, with less cost, and higher production, they tend to abandon their monocultures and associated conventional agriculture practices. Another way is to use a methodology featuring participatory and interactive techniques to facilitate awareness of the consequences of adopting a technology. For example, one activity consists of giving farmers different colors to the resources they need for their agriculture projects: in green, the resources of the farmers, in red, what comes from the industry, in blue, what comes from the government. So if farmers want to adopt monocultures of cassava, they need to identify what resources they need and where will they come from. Improved seeds? Red or blue, as they come from the industry or government. Labour? Green, if it is family labor, if hired, red. Pesticides and fertilizers – red, information about agrochemical use and dosage blue, as this information is usually provided by government extension agents, or red, if provided by pesticide salesmen. This exercise can be very useful for a community, as they can visualize that if their chosen approach to agriculture required more than 50 per cent red and blue cards, they can easily realize that they are losing control over their production process, and becoming dependent on outside forces. This is when farmers usually propose approaches that enable more 'green-colored' solutions, as they realize they can be more independent and autonomous.

On the second question, well, I think that the issue is that, at least in Latin America, 80 per cent of the small farmers live in marginal areas and they are very poor. So the main problem is access to the technology and also who controls the technology. Many people say that these poor farmers should be using drones and geographic information system, which would be fantastic because it would provide them with key information to increase crop production. But if they don't have access to technology, then how can they do that? The point is democratizing GIS technology. We have to find a way to provide farmers with digital tools and drones that are owned by the community, allowing this technology to be more accessible and user-friendly. But, we have also to consider that if they have access to this technology, what will they use it for? How will the technology provide them with more information to make decisions but without bypassing their own rationale? We rationalize things from our Western perspective, but most peasants do not make decisions based on the same parameters and indicators that we use. For example, in Mexico, most farmers practice the 'milpa', which is a system where maize, common beans, and squash are grown in association (Altieri, Nicholls, and Montalba, 2017). Many economists have done studies of this system and have shown that the milpa, from a neo-liberal economic perspective, doesn't make sense. But the milpa persists and it is used by thousands of farmers because there are other factors – cultural and ecological – at the basis of its use, despite economic studies affirming they are not viable economically. I want to repeat it: democratizing technology is very important. We have to find out a way that these fantastic innovations are accessible to the community. But also, we need to be open to the fact that some communities may want to reject this technology. Why do they have to accept drones? Just because we say that is good for them? They need to make an informed decision, understanding what it means for them; why is it useful for them; what is going to be the impact on their culture and their social relations; who is going to have access to the technology. Because sometimes what happens in communities is that some people have access to the innovations and some don't, thus creating social gaps within the community. For example, fair-trade coffee is a great idea, but it turns out that it is promoting big inequalities in many communities of Latin America. Why? Because only a few farmers have the quality demanded by the market and are part of the network of this so-called 'fair-trade system'; other farmers are left out and do not receive a premium price, creating a social stratification within the communities. So we need to make sure that you are not going to exacerbate inequalities with GIS technology. Another aspect in relation to this issue is that, if you provide technology to a community, you will find that some people learn quickly and get ahead easily, leaving behind those who learn more slowly, and may not even benefit from the technology at all. It is important to utilize the information and the indicators that the farmers use. For example, if we go to a community and we want to measure how good is the soil, we use analytic methods that measure pH, nitrogen, phosphorus, etc. Conversely, farmers simply taste or smell the soil to assess its quality. Perhaps the chemical parameters measured may not be useful for farmers. We need to figure out a pedagogical way of involving the people in a participatory manner from the beginning to see if they need the technology, how accessible it is, and if they want to use it. At best, we can try to combine both sets of indicators.

12.5. Changing the System

Question: As you said, the current food system is not sustainable and has to be changed together with the global capitalistic system, which rules the economy, including food production. How can we do it?

Altieri, M.A.: Well, first of all, we can try to change the world in two ways, using reformist or transformative strategies. Reformists don't question the capitalist system; thus alternatives are proposed to align with the logic of the market economy. For example, more than 80 per cent of certified organic farming in the world maintains monoculture, using input-substitution approaches. Most of the production is for export, so it does not contribute to national food security, and only wealthy people benefit from the food as only a few people can pay for the high prices of organic food. Why is organic food more expensive than conventional? It is because organic farming is playing the game of the market economy, which it is part of. It simply takes advantage of the windows left in the capitalist system but it does not attempt to change it. Of course, organic agriculture is better for the environment and generates cleaner food. On the contrary, agroecology is transformative, as it wants to change the system by changing the way we produce, distribute, and consume food.

Changing the structure of the dominant food system is very difficult. It is more practical to start by creating autonomous territories with markets that are based more on an economy ruled by solidarity principles between producers and consumers, rather than the principles of the capitalistic market economy. We need to democratize food, so that the vulnerable and poor people may have access to healthy food. For example, in Brasil, there is a network called Rede de Agroecologia Ecovida, which is a cooperative between consumers and producers where they agree on the price that has to be fair for both the farmers and the consumers. These are market rules based on solidarity.

With the Covid-19, there are a lot of interesting experiences that are happening to make the food accessible to people at fair prices. There are many people who have lost their jobs and don't have anything to eat. Therefore, there are communities that are mobilizing to develop new production and food distribution systems, like kitchen soups.

We need to create new networks of food production and consumption that reduce the distance between producers and consumers while ensuring that the food is accessible and healthy to everybody. I think an important lesson from

Covid-19 is that we need to put food production in the hands of small farmers and urban farmers. This is the only way to ensure the supply of fresh food at affordable prices in local markets.

This re-design of the food system, based on short supply chains, will require some profound changes. We need to provide small farmers with access to land, seeds, water, and equitable markets. There is a need for training in agroecology and research on the agroecological systems, which is the role of the university.

Anyway, we cannot put the weight of the change of the food system only on farmers. A big difficulty is that the big corporations are controlling the food system, determining what farmers should grow, for whom and the technologies they are going to use. They also control the supermarkets and what people are going to eat, the quality of the food, and its price. In fact, every time we go to the supermarket, we support the capitalist food-chain, but if, instead, we support local farmers' markets instead of the corporate food-chain, we promote socio-ecological sustainability and resilience in our communities. So those of us who have jobs, have a huge responsibility with our wallets in terms of deciding what we consume. Profound changes are needed, but substituting the industrial monocultures with ecological practices is not enough. We need to dismantle the control of the multinationals on the food system and the neo-liberal policies that maintain this structure. This is not a matter of painting capitalism a bit more green or making it a little bit more sustainable with reformist practices; it requires a complete transformation, a full shift from the market economy to a solidarity economy, from fossil fuel dependence to renewable energy, from big corporations controlling the food system to cooperatives between producers and consumers. Such a new world should be led by allied social, urban, and rural movements.

Covid-19 has exposed the tragedy of animal farming and industrial agriculture that has led to a dramatic loss of biodiversity and caused obesity, malnutrition, food waste, bad conditions for the workers, while undermining the livelihoods of small farmers, who are the ones that produce the food we eat. Now that the global supply chains are in a disarray, it is time to enhance regional food systems in order to feed the people in a more equitable way, with food produced through agroecological production practices. In this light, agroecology is today positioning itself as a key agricultural path for the future.

Question: What are the drivers of the change we need?

Altieri M.A.: Well, one of the drivers is a crisis, something that usually motivates changes. I really hope that this crisis caused by Covid-19 and which is linked to other crises we face (climate change, social inequality, etc.) is going to motivate a transformative change, which goes beyond mere reforms. The problems unfolded by the current pandemic can be a key driver to change industrial agriculture for a transition towards agroecological-based food systems. The second driver is social movements. Social movements have been behind most changes in history. If you do not have social movements, pushing agriculture ahead, no change is going to happen. We cannot just depend on technological changes in agriculture; innovations must run parallel to social and economic changes.

In many rural areas of Latin America, you find that farmers are doing agroecology, promoting changes at the local level, and showing the way of how we can do things differently. This movement, called the *campesino to campesino* (CAC) movement, is basically a grassroots movement using pedagogical tools that allow for the horizontal exchange of information between farmers. A member of a community that knows about agroecology (a promoter) shares his knowledge with the rest through field days and demonstration activities. If a farmer trains ten to fifteen other farmers, then each one of these farmers can become promoter of the agroecological principles and train other ten or fifteen people. This is how agroecology is scaled up.

In Cuba, for example, right after the collapse, only 216 farmers were managing their farms based on agroecology. By adopting CAC methodology in less than ten years, more than 130 thousand farmers adopted agroecological practices. Another grassroots movement is also the *via campesina,* to which millions of farmers in the world belong. They have their voice heard in international fora and can make alliances with other movements, enriching the political discourse for changing agriculture on a global scale. Clearly, agroecological innovations do not emerge from the universities or research institutes, but from farmers in rural areas. Social movements can spread the agroecological principles towards societies no more embedded in the market economy but in alternative sustainable and equitable food systems.

Another driver is to spread agroecological practices that really work and provide solutions to problems affecting agroecosystems. There is too much social and political discourse about agroecology which is good, but we need effective agroecological practices that really work, that are effective in regulating pests, in providing soil fertility, in increasing productivity.

A fourth driver is the political will, which is the support from local politicians that promote enabling policies to scale up agroecology. For example, there are many communities which elected mayors (many of them women) who are agroecological farmers, who, now, from a position of power at the municipality level, are promoting important agroecological initiatives.

Question: You considered many examples of agroecology based in Latin America. But is it possible to change the system through agroecology also in Western countries?

Altieri M.A.: Many examples of agroecology are coming from Latin America because in this region, agriculture is deeply rooted in traditional farming which has been developed for thousands of years. But I would like to stress that agroecological principles are universal and can be applied in Europe, in the USA, or wherever in the world. It is just that the principles take different technological forms depending on the social, cultural, economic, and political conditions prevailing in each region. In California, for example, there are large-scale farms of more than 200 hectares of vineyards which use agroecological principles, but their practices are different from those of a Central American peasant

One of the big differences between geographical areas, such as Latin America, Africa, and Asia and the more industrialized regions, is that in developing countries rural populations constitute a high proportion of the total. Here rural social movements are very strong. In Europe and the USA, you almost have no farmers left compared to the urban population and thus needed changes in agriculture must emerge from the urban movements. In this light, consumers in Western countries have to become very active and aware of their weight as a big social movement that can support small farmers, locally-based food systems, and promote much-needed socio-ecological change.

Consumers must understand that local agriculture's role is more than provisioning healthy and accessible food. For example, in Brazil, case studies demonstrate that towns surrounded by sugarcane cultivations are 10 degrees hotter than towns surrounded by small farms with diversified ecosystems because of the albedo effect. Other studies show that towns surrounded by industrial large-scale farms have more crime and violence episodes, compared to towns surrounded by agroecological small farmers, where social relations between farmers and consumers are more intimate, thus reflecting a more developed social network that creates conditions of harmony as opposed to towns surrounded by big farms where inequities are huge.

Another important aspect to spread agroecology in Western countries is the existence of an adequate policy framework supporting the adoption and amplification of agroecological principles in agriculture. There are countries where, i.e. Brazil, there is a national law on agroecology (created due to the pressure of social movements) that boosts agroecological practices and alternative marketing schemes. Another case is Uruguay, where the national plan for agroecology represents a tremendous opportunity to promote the agroecological changes in agriculture. Many of these laws contemplate school lunch programs where 30 per cent of the food for school lunches is required to come from small farmers who practice agroecology. So, imagine that in Italy, you were to create a law that requires that all the food consumed in schools, universities, and hospitals has to come from small farmers to nearby cities, this would catapult the promotion of agroecology, where small-scale farmers are actively supported by the government.

Overall, agroecology has developed as a global movement pushed by farmers, peasants, and activists within their pursuits for food sovereignty, biodiversity protection, and promotion, ecological restoration, and a transition to more socially and sustainable rural societies. The agroecological principles work worldwide; the only limits are the imperatives imposed by the globalizing market economy ruling which produces the food and what we eat, and its cost. We need to understand that breaking this system represents an ecological, economic, and political rupture. The choice to change is in our hands.

De Marchi M.: The Covid pandemic, arriving at the end of six decades of uneven development, highlights the global predatory capitalism embodied in many development discourses consolidating social exclusion, resource extraction, environmental injustice, and accumulation by dispossession. Agroecology, as a place-based approach to healing people and ecosystems, offers a rich texture of reflections and practices, challenging the menu of globalizing universalizing development theories and initiatives to propose a pluriverse of words and worlds.

Freire (1992) reminds us that we can create possible futures: the unprecedented achievable. The future is not inevitable and not even given; the world itself is not given, but it is giving itself in a dialectical and conflictual way. Men and women, not only live, they also exist conditioned, but not determined; they can experience oppression, but also liberation. History, seen as a possibility, opens up spaces for responsibility, in which the dream has a fundamental function in a tension between denunciation of the present and announcement of the future. The future must be done and produced, otherwise, it will not arrive as individuals want it (Freire, 1992, pp. 91-102).

Bibliography

Altieri, M.A. (1999). Applying agroecology to enhance productivity of peasant farming systems in Latin America, *Environment, Development and Sustainability*, 1: 197-217.

Altieri, M.A. and C.I. Nicholls (2014). *Manage Insects in Your Farm: A Guide to Ecological Strategies*, Sustainable Agriculture Research and Education Handbook Series, Sustainable Agriculture Research and Education (SARE), College Park, USA.

Altieri, M.A., C.I. Nicholls, A. Henao and M.A. Lana (2015). Agroecology and the design of climate change – Resilient farming systems, *Agronomy for Sustainable Development*, 35: 869-890.

Altieri, M.A., C.I. Nicholls and R. Montalba (2017). Technological approaches to sustainable agriculture at a crossroads: An agroecological perspective, *Sustainability*, **9**(3): 349.

Altieri, M.A., and C.I. Nicholls (2020). Agroecology and the reconstruction of a post-COVID-19 agriculture, *The Journal of Peasant Studies*, 47: 5, 881-898.

Clouse, C. (2014). *Farming Cuba: Urban Agriculture from the Ground Up*, Princeton Architectural Press, New York, USA.

Constanza, R., R. de Groot, P. Sutton, S. van der Ploeg, S.J. Anderson, I. Kubiszewski, S. Faber and K. Turner (2014). Changes in the Global Value of Ecosystem Services, *Global Environmental Change*, 26: 125-156.

de Bon, H., L. Parrot and P. Moustier (2009). Sustainable Urban Agriculture in Developing Countries: A Review, *Agronomy for Sustainable Development*, 30: 21-32.

de Souza Santo, B. (2020). *La Cruel Pedagogia del Virus, Ciudad Autónoma de Buenos Aires*, CLACSO.

ETC (Action Group on Erosion, Technology and Concentration). (2017). *Who will Feed Us? The Peasant Food Web vs. The Industrial Food Chain*; retrieved from: https://www.etcgroup.org/whowillfeedus; accessed on 27 April, 2021.

Francis, C., G. Lieblein, S. Gliessman, T.A. Breland, N. Creamer, R. Harwood, L. Salomonsson, J. Helenius, D. Rickerl, R. Salvador, M. Wiedenhoeft, S. Simmons, P. Allen, M.A. Altieri, C. Flora and R. Poincelot (2003). Agroecology: The ecology of food systems, *Journal of Sustainable Agriculture*, **22**(3): 99-118.

Freire, P. (1992). *Pedagogia da esperança, um rencontro com a pedagogia do oprimido, Paz e Terra*, Rio de Janeiro, Br.

Funes, F.A. and L.M. Vazquez (2016). *Avances de la agroecología en Cuba, Estación Experimental de Pastos y Forrajes Indio Hatuey, Editora*, Matanzas, Cu.

Jackson, L.E., U. Pascual and T. Hodgkin (2007). Utilizing and conserving agrobiodiversity in agricultural landscapes, *Agriculture, Ecosystems & Environment*, 1(21): 196-210.

Koohafkan, P. and M.A. Altieri (2016). *Forgotten Agricultural Heritage: Reconnecting Food Systems and Sustainable Development*, Routledge, London, UK.

Murgueitio, E., M. Flores, Z. Calle, J. Chará, R. Barahona, C. Molina and F. Uribe (2015). Productividaden Sistemas Silvopastoriles Intensivos en América Latina. *In:* F. Montagnini, E. Somarriba, E. Murgueitio, H. Fassola and B. Eibl (Eds.). *Sistemas Agroforestales, Funciones Productivas, Socioeconómicas y Ambientales*, CATIE, Cali, Co: 59-101.

Nicholls, C.I., M.A. Altieri and L. Vazquez (2016). Principles for the conversion and redesign of farming systems, *Journal of Ecology and Ecography*, 5: 1-18.

Oliveira, G.L.T. and S.B. Hecht (2017). Soy, globalization, and environmental politics in South America, *Critical Agrarian Studies*, Routledge, New York, USA.

Pimentel, D., D. Andow, R. Dyson Hudson, D. Gallahan, S. Jacobson, M. Irish, S. Kroop, A. Moss, I. Schreiner, M. Shepard, T. Thompson and B. Vinzant (1980). Environmental and Social Costs of Pesticides: A Preliminary Assessment, *Oikos*, 34: 127-140.

Purdy, C. (2020). Covid-19 is about to Reach US Farms in a Major Test for Food Supply Chains; Retrieved from https://qz.com/1829558/covid-19-is-about-to-reach-us-farms/; accessed on 27 April, 2021.

Repetto, R. and S.S. Baliga (1996). Pesticides and immunosuppression: The risks to public health, *Health Policy and Planning*, 12: 97-106.

Rosset, P.M. and M.A. Altieri (2017). *Agroecology: Science and Politics*, Fernwood Publishing, Nova Scotia, USA.

Rosset, P.M., B. Machin-Sosa, A.M. Roque-Jaime and D.M. Avila-Lozano (2011). The Campesino a Campesino Agroecology Movement of ANAP in Cuba, *Journal of Peasant Studies*, **38**(1): 161-191.

UNSCN (United Nations System Standing Committee on Nutrition). (2020). The COVID-19 Pandemic is Disrupting People's Food Environments; Retrieved from https://www.unscn.org/en/news-events/recent-news?idnews=2039; accessed on 27 April, 2021.

Weiss, T. (2013). *The Ecological Hoofprint: The Global Burden of Industrial Livestock*, Zed Books, London, UK.

Zheng, Y. and G. Deng (1998). Benefits analysis and comprehensive evaluation of rice-fish-duck symbiotic model, *Chin. J. Eco-Agric.*, 6: 48-51.

Index

For Product Safety Concerns and Information please contact our EU representative GPSR@taylorandfrancis.com Taylor & Francis Verlag GmbH, Kaufingerstraße 24, 80331 München, Germany